Application of Artificial Intelligence Methods in Processing of Emotions, Decisions and Opinions

Application of Artificial Intelligence Methods in Processing of Emotions, Decisions and Opinions

Editors

Pawel Dybala
Rafal Rzepka
Michal Ptaszynski

Basel • Beijing • Wuhan • Barcelona • Belgrade • Novi Sad • Cluj • Manchester

Editors

Pawel Dybala
Institute of the Middle and Far East
Faculty of International and Political Studies
Jagiellonian University
Krakow
Poland

Rafal Rzepka
School of Information Science and Technology
Hokkaido University
Hokkaido
Japan

Michal Ptaszynski
Text Information Processing Laboratory
Kitami Institute of Technology
Kitami
Japan

Editorial Office
MDPI AG
Grosspeteranlage 5
4052 Basel, Switzerland

This is a reprint of articles from the Special Issue published online in the open access journal *Applied Sciences* (ISSN 2076-3417) (available at: https://www.mdpi.com/journal/applsci/special_issues/NTMDOE41MY).

For citation purposes, cite each article independently as indicated on the article page online and as indicated below:

Lastname, A.A.; Lastname, B.B. Article Title. *Journal Name* **Year**, *Volume Number*, Page Range.

ISBN 978-3-7258-1715-3 (Hbk)
ISBN 978-3-7258-1716-0 (PDF)
doi.org/10.3390/books978-3-7258-1716-0

Cover image courtesy of Michal Ptaszynski

© 2024 by the authors. Articles in this book are Open Access and distributed under the Creative Commons Attribution (CC BY) license. The book as a whole is distributed by MDPI under the terms and conditions of the Creative Commons Attribution-NonCommercial-NoDerivs (CC BY-NC-ND) license.

Contents

About the Editors vii

Michal Ptaszynski, Pawel Dybala and Rafal Rzepka
Application of Artificial Intelligence Methods in Processing of Emotions, Decisions, and Opinions
Reprinted from: *Appl. Sci.* **2024**, *14*, 5912, doi:10.3390/app14135912 1

Yuan Zhou and Siamak Khatibi
A New AI Approach by Acquisition of Characteristics in Human Decision-Making Process
Reprinted from: *Appl. Sci.* **2024**, *14*, 5469, doi:10.3390/app14135469 4

Maresha Caroline Wijanto and Hwan-Seung Yong
Combining Balancing Dataset and SentenceTransformers to Improve Short Answer Grading Performance
Reprinted from: *Appl. Sci.* **2024**, *14*, 4532, doi:10.3390/app14114532 22

Pei Yang, Ying Liu, Yuyan Luo, Zhong Wang and Xiaoli Cai
Text Mining and Multi-Attribute Decision-Making-Based Course Improvement in Massive Open Online Courses
Reprinted from: *Appl. Sci.* **2024**, *14*, 3654, doi:10.3390/app14093654 42

Tanjim Mahmud, Michal Ptaszynski and Fumito Masui
Automatic Vulgar Word Extraction Method with Application to Vulgar RemarkDetection in Chittagonian Dialect of Bangla
Reprinted from: *Appl. Sci.* **2023**, *13*, 11875, doi:10.3390/app132111875 67

Jie Long, Zihan Li, Qi Xuan, Chenbo Fu, Songtao Peng and Yong Min
Social Media Opinion Analysis Model Based on Fusion of Text and Structural Features
Reprinted from: *Appl. Sci.* **2023**, *13*, 7221, doi:10.3390/app13127221 99

Rosa A. García-Hernández, José M. Celaya-Padilla, Huizilopoztli Luna-García, Alejandra García-Hernández, Carlos E. Galván-Tejada, Jorge I. Galván-Tejada, et al.
Emotional State Detection Using Electroencephalogram Signals: A Genetic Algorithm Approach
Reprinted from: *Appl. Sci.* **2023**, *13*, 6394, doi:10.3390/app13116394 117

Wafa Alshehri, Nora Al-Twairesh and Abdulrahman Alothaim
Affect Analysis in Arabic Text: Further Pre-Training Language Models for Sentiment and Emotion
Reprinted from: *Appl. Sci.* **2023**, *13*, 5609, doi:10.3390/app13095609 132

Dušan Radisavljević, Rafal Rzepka and Kenji Araki
Personality Types and Traits—Examining and Leveraging the Relationship between Different Personality Models for Mutual Prediction
Reprinted from: *Appl. Sci.* **2023**, *13*, 4506, doi:10.3390/app13074506 158

Yeo-Gyeong Noh, Junryeol Jeon and Jin-Hyuk Hong
Understanding of Customer Decision-Making Behaviors Depending on Online Reviews
Reprinted from: *Appl. Sci.* **2023**, *13*, 3949, doi:10.3390/app13063949 193

Yanping Shen, Qingjie Liu, Na Guo, Jing Yuan and Yanqing Yang
Fake News Detection on Social Networks: A Survey
Reprinted from: *Appl. Sci.* **2023**, *13*, 11877, doi:10.3390/app132111877 209

About the Editors

Pawel Dybala

Pawel Dybala received an MA degree from Jagiellonian University, Krakow, Poland, in 2006. He was a research student at Hokkaido University, Japan. Since 2007, he has been working toward a Ph.D. degree at the Graduate School of Information Science and Technology, Hokkaido University, Japan. At present, he is an Assistant Professor at Jagiellonian University in Krakow, Poland, Faculty of International and Political Studies. His research interests include natural language processing, dialog processing, humor processing, HCI, and information retrieval.

Rafal Rzepka

Rafal Rzepka received a master's degree from the University of Adam Mickiewicz, Poznan, Poland in 1999 and a Ph.D. from Hokkaido University, Japan, in 2004. Currently, he is an associate professor in the Graduate School of Information Science and Technology at Hokkaido University. His research interests include natural language processing, common sense knowledge retrieval, dialog processing, artificial general intelligence, affect and sentiment analysis, and machine ethics. He is a member of the AAAI, JSAI, JCSS, and ANLP.

Michal Ptaszynski

Michal Ptaszynski received his Master's degree from the University of Adam Mickiewicz, Poznan, Poland, in 2006, and his Ph.D. in information science and technology from Hokkaido University, Japan in 2010. From 2010 to 2012, he was a JSPS post-doctoral research fellow at the High-Tech Research Center, Hokkai-Gakuen University, Japan. Currently, he is an associate professor at the Kitami Institute of Technology. His research interests include natural language processing, artificial intelligence, and affective computing, and he is a founder of the fields of automatic cyberbullying detection and Ainu language processing. He is a senior member of the IEEE and a member of the AAAI and ACL.

Editorial

Application of Artificial Intelligence Methods in Processing of Emotions, Decisions, and Opinions

Michal Ptaszynski [1,*], Pawel Dybala [2] and Rafal Rzepka [3]

[1] Text Information Processing Laboratory, Kitami Institute of Technology, Kitami 090-8507, Japan
[2] Faculty of International and Political Studies, Institute of Middle and Far Eastern Studies, Jagiellonian University, 30-387 Krakow, Poland; pawel.dybala@uj.edu.pl
[3] Langauge Media Laboratory, Hokkaido University, Sapporo 060-0808, Japan; rzepka@ist.hokudai.ac.jp
* Correspondence: michal@mail.kitami-it.ac.jp

1. Introduction

The rapid advancement of artificial intelligence (AI) and natural language processing (NLP) has profoundly impacted our understanding of emotions, decision-making, and opinions, particularly within the context of the Internet and social media. This special issue of Applied Sciences, titled "Application of Artificial Intelligence Methods in Processing of Emotions, Decisions, and Opinions", brings together a collection of pioneering research articles that study these intertwined domains. This issue aims to highlight innovative AI methodologies and their application in analyzing and understanding the intricate human experiences reflected in online data. We discuss the accepted papers in their respective thematic groups.

2. Understanding Human Decision-Making through AI

Yuan Zhou and Siamak Khatibi [1], in their paper A New AI Approach by Acquisition of Characteristics in Human Decision-Making Process, challenge the conventional Markovian decision-making models by proposing an ambiguity probability model that captures the dynamic nature of human decision strategies. Their innovative decision map approach offers a nuanced understanding of human decision dependencies, presenting a significant advancement in modeling complex decision-making processes.

3. Enhancing Educational Assessments and Course Improvement

Maresha Caroline Wijanto and Hwan-Seung Yong [2] address the challenges of automatic short-answer grading in their paper Combining Balancing Dataset and Sentence-Transformers to Improve Short Answer Grading Performance. By integrating a balanced dataset with a simplified SentenceTransformers model, they achieve high grading accuracy, demonstrating that efficient and cost-effective AI models can rival more resource-intensive techniques.

In the realm of online education, Pei Yang Ying Liu, Yuyan Luo, Zhong Wang and Xiaoli Cai [3] present Text Mining and Multi-Attribute Decision-Making-Based Course Improvement in Massive Open Online Courses. Their method combines text mining and decision-making techniques to extract and analyze learner feedback, providing actionable insights for enhancing course quality and learner satisfaction.

4. Detecting and Analyzing Offensive Language

The proliferation of social media has necessitated robust methods for detecting harmful content. Tanjim Mahmud, Michal Ptaszynski and Fumito Masui [4] tackle this issue in Automatic Vulgar Word Extraction Method with Application to Vulgar Remark Detection in Chittagonian Dialect of Bangla. By comparing keyword matching with advanced machine

learning techniques, they highlight the efficacy of deep learning in capturing nuanced vulgar language, particularly in low-resource languages and dialects.

5. Opinion and Emotion Analysis in Social Media

Jie Long, Zihan Li, Qi Xuan, Chenbo Fu, Songtao Peng and Yong Min [5] introduce an innovative model in their paper Social Media Opinion Analysis Model Based on Fusion of Text and Structural Features. Their heterogeneous graph attention network (HGAT) effectively integrates text and contextual data, enhancing the accuracy of opinion recognition in complex Chinese texts, thereby contributing significantly to the field of sentiment analysis.

Wafa Alshehri, Nora Al-Twairesh and Abdulrahman Alothaim [6] study emotion detection in Arabic texts with their work Affect Analysis in Arabic Text: Further Pre-Training Language Models for Sentiment and Emotion. By adapting BERT-based models for Arabic sentiment and emotion tasks, they achieve notable improvements, showcasing the potential of pre-trained language models in underrepresented languages.

6. Integrating Cognitive Science and AI

Rosa A. García-Hernández and her research team [7] explore the intersection of cognitive science and AI in their paper Emotional State Detection Using Electroencephalogram Signals: A Genetic Algorithm Approach. Utilizing genetic algorithms to optimize EEG data features, they achieve impressive accuracy in emotion classification, paving the way for real-time emotion detection using portable sensors.

7. Understanding Consumer Behavior

In Understanding of Customer Decision-Making Behaviors Depending on Online Reviews, Yeo-Gyeong Noh, Junryeol Jeon and Jin-Hyuk Hong [8] study how consumers process online reviews. Their findings reveal the differential impact of star ratings and comments on consumer decisions, providing valuable insights for businesses aiming to leverage online reviews to influence purchasing behavior.

8. Personality Analysis in Social Media

Dušan Radisavljević, Rafal Rzepka, and Kenji Araki [9] investigate the relationship between various personality models in their paper Personality Types and Traits—Examining and Leveraging the Relationship between Different Personality Models for Mutual Prediction. By bridging the gap between MBTI, Big Five, and Enneagram models, they enhance the resources available for personality research, offering new methods for personality recognition.

9. Comprehensive Review on Fake News Detection

Finally, the review article Fake News Detection on Social Networks: A Survey by Yanping Shen, Qingjie Liu, Na Guo, Jing Yuan and Yanqing Yang [10] provides a comprehensive overview of the current state of fake news detection. Their detailed classification of detection techniques and analysis of datasets serves as a valuable resource for researchers dedicated to combating the spread of misinformation.

10. Conclusions

This special issue of Applied Sciences showcases the cutting-edge advancements in AI methods for processing emotions, decisions, and opinions. The diverse range of studies presented here not only underscores the interdisciplinary nature of this field but also highlights the transformative potential of AI in understanding and addressing complex human behaviors in the digital age. We hope these contributions will inspire further research and innovation, driving the continued evolution of AI and its applications in society.

Conflicts of Interest: The authors declare no conflicts of interest.

References

1. Zhou, Y.; Khatibi, S. A New AI Approach by Acquisition of Characteristics in Human Decision-Making Process. *Appl. Sci.* **2024**, *14*, 5469. [CrossRef]
2. Wijanto, M.C.; Yong, H.S. Combining Balancing Dataset and SentenceTransformers to Improve Short Answer Grading Performance. *Appl. Sci.* **2024**, *14*, 4532. [CrossRef]
3. Yang, P.; Liu, Y.; Luo, Y.; Wang, Z.; Cai, X. Text Mining and Multi-Attribute Decision-Making-Based Course Improvement in Massive Open Online Courses. *Appl. Sci.* **2024**, *14*, 3654. [CrossRef]
4. Mahmud, T.; Ptaszynski, M.; Masui, F. Automatic Vulgar Word Extraction Method with Application to Vulgar Remark Detection in Chittagonian Dialect of Bangla. *Appl. Sci.* **2023**, *13*, 11875. [CrossRef]
5. Long, J.; Li, Z.; Xuan, Q.; Fu, C.; Peng, S.; Min, Y. Social Media Opinion Analysis Model Based on Fusion of Text and Structural Features. *Appl. Sci.* **2023**, *13*, 7221. [CrossRef]
6. Alshehri, W.; Al-Twairesh, N.; Alothaim, A. Affect Analysis in Arabic Text: Further Pre-Training Language Models for Sentiment and Emotion. *Appl. Sci.* **2023**, *13*, 5609. [CrossRef]
7. García-Hernández, R.A.; Celaya-Padilla, J.M.; Luna-García, H.; García-Hernández, A.; Galván-Tejada, C.E.; Galván-Tejada, J.I.; Gamboa-Rosales, H.; Rondon, D.; Villalba-Condori, K.O. Emotional State Detection Using Electroencephalogram Signals: A Genetic Algorithm Approach. *Appl. Sci.* **2023**, *13*, 6394. [CrossRef]
8. Noh, Y.G.; Jeon, J.; Hong, J.H. Understanding of Customer Decision-Making Behaviors Depending on Online Reviews. *Appl. Sci.* **2023**, *13*, 3949. [CrossRef]
9. Radisavljević, D.; Rzepka, R.; Araki, K. Personality Types and Traits—Examining and Leveraging the Relationship between Different Personality Models for Mutual Prediction. *Appl. Sci.* **2023**, *13*, 4506. [CrossRef]
10. Shen, Y.; Liu, Q.; Guo, N.; Yuan, J.; Yang, Y. Fake News Detection on Social Networks: A Survey. *Appl. Sci.* **2023**, *13*, 11877. [CrossRef]

Disclaimer/Publisher's Note: The statements, opinions and data contained in all publications are solely those of the individual author(s) and contributor(s) and not of MDPI and/or the editor(s). MDPI and/or the editor(s) disclaim responsibility for any injury to people or property resulting from any ideas, methods, instructions or products referred to in the content.

Article

A New AI Approach by Acquisition of Characteristics in Human Decision-Making Process

Yuan Zhou * and Siamak Khatibi

Department of Technology and Aesthetic, Blekinge Institute of Technology, 371 79 Karlskrona, Sweden; siamak.khatibi@bth.se
* Correspondence: yuan.zhou@bth.se

Abstract: Planning and decision making are closely interconnected processes that often occur in tandem, influence and informing each other. Planning usually precedes decision making in the chronological sequence, and it can be viewed as a strategy to make decisions. A comprehensive planning or decision strategy can facilitate effective decisions. Thus, understanding and learning human decision-making strategies has drawn intensive attention from the AI community. For example, applying planning algorithms into reinforcement leaning (RL) can simulate the consequence of different actions and select optimal decisions based on learned models, while inverse reinforcement learning (IRL) learns a reward function and policy from expert demonstration and applies them into new scenarios. Most of these methods work based on learning human decision strategies by using modeling of a Markovian decision-making process (MDP). In this paper, we argue that the property of MDP is not fit for human decision-making processes in the real-world and it is insufficient to capture human decision strategies. To tackle this challenge, we propose a new approach to identify the characteristics of human decision-making processes as a decision map, where the decision strategy is defined by the probability distribution of human decisions that are adaptive to the dynamic changes in the environment. The proposed approach was inspired by imitation learning (IL) but with fundamental differences: (a) Instead of aiming to learn an optimal policy based on expert's demonstrations, we aimed to estimate the distribution of decisions of any group of people. (b) Instead of modeling the environment by an MDP, we used an ambiguity probability model to consider the uncertainty of each decision. (c) The participant trajectory was obtained by categorizing each decision of a participant to a certain cluster based on the commonness in the distribution of decisions. The result shows a feasible way to capture human long-term decision dependency, which provides a complement to the existing machine learning methods for understanding and learning human decision strategies.

Keywords: decision-making process; decision strategy; knowledge representation; artificial intelligence; data clustering

Citation: Zhou, Y.; Khatibi, S. A New AI Approach by Acquisition of Characteristics in Human Decision-Making Process. *Appl. Sci.* **2024**, *14*, 5469. https://doi.org/10.3390/app14135469

Academic Editors: Pawel Dybala, Rafal Rzepka and Michal Ptaszynski

Received: 13 May 2024
Revised: 10 June 2024
Accepted: 12 June 2024
Published: 24 June 2024

Copyright: © 2024 by the authors. Licensee MDPI, Basel, Switzerland. This article is an open access article distributed under the terms and conditions of the Creative Commons Attribution (CC BY) license (https://creativecommons.org/licenses/by/4.0/).

1. Introduction

Planning and decision making have been extensively studied in the AI community, with the aim to design and develop an intelligent agent that can interact and act in a variety of environments involving individuals or agents. Planning and decision making are closely related processes. Effective decision making often relies on some form of planning, where individuals or agents consider different options, anticipate potential consequences, and evaluate trade-offs before making a choice. In other words, planning can be viewed as a higher-level cognitive process that supports decision making by providing a structured framework for considering alternatives, setting goals, and organizing actions in pursuit of desired outcomes.

Planning or decision strategies refer to the approaches that individuals use to make decisions [1]. These strategies can vary based on factors such as the decision context, available information, preferences, and cognitive biases [2]. Common decision strategies include heuristic-based approaches (e.g., using rules of thumb or shortcuts) [3], analytical approaches (e.g., systematic analysis of options) [4], and intuitive approaches (e.g., relying on gut feelings) [5]. In other hand, decision making involves various cognitive processes, including perception, attention, memory, reasoning, and judgment. Decision making is also influenced by environmental factors, such as social norms, situational context, available resources, and external constraints. IL is a machine learning approach where an agent learns to perform tasks by observing and mimicking demonstrations provided by an expert. The goal is to replicate the expert's behavior based on observed actions and outcomes [6]. In IL, the state space and action space are critical components, where the state space encapsulates the environment's current condition, and the action space defines the agent's possible decisions or behaviors in response to different states. Imitation learning methods often incorporate planning algorithms to enhance decision-making capabilities. For example, RL techniques combined with planning (such as model-based RL) can enable agents to simulate outcomes of different actions and select optimal decisions based on learned models [7,8]. Planning can also aid in generalization and adaptation in imitation learning settings. Agents can use planning to infer strategies or policies underlying expert demonstrations and apply similar decision-making principles to new scenarios [9]. By integrating planning with imitation learning, agents can exhibit more flexible and robust decision-making behavior, allowing them to handle complex tasks and adapt to changing environments beyond the specific demonstrations they were trained on.

Regarding the different methodologies for reproducing the expert behavior, IL can be generally divided into two major approaches: behavior cloning (BC) and IRL. BC simply tackles the IL problem as a supervised learning problem that aims to build a straightforward mapping from the state space to the action space under the learned policy. But this method can only provide relatively acceptable performance in environments that are the same or similar to the training dataset due to the property of an MDP. However, minor errors can quickly compound when the learned policy departs from the observed states in the demonstration [10]. To increase the awareness of the environment dynamics, an IRL approach uses reinforcement learning methods to extract a reward function based on the optimal expert trajectory and learning a policy from the inferred reward function. However, the IRL method is hardly implemented into real-world tasks since recovering a reward function requires intensive interactions with an external environment, which can be costly in terms of computation and safety [11,12]. Decision making is performed as the center of human interaction with the real world and the decision-making process has a tight relationship with individual characteristics, which was mentioned in our previous work [13]. In this paper, we propose an approach to characterize human decisions. Generally defining the characteristics of decision making may involve understanding the decision strategies employed by individuals or groups in different contexts and examining how the cognitive processes interact to influence decision outcomes. Additionally understanding how the environmental factors shape decision-making behavior can contribute to defining its characteristics. The characteristics of decision making can also be defined in terms of decision outcomes, such as the quality of decisions, decision confidence, decision speed, and decision consistency. Analyzing these outcomes can provide insights into the underlying characteristics of decision-making processes.

The characteristics of decision making can be inferred from analyzing the probability distribution of a group of people's decisions. Let us break down how this can be achieved:

- Tendency and preference: Analyzing the probability distribution allows us to identify tendencies and preferences within a group. For example, if a certain option is chosen more frequently than others, it suggests a preference for that option within the group [14,15].
- Variability and consensus: The spread or variability of decisions within the probability distribution provides insights into the level of agreement or disagreement between

decision makers. A narrow distribution indicates a high level of consensus, whereas a broader distribution suggests greater variability in decision preferences [16,17].
- Decision biases: Certain decision biases may manifest as skewed or asymmetric probability distributions. For instance, if decisions tend to cluster around a particular option due to an anchoring bias or status quo bias, it will be reflected in the shape of the probability distribution [18].
- Risk preferences: Decision making often involves trade-offs between risks and rewards. By analyzing the probability distribution of decisions, we can infer the risk preferences of the group. A risk-averse group may exhibit a probability distribution skewed toward safer options, while a risk-seeking group may display a distribution skewed toward riskier alternatives [19,20].
- Decision stability: The stability of decision making over time can also be assessed through changes in the probability distribution. A consistent probability distribution indicates stable decision-making behavior, whereas fluctuations or shifts in the distribution may signify changes in preferences or external influences [21,22].

While analyzing the probability distribution of decisions provides valuable insights into the characteristics of decision making within a group, it is essential to complement this analysis with other approaches to gain a comprehensive understanding. Factors such as decision strategies, cognitive processes, individual differences, and environmental influences also play significant roles in shaping decision-making behavior. Therefore, a holistic approach that integrates various methods and perspectives is often necessary to fully define the characteristics of decision making.

Our insight is that the difficulty of environment dynamics with previous IL approaches arises from the method for the extraction of human knowledge or experience, i.e., the dynamics of the environment are modeled by an MDP that assumes the next state only depends on the decision (or action) in the current state in the expert trajectory. Thereby, we argue that the extracted human knowledge or experience under MDP settings can only perform well in the low-level task, e.g., object manipulations, where the environmental dynamics are limited. Meanwhile, it is prone to failing in a high-level task, e.g., driving on the highway, which requires high-level knowledge that refers to planning, causal reasoning, or analogical reasoning to handle the environment dynamics. In reality, when human agents use this high-level knowledge to make decisions, they usually consider a long-term relationship, such as what they experienced and what decisions were made. However, MDP settings usually fail to extract such long-term relationships lying in the demonstration data [23,24].

In this paper, we propose a method to extract human knowledge based on the characteristics of decision-making processes from a group of people. This method was inspired by IL but with some fundamental differences: (a) Instead of aiming to learn an optimal policy based on demonstrations, we aimed to estimate the distribution of decisions of a group of people. (b) Instead of modeling the environment by an MDP, we used an ambiguity probability model to consider the uncertainty of each decision. (c) The participant trajectory was obtained by categorizing each decision into a certain cluster based on the commonness compared with the distribution of decisions.

The rest of the paper is organized as follows: the related works are described in Section 2; the challenges in imitation learning are elaborated in Section 3; the proposed method is explained with an application example in Section 4; the results of application example are discussed in Section 5; conclusion and future work are given in Section 6.

2. Related Works

In past decades, the classical AI, known as the good old-fashioned artificial intelligence (GOFAI), has shown itself as powerful and efficient when considering numerous applications [25–27]. The GOFAI-based systems supported or even outperformed humans in specific tasks that used to be dominated by human intelligence, for example, IBM's Deep Blue first defeated the world champion player in the epic chess battle rematch in

1997 [28], and more recently, Google's AlphaGo system was reported to produce remarkable results in the series of competition with professional Go players from 2015 to 2017 [29]. However, these complicated and challenging problems are considered to be just a fragment of human intelligence in dealing with finite, deterministic, and constrained problems that are a part of the complex and multidimensional intelligence problem spaces [30]. Let us take the board game of checkers as an example. Despite the complexity of the game, which can contain nearly 500 billion possible position moves, it still has a finite number of combinations and a certain number of parameters (i.e., pieces on the board). Besides this, the most important thing is that the board game has fixed rules and perfect information, which means players have insight into everything that has happened before they decide to move a piece. However, unlike these finite, deterministic, constrained board gaming spaces, the real world where humans live are full of uncertainties and usually with limited availability of information for making decisions. Unfortunately, the classical AI tends to fail in pursuing human-level intelligence to deal with problems that are highly dynamic and uncertain. Generally speaking, classical AI has been successful in those tasks that are normally considered difficult, such as playing chess, applying rules of logic, proving a mathematical theorem, or solving certain abstract problems, but fail in those tasks where human actions are experienced as natural, effortless, and dynamic, such as seeing, hearing, talking, walking, and driving; all these skills require common sense [31]. Thus, building systems that incorporate the common sense knowledge became the goal for many classical problem-solving systems, such as CYC (stands for encyclopedia) [32], with the aim to make inferences based on a database of common sense, or the DARPA (Defense Advanced Research Project Agency) project in [33], which is trying to build a machine that can learn by reading a text. However, the successes of these projects are disputed because they are incapable of dealing with common sense in a flexible and adaptive way. One of main reasons is that such systems represent and manipulate the common sense knowledge in an explicit way. For example, in the CYC system, the building blocks of common sense knowledge are propositional statements, such as "cars cannot fly", "viruses cause infection", "smoking is harmful to health", etc.; it is believed that the large collection of such logic-based statements in conjunction with a set of inference rules is all that is needed to represent common sense knowledge. However, common sense is not rigid but highly dynamic and varies in relation to people's characteristics and background, especially when it refers to the trivial matters in our daily lives [34,35]. For example, we all have an intuitive understanding of the word "driving" that will come to mind if we freely associate with "driving". It might be turning the wheel, stepping on the gas, a luxury car, a road journey, a car race, the jammed traffic, etc. It is these kinds of variational experiences that form the basis of our common sense, which can never be captured or modeled by such a set of logical statements and rules but requires interactions with the real world.

Common sense reasoning presents challenges in formalization and computational modeling due to its inherent complexity and context-dependent nature. Incorporating common sense into machine learning algorithms, including imitation learning, requires effective representation and integration with data-driven approaches. On the other hand, leveraging common sense can enhance the capabilities of imitation learning agents in addressing real-world tasks that involve rich contextual understanding, intuitive decision making, and adaptive behavior. Integrating common sense reasoning with imitation learning opens up new avenues for developing human-like intelligent systems. Many studies were conducted to explore the relationship between common sense and decision making in different fields, such as managerial decisions in organization [36], integrating common sense as prior knowledge in reinforcement learning [37], and medical decisions in treatment [38]. However, that research put more focus on how common sense as "practical judgement" influences a decision rather than how common sense is captured as knowledge from lived experience. Despite the fact that in our daily lives, most decision problems are trivial without general agreed-upon criteria, individuals make decisions based on their own perceptions and personality, which are formed by their life experience. This kind of common sense is highly dynamic to different groups of people, and it cannot be simply formalized by a set of rules but requires observations of how individuals interact with the real world and what are the characteristics in those interactions.

3. Challenges in Imitation Learning

IL as learning from demonstration or apprenticeship learning is generally mathematically formulated within the framework of reinforcement learning [39]. The goal of imitation learning is to enable an agent to mimic expert behavior by learning from observed demonstrations. The following notations are used:

- State space: S, set of all possible states.
- Action space: A, set of all possible actions.
- Reward function: $R : S \times A \to \mathbb{R}$, defines the immediate reward for taking an action in a state.
- Demonstration dataset: $D = \{(s_1, a_1), (s_2, a_2), \ldots, (s_t, a_t)\}$, set of state–action pairs from expert demonstrations.
- Policy: $\pi : S \to A$, agent's policy mapping states to actions.

The objective of IL is to learn a policy π that mimics the expert's behavior, as demonstrated in D, where the aim is to find a policy π that maximizes the expected cumulative reward, like RL but without direct interaction with the environment. In the BC approach, a policy π is directly learned that maps state s to action a based on the observed demonstrations [40]. The learning objective of BC can be formulated as minimizing the expected loss between the actions predicted by π and the actions in the demonstration dataset D according to

$$min_\pi \mathbb{E}_{s \sim D}[loss(\pi(s), a)]. \tag{1}$$

Here, $loss$ is a suitable loss function (e.g., mean squared error or cross-entropy) that measures the discrepancy between the predicted action $\pi(s)$ and demonstrated action a. In the inverse reinforcement learning (IRL) approach, the agent infers the underlying reward function R from the expert's behavior and then learns a policy π that maximizes this inferred reward [41]. The learning process involves estimating the reward function R based on the observed demonstrations and then optimizing the policy π to maximize the expected cumulative reward under this reward function according to

$$max_\pi \sum_{t=1}^{T} \mathbb{E}_{S_t \sim p(.)}[R(s_t, \pi(s_t)], \tag{2}$$

where $p(.)$ is the distribution over states encountered during the execution of policy π.

Considering the training and optimization in IL methods, two major trends can be mentioned: the supervised learning approach and iterative improvement. BC can be framed as a supervised learning problem, where the policy π is learned using supervised learning techniques (e.g., neural networks) to minimize the prediction error between $\pi(s)$ and a from D. Iterative approaches, such as policy iteration or gradient-based optimization methods (e.g., policy gradient), can be employed to iteratively update the policy π to maximize the expected reward based on the observed demonstrations.

Three major challenges can be considered in relation to IL, which are as follows [6,10,42]:

- Generalization: one key challenge in imitation learning is generalizing the learned policy π to new, unseen situations that may differ from the demonstrations;
- Distribution mismatch: the distribution of states encountered during the policy execution by the agent may differ from the distribution of states in the demonstration dataset D, leading to challenges in effective learning;
- Bias and variance trade-off: balancing bias (due to approximation error) and variance (due to noisy demonstrations) is critical for successful imitation learning.

Various techniques and algorithms can be applied to address challenges and achieve effective imitation learning. However, we believe the difficulty of environment dynamics that can be expressed with common sense problems is the major challenge with IL approaches, which cannot be solved due to their fundamental limitation of using an MDP.

4. Method

In this section, we explain the proposed method and elaborate upon it with a game example as a one-dimensional application of the method. The proposed method was inspired by IL approaches.

Figure 1 illustrates the differences between IL approaches and the proposed method to extract human knowledge or experiences. As shown in the left part of Figure 1, an agent learns a task by observing demonstrations performed by an expert. Generally, it extracts expert knowledge through the three processes of a learning algorithm, policy optimization, and evaluation-refinement process. In the demonstration collection procedure, a set of demonstrations is collected from an expert. These demonstrations typically consist of sequences of states and actions that the expert performs while completing the task [43]. The applied learning algorithm process for different methods, such as BC, IRL, or adversarial structured IL, varies, where the goal is to learn a policy or a mapping from states to actions that mimics the behavior of the expert. In the policy optimization process, the learned policy is optimized to minimize the discrepancy between its behavior and that of the expert. This optimization process may involve techniques such as reinforcement learning or supervised learning, depending on the learning algorithm. In the evaluation-refinement process, the learned policy is evaluated on new data to assess its performance. If the performance is not satisfactory, the process may be iterated by collecting more demonstrations, refining the representation, or adjusting the learning algorithm parameters. As shown in the right part of Figure 1, instead of collecting the expert knowledge decisions of a group of people, our proposed method collects decisions to extract the probability distribution of optional decisions through the three processes of decision observation, uncertainty estimation, and the decision characteristics process. In the following, these three processes are elaborated in detail.

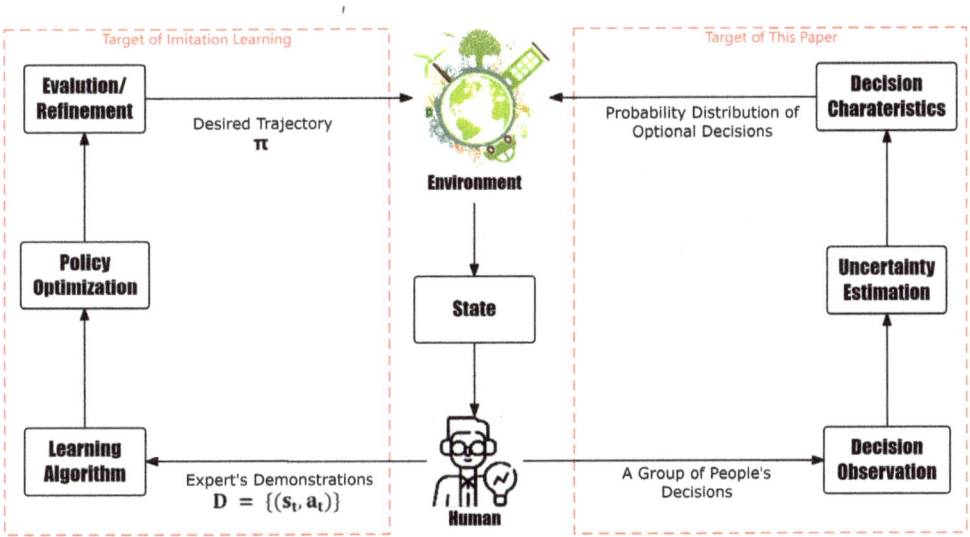

Figure 1. Comparison of task event processing in IL approaches and the proposed method.

As mentioned in the introduction, decision making is the fundamental human activity for interaction with the real world. Figure 2 shows how individuals generally interact with the outside world through decision making. Let us assume the world consists of a set of infinite task events S, where each task event is a subset of S and is denoted as $S' = s_i, i = 0 \ldots n$, where s_i is a state of the task event and n is the number of states, which is determined by the sequence of actions (see Figure 2 bottom). The impact of an action on the state, i.e., the state of a task event, is represented by the state transition function $T(s_{i+1} s_i, a_i)$, where the a_i represents the action and occurs according to decision making (see Figure 1 top). The decision making contains three processes: perception, reasoning, and acting. In perception, the state is perceived through the measurements of the current state, such as temperature, velocity, and position, so that the state is perceived by a set of parameters and their values. Note that the perception of state is task-oriented, which means the choices of parameters are related to the task under the decision-making process. For example, normally a driver does not look at the back mirror when the car is driven straight forward, but the back mirror is used when driver needs to change lane or drive in reverse. For succinctness, the "state" is used interchangeably with "perceived state" throughout the text and is denoted by s_i. The decision d_i is selected after the perception through the reasoning of alternatives, denoted as the function $R(s_i)$; this process is mostly based on personal experience and other personal factors, such as preference and emotions [44]. A physical action, denoted by the function $\phi(d_i)$, happens after a decision. The states during the task execution keep changing until the task event is accomplished. The number of states and their order within the task event are determined by the sequence of actions. It should be noted that Figure 1 illustrates two types of common sense in conjunction with the relation between the decision making and task execution. Implementing the flow chart in the figure from the top to bottom, i.e., using the function $\phi(R(s_i))$, the figure represents the first type of common sense, i.e., "the knack for seeing things as they are, or doing things as they ought to be done". While, implementing the flow chart from the bottom to top, i.e., estimation of the function $\phi(R(s_i))$, the figure represents the second type of common sense, i.e., the commonsensical experiences of a group for the execution of a task.

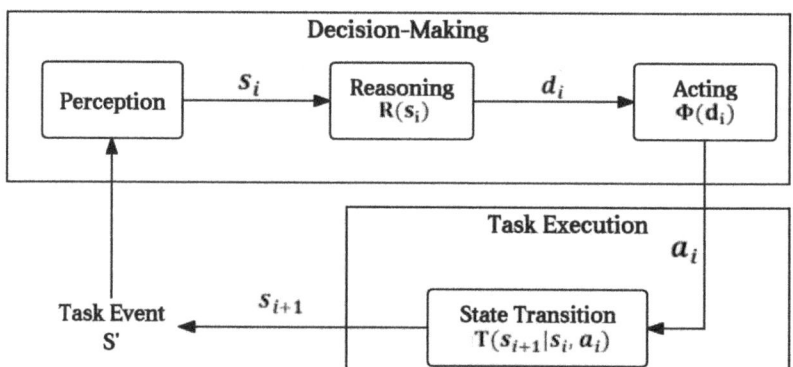

Figure 2. Detailed task event processing in the proposed method.

4.1. Application Example

A number-guessing game as a one-dimensional application of the method was designed. The collected data from the game was used to estimate the function ϕ through its inputs s_i and outputs a_i, as shown from the bottom to top in Figure 2.

In the game, participants were asked to find out an integer number that was randomly predefined within the range of 1 to 1000. During the task, each participant had unlimited attempts to ask questions to achieve the goal. In each attempt, participants needed to select one of two types of question: (1) Is the next targeted number bigger than participant's previously chosen number? (2) Is the next targeted number equal to participant's previous chosen number? Then, an answer of "Yes" or "No" was given to participants based on their questions and chosen number. The task event did not end until the targeted number is identified. The flowchart of the task event is shown in Figure 3. In this task event, the ranges of the targeted number represented states; at the beginning, all participants faced the same state, where the targeted range was from 1 to 1000; after each attempt, the states were changed based on each participant's actions (i.e., the question and number they chose).

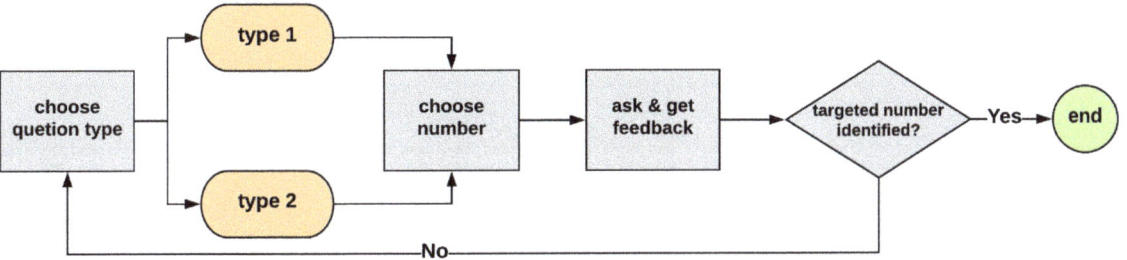

Figure 3. Flow chart of number guessing game.

A web-based application was deployed to collect data, and 54 people participated in the game. The decision data of the participants, i.e., the chosen number and type of question, were collected, but only bigger-type question and its chosen number were used in the later data processing since this type of question changed the targeted range more significantly than the other one, especially when the targeted range remained wide in the early phase of task event. In addition, demographic information of the participants was collected as well.

4.2. Method Implementation in Game Application

In this subsection, the usage of the proposed method for the task event processing in the game is explained, where the chosen number a_i and relevant state s_i are processed to estimate the function ϕ. The process is achieved in the five steps of normalization, clustering, generation of an ambiguous cluster, and association accumulation.

4.2.1. Feature Space Mapping

Throughout the execution of task, the decision sequence of each participant was recorded as

$$d_i^j = \{s_i^j, a_i^j\}, s_i^j \in S, a_i^j \in A^j \tag{3}$$

where j is the index of participant; pairs of states and actions were used to represent the decision sequence of each participant. A^j is the set of possible actions of participant j and i is the index of each attempt. The decision sequence of each participant was a consequence of the participant actions, which was executed as

$$d_i^j = \{s_i^j, a_i^j\} \xrightarrow{T_i^j} \{s_{i+1}^j, a_i^j\} \tag{4}$$

which resulted in a new situation of s_{i+1}^j. Using the action of each decision as a comparable measurement factor among all the decisions and all the participants created a multiple-measurement space where the decisions of each participant could be considered as one dimension that had a special measurement unit depending on each participant. To be able to compare all the decisions of all the participants a feature space was considered, where the multiple-measurement space was mapped to a one-dimensional feature space according to

$$d_i^j = \{s_i^j, a_i^j\} \xrightarrow{T_i^j} \{s_{i+1}^j, a_i^j\} \xrightarrow{F_i^j} df_i^j = \{s_i^j, \frac{s_{i+1}^j}{s_i^j}, a_i^j s_{i+1}^j < s_i^j\} \tag{5}$$

where F is the feature-mapping operation and df_i^j is the decision in feature space with respect to each decision of a participant given as i and each participant given as j. According to the defined decision in the feature space, the measurable parameter in the feature space became

$$d_i^j = \{s_i^j, a_i^j\} \xrightarrow{F_i^j} \{p_i^j\} \qquad (6)$$

The process of feature mapping is shown in Figure 4, where each participant's decision sequence was mapped as $df_i^j = \{p_i^j\}$.

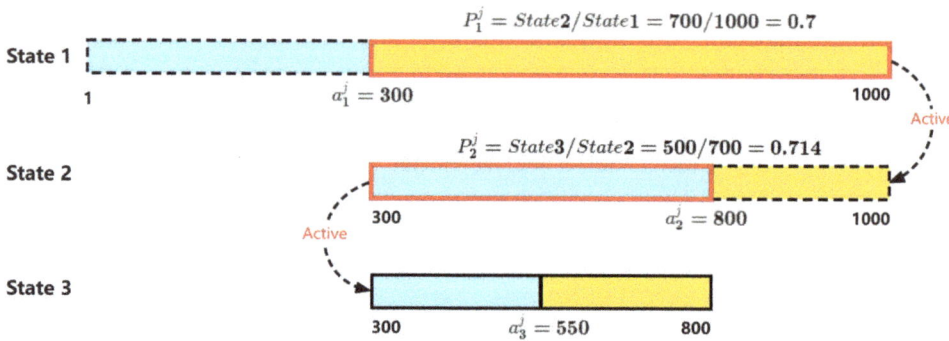

Figure 4. Mapping decisions to feature space.

4.2.2. Clustering

The decision sequence of each participant, i.e., $df^j = \{p_i^j\}$, was clustered as $df^j = \{p_{ik}^j\}$, where k is the index of the cluster. The clustering method was proposed in two steps. In the first step, the number of clusters for each decision sequence was identified; the result of the clustering was obtained as a one- or multi-element cluster. A one-element cluster is a cluster that includes either only one decision or several decisions but with the same df value, i.e., $C_{one} = \{p_{ik}^j\}, p_{1k}^j = \ldots = p_{ik}^j, i = 1, \ldots, n$. A multi-element cluster is a cluster that includes at least two varied df_i^j values, i.e., $C_{mul} = \{p_{ik}^j\}, p_{ik}^j \in [0,1], \nexists p_{1k}^j = \ldots = p_{ik}^j, i = 1, \ldots, n, n >= 2$. In the second step, all one-element clusters were combined and re-clustered. Details of the two steps are described as follows:

Step 1: K-medoids method [45] was used as core of clustering. Algorithm 1 shows the number of clusters was identified in two sub-steps: (a) A condition was defined to perform k-medoids for constraining the span of each cluster (SEC) less than a threshold, i.e., the value of error tolerance (VET). The SEC was considered to be the distance in each cluster, denoted as $each\_span = p_k^{max} - p_k^{min}$ where $p_k^{max} = max\{p_{ik}^j\}$ and $p_k^{min} = min\{p_{ik}^j\}$. VET was increased iteratively from 0.01 to 0.15 by steps of 0.01. Meanwhile, in each iteration, the number of clusters was also increased from one by a one-step manner and continued until the condition $SEC < VET$ was satisfied. Then, the number of clusters regarding each VET was recorded as candidates, denoted as $candi\_num$. (b) The difference between each two sequential candidates was calculated, denoted as $difference = candi\_num_i - candi\_num_{i+1}$; the final number of clusters was identical to the first two candidates whose difference was zero. Figure 5a,b shows two examples where the two clustering results in step 1 are presented. On the left of Figure 5, the candidate number of clusters for each VET are plotted and the final number of clusters is marked by the red circle; on the right, the results of clustering according to final numbers of clusters are plotted, where different clusters are distinguished by colors, the one-element clusters are marked by dash lines, and the multi-elements clusters are marked by solid lines.

Step 2: All one-element clusters from all decision sequences of participants were gathered and considered as a new decision sequence of a virtual participant. Then, the ensemble of one-element clusters was re-clustered by using step 1. The result of re-clustering is shown in Figure 5c right.

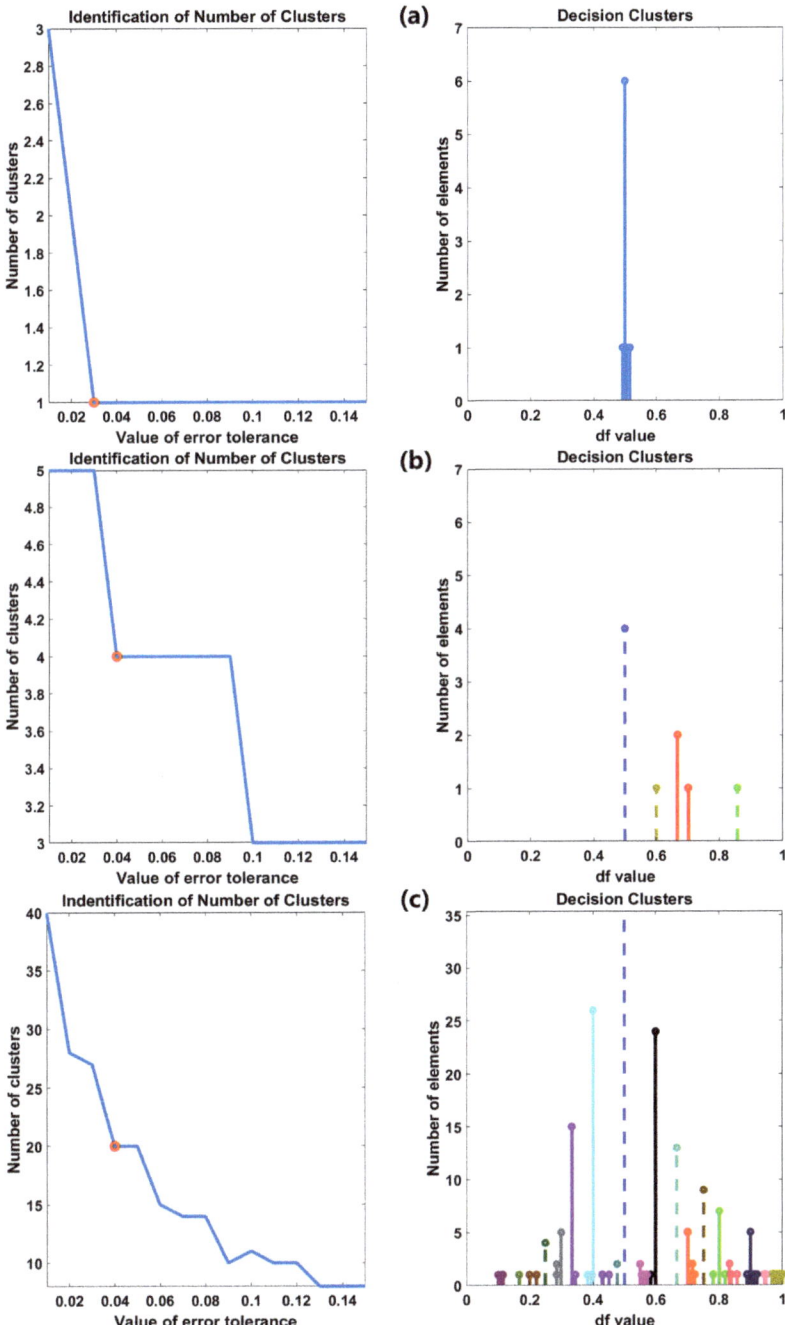

Figure 5. Examples of clustering result, on the right figures, different clusters of decisions are presented by different colors, one-element clusters are presented by dash lines, and multi-element cluster are presented by solid lines: (**a**) a sequence of decisions in one cluster; (**b**) a sequence of decisions in several clusters; (**c**) re-clustering of one-element clusters.

Algorithm 1: Clustering

```
scale = 1 Result: final_number_of_clusters
breakflag = 0;
index = 1;
candi_num;       /* records of all candidates regarding each VET for the final number of clusters */
for VET = 0.01 : 0.01 : 0.15 do
    temp_num = 1;
    while (1) do
        [k, decision_sequence] = kmedoids(decision_sequence, temp_num);
        SEC = p_k^max - p_k^min;
        if SEC < VET then
            | break;
        end
        temp_num = temp_num + 1;
    end
    candi_num(index) = temp_num + 1;              /* save each candidate number of clusters */
    index = index + 1;
end
for i = 1 : length(candi_num) do
    difference = candi_num(i) - candi_num(i + 1);
    if differnce == 0 then
        | break;
    end
end
final_number_of_clusters = candi_num(i);
```

4.2.3. Generation of Ambiguous Clusters

Each cluster was improved to be a more general and analog cluster based on its discrete elements in this process. The improved cluster is called an ambiguous cluster, denoted as ψ_i, which was achieved by modeling one-element and multi-element clusters as a Gaussian distribution and Student's t distribution, respectively. For the Gaussian distribution, the mean was estimated as the value of the element in each one-element cluster; the variance was estimated as half of the relevant VET. For the Student's t-distribution, the mean and variance were estimated by using the modeling function in Matlab (2021a), i.e., fitdist(x, 'tLocationScale'), to each multi-element cluster. Due to the limited number of elements (less than 30) in each multi-element cluster, the Student's t-distribution was used for the estimation rather than using a Gaussian distribution directly. The amplitude of estimated distribution was normalized and weighted by a multiplier factor denoted as λ_i. The factor was determined by the number of discrete elements in each cluster. Examples of generating an ambiguous cluster are shown in Figure 5 left.

4.2.4. Association and Accumulation

Each two ambiguous clusters were associated by $\Psi_{ij} = W_i \psi_i + W_j \psi_j$, where Ψ_{ij} is the association of each two ambiguous clusters; i and j were the indexes of the ambiguous clusters; $W_i = \lambda_i / (\lambda_i + \lambda_j)$, $W_j = 1 - W_i$, W_i, and W_j are the relative weights of two ambiguous clusters in the association. An example of association is shown in Figure 6. Finally, the function ϕ was estimated by accumulating all associations of each two ambiguous clusters, denoted as $\phi = \sum_1^M \Psi_{ij}$, where Ψ_{ij} is the association of each two ambiguous clusters, and $M = n(n-1)/2$ and n are the total numbers of ambiguous clusters.

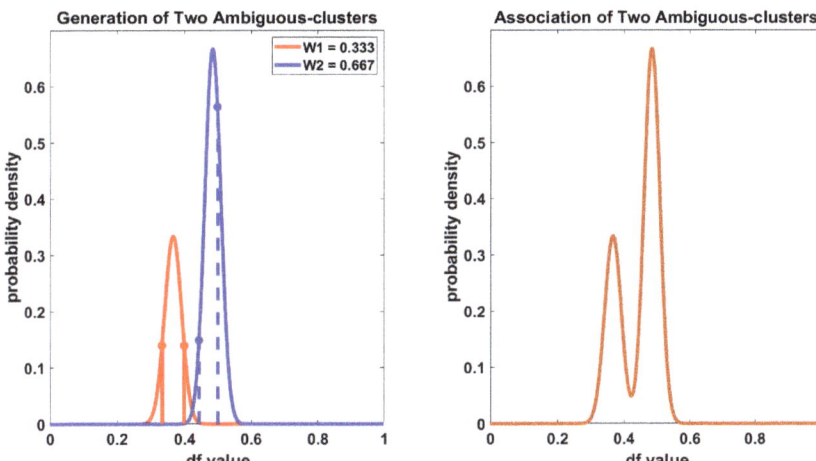

Figure 6. Example of generating two ambiguous clusters and operation of their association.

5. Results and Discussion

The application example and its data processing demonstrated the process of estimating the function ϕ based on the initial decisions made by a group of people. The result of estimation was called a "Decision Map", where all possible decisions, i.e., initial decisions and their variations, were characterized by a probability distribution, as presented in Figure 7. The amplitude of the probability distribution reveals the commonness among all possible decisions, i.e., a higher amplitude corresponds to more commonness of the decision. The meanings of each process are explained as following: (1) The process of the feature space mapping interpreted the sequence of each participant's decisions (the chosen number) into a df value in the feature space. The df value indicates how each participant decided to partition the current range to narrow down the searching area of the targeted number for the next attempt. Moreover, the df value is a comparable measurement among all decisions regardless of the dynamic range. (2) To find the common characteristics between those decisions made by each participant, their decision sequence was clustered and each cluster was considered as a decision strategy. For instance, in Figure 5a right, the df values were around 0.5, meaning the participant always used the strategy that partitioned the targeted range at around middle point in each attempt. However, each cluster was just a preliminary representation of a decision strategy that only contained limited decisions, but other possible variations of decision in this strategy might not be observed in the limited attempts during the application example. (3) To include those possible variations of decision into the decision strategies, ambiguous clusters were generated. The generation of an ambiguous cluster was carried out by modeling the uncertainty of initial clusters as a Gaussian distribution or Student's t-distribution. Note that the choices of modeling type depended on the attribute of initial data in the cluster and subjective preference that had impact on the decision-making process. In this study, the Gaussian distribution and Student's t-distribution were used for the generation of ambiguous clusters by assuming that possible decisions, i.e., the initial decision and its variations, were normally distributed within a strategy. By generation of an ambiguous cluster, each decision strategy was represented by a probability distribution, where the characteristics of possible decisions within the strategy was reflected by the probability density. Furthermore, the weighting of the normalized ambiguous cluster reflected the commonness of each decision strategy; in other words, the higher-weighted ambiguous cluster meant more inclusions of initial decisions, thereby, the more commonness among decisions was integrated into the relevant strategy. However, each ambiguous cluster only revealed the characteristics of possible decisions within the decision strategy. (4) To understand the characteristics of all possible decisions among the different decision strategies, first, each two ambiguous clusters were associated with their relative weights. As Figure 6 presents, the overlapping part shows the common decisions between two strategies, their amplitudes in the corresponding positions are increased after association. Then, after accumulating all associations of two ambiguous clusters, the commonness of decisions was displayed by amplitudes in the decision map, and the curve also indicated the characteristics of the decision making in this group of people. For example, in Figure 7, the most common decisions were around 0.5; according to this, we can conclude a characteristic of decision making that most participants in this group made decisions cautiously since they preferred to control the consequent risk of decisions (results in the searching ranges in the next attempt) equally.

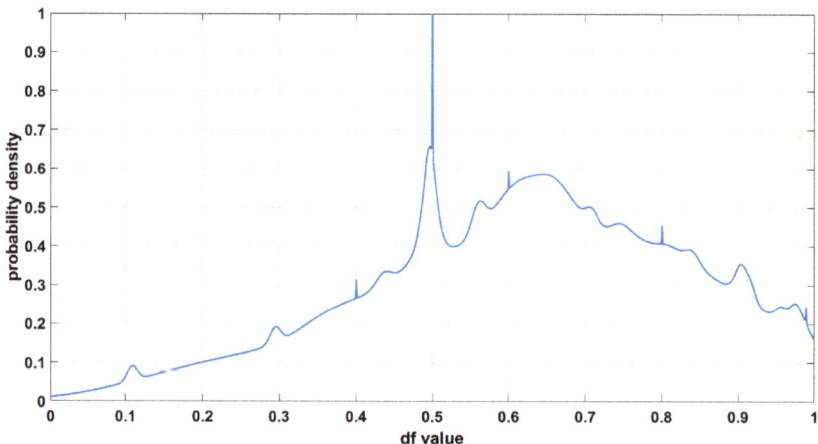

Figure 7. Decision map of 54 participants.

In addition, the decision map is adaptable to the different groups of people, and it can also indicate the tendency of common decisions as the number of people increases. To investigate this, we divided the dataset of all participants into sub-groups and generated decision maps based on the sub-groups. In each sub-group, participants and their decision data were randomly selected; the number of people increased from 4 to 52 by adding four people each time; and this process of division was repeated 30 times. Figure 8 shows several examples of decision maps based on different sub-groups of participants; each row presents decision maps of sub-groups with a different number of participants; each column presents decision maps of sub-groups with the same number of participants but different individuals. As we can see, the commonness of decisions in the decision maps varied with different numbers of people and different individuals in the sub-groups, but those variations tended to be limited as the number of people grew. Furthermore, the variation in the decision map was measured by $\alpha = \frac{RSS}{TPDM}$, where $TPDM$ stands for the sum of the total population's decision map (as in Figure 7), and RSS stands for the residual sum of the square between the sub-group's and total population's decision maps. The value of α indicates the percentage of difference from the sub-group's decision map to the TPDM. Figure 9 shows the variations in the decision map with different numbers of people. As shown in the figure, the α values decreased as the number of people grew, especially after 28 people, where an increased number of people made less of a difference on the decision map, which indicates that the commonness contributions of decisions were limited. Moreover, the error bars of the α value reduced, which indicates the random selection of participants had less of an effect on the decision map as the number of people grew.

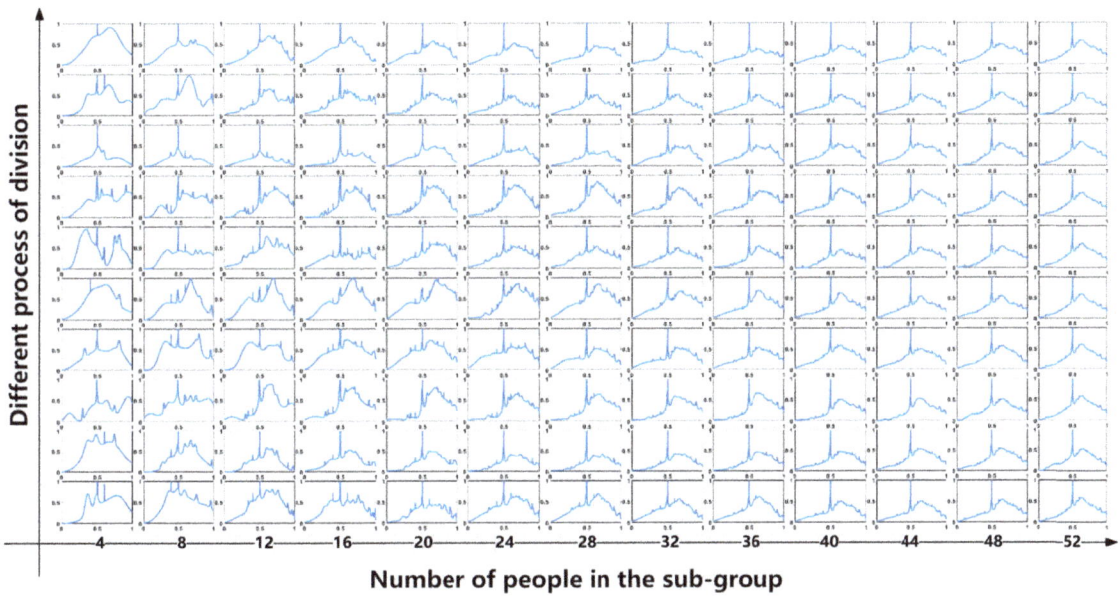

Figure 8. Ten examples of derived decision maps based on different sub-groups of participants.

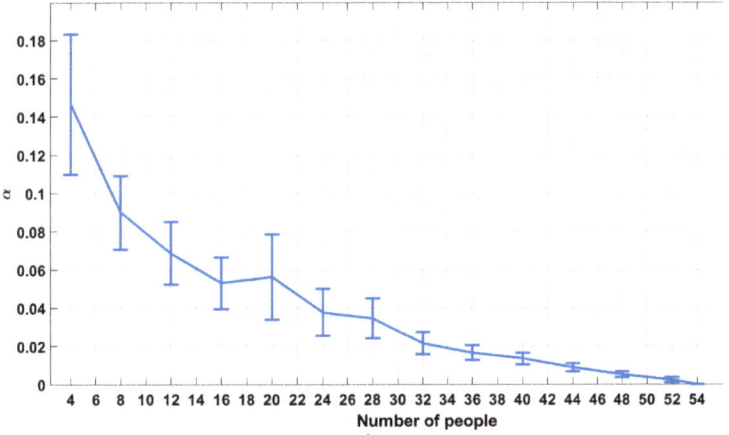

Figure 9. The variation in decision map with different number of participants.

Another angle of looking at a decision map is its application. Based on the decision map, a one-dimensional ontological structure of decisions (1-DOSE) was applied. Ontology is known as a method to study a concept regarding where and what entities exist and how such entities may be grouped and related to each other [46]. In the 1-DOSE, the concept was regarded as a group of people's common sense regarding dealing with a specific problem or task; their decisions to solve the problem were regarded as the entities of the concept; the decision strategy was regarded as a groups of entities; and the relationship between strategies was hierarchically connected by ranks. Table 1 shows an example of a 1-DOSE with 30 decision strategies. The rank was calculated in three steps: (1) each strategy's initial decisions were projected to the corresponding positions in the decision map; (2) the corresponding amplitudes at those positions were accumulated and the accumulated value was assigned to each decision strategy; (3) the ranks of decision strategies were identified based on the sorting of their accumulated values of amplitude in descending order. This rank revealed the hierarchical similarities of decision strategies since the decision strategy with a higher rank meant it had a higher accumulated value of amplitude and its initial decisions were shared by more participants in their decision-making process. Moreover, a 1-DOSE provides a way to quantitatively analyze the characteristics of the decisions made by a person regarding a specific problem or task. Table 2 shows an example of such an analysis between two participants. Each df value in their decision sequences was assigned with a rank number; the rank number corresponded to the rank of a relevant decision strategy in the 1-DOSE; the sum of ranks reflects the commonness of the participant's decision sequence, i.e., a lower sum of the ranks number means more common decisions were made by the participant in his or her decision sequence. In the analysis of Table 2, the participants' personalities were exposed somehow by how their decisions were close to the common strategy applied by other participants. By using "somehow", we mean that only one decision sequence was not good enough to evaluate the whole personality of the participant, but it could be considered as one piece of personality itself in a decision-making process. This provides a new perspective for current research of an autonomous decision-making system, where the absence of situation data in reconstructions or modeling used leads to poor or no decision making. By implementation of the 1-DOSE, the dependence between situation reconstruction and decision making can be reduced since humans have contributed to the hard part in the decision-making process, i.e., the process from the perception of the situation to the decision making. The machine only needs to learn how to apply one characteristic of making decisions to a specific problem or task, which can refer to the 1-DOSE. In addition, the 1-DOSE does not necessarily show the optimal decisions, but it always shows an overview of how decisions are made by people related to their characteristics. However, the decision map can be optimized by characterizing a group of experts' decisions regarding a given problem or task and then the 1-DOSE based on the experts' decision map can be implemented, where the optimal decisions are aligned with the most common decisions.

Table 1. 1-DOSE with 30 decision strategies (identified by cluster index).

Rank	Cluster Index	Rank	Cluster Index	Rank	Cluster Index
1	2	11	9	21	10
2	30	12	28	22	24
3	7	13	20	23	16
4	23	14	17	24	12
5	3	15	22	25	5
6	4	16	29	26	25
7	14	17	15	27	18
8	27	18	1	28	26
9	8	19	13	29	11
10	21	20	6	30	19

Table 2. The commonness analysis with two participants' decisions.

Participant 1			Participant 2		
Decision Sequence	D.S. Index	Rank	Decision Sequence	D.S. Index	Rank
0.5000	2	1	0.5000	2	1
0.4000	7	3	0.6000	49	57
0.7500	8	9	0.3333	38	91
0.6667	23	4	0.4000	38	91
0.5000	2	1	0.5000	2	1
0.6000	30	2	0.7500	60	45
0.8333	53	75	0.6667	49	57
0.8000	53	75	0.5000	2	1
0.5000	2	1	0.8000	60	45
0.5000	2	1	0.7500	60	45
Sum of ranks		172	Sum of ranks		434

6. Conclusions

In this paper, we propose a approach to acquire characteristics of decision making based on a group of people's decisions regarding a specific task. The idea of this method was inspired by IL, but with fundamental differences: (a) Instead of aiming to learn an optimal policy based on expert's demonstrations, we aimed to estimate the distribution of decisions of a group of people. (b) Instead of modeling the environment by an MDP, we used an ambiguity probability model to consider the uncertainty of each decision. (c) The participant trajectory was obtained by categorizing each decision of the participant to a certain cluster based on the commonness in the distribution of decisions. The feasibility of the approach was addressed by an application example, where a one-dimensional intelligence problem was designed for data collection. In the processes of applying example data, we used statistical methods to estimate and characterize all possible decisions that could be made by this group of participants. The characteristics of decision making was estimated and represented by a probability distribution called a decision map, where the curve of the amplitude indicates the commonness of decisions within the group of participants. We also discussed the variations in the decision map, which shows the decision map is adaptable to unveil the characteristic of decision making based on different groups of participants. In the end, we discussed a decision-map-based application, named 1-DOSE, which used a group of peoples' decision map as a reference to measure the commonness of an individual's decision trajectory. This application provides a new perspective to the research of learning and mimicking human decision making in autonomous systems. However, in this study, the lack of a comparison of the proposed approach with other existing approaches is a shortcoming, which is planned to be investigated in future research work.

Author Contributions: Conceptualization, Y.Z. and S.K.; methodology, Y.Z. and S.K.; software, Y.Z.; validation, Y.Z. and S.K.; formal analysis, Y.Z. and S.K.; investigation, Y.Z.; resources, Y.Z.; data curation, Y.Z.; writing—original draft preparation, Y.Z.; writing—review and editing, Y.Z. and S.K.; visualization, Y.Z.; supervision, S.K.; project administration, Y.Z.; funding acquisition, Y.Z. and S.K. All authors read and agreed to the published version of this manuscript.

Funding: This research received no external funding.

Institutional Review Board Statement: Not applicable.

Informed Consent Statement: Informed consent was obtained from all subjects involved in this study.

Data Availability Statement: Publicly available datasets were analyzed in this study. This data can be found here: https://www.dropbox.com/scl/fo/d54zqkoxo7zziq8stmq50/AEc_WiNrWh-97N1pNw4Oe5Q?rlkey=isj3yapigdykaqdfmvqlj10tt&st=8f9u70lu&dl=0, accessed on 12 May 2024.

Conflicts of Interest: The authors declare no conflicts of interest.

References

1. Daniel, K. *Thinking, Fast and Slow*; Macmillan: New York, NY, USA, 2011.
2. Mousavi, S.; Gigerenzer, G. Risk, uncertainty, and heuristics. *J. Bus. Res.* **2014**, *67*, 1671–1678. [CrossRef]
3. Gilovich, T.; Griffin, D.; Kahneman, D. *Heuristics and Biases: The Psychology of Intuitive Judgment*; Cambridge University Press: Cambridge, UK, 2002.
4. Bazerman, M.H.; Moore, D.A. *Judgment in Managerial Decision Making*; John Wiley & Sons: Hoboken, NJ, USA, 2012.
5. Chen, V.; Liao, Q.V.; Wortman Vaughan, J.; Bansal, G. Understanding the role of human intuition on reliance in human-AI decision-making with explanations. *Proc. ACM Hum. Comput. Interact.* **2023**, *7*, 1–32.
6. Zheng, B.; Verma, S.; Zhou, J.; Tsang, I.W.; Chen, F. Imitation Learning: Progress, Taxonomies and Challenges. *IEEE Trans. Neural Netw. Learn. Syst.* **2022**, *35*, 6322–6337. [CrossRef]
7. Edwards, A.; Sahni, H.; Schroecker, Y.; Isbell, C. Imitating latent policies from observation. In Proceedings of the International Conference on Machine Learning, Long Beach, CA, USA, 10–15 June 20109; PMLR: New York, NY, USA, 2019; pp. 1755–1763.
8. Nair, A.; Chen, D.; Agrawal, P.; Isola, P.; Abbeel, P.; Malik, J.; Levine, S. Combining self-supervised learning and imitation for vision-based rope manipulation. In Proceedings of the 2017 IEEE International Conference on Robotics and Automation (ICRA), Singapore, 29 May–3 June 2017; IEEE: New York, NY, USA, 2017; pp. 2146–2153.
9. Garg, D.; Chakraborty, S.; Cundy, C.; Song, J.; Ermon, S. Iq-learn: Inverse soft-q learning for imitation. *Adv. Neural Inf. Process. Syst.* **2021**, *34*, 4028–4039.

10. Ding, Z. Imitation learning. In *Deep Reinforcement Learning: Fundamentals, Research and Applications*; Springer: Berlin/Heidelberg, Germany, 2020; pp. 273–306.
11. Finn, C.; Levine, S.; Abbeel, P. Guided cost learning: Deep inverse optimal control via policy optimization. In Proceedings of the International Conference on Machine Learning, New York, NY, USA, 19–24 June 2016; PMLR: New York, NY, USA, 2016; pp. 49–58.
12. Ho, J.; Ermon, S. Generative adversarial imitation learning. *Adv. Neural Inf. Process. Syst.* **2016**, *29*.
13. Zhou, Y.; Khatibi, S. Mapping and Generating Adaptive Ontology of Decision Experiences. In Proceedings of the 3rd International Conference on Information Science and Systems, Cambridge, UK, 19–22 March 2020; pp. 138–143.
14. Kuehnhanss, C.R. The challenges of behavioural insights for effective policy design. *Policy Soc.* **2019**, *38*, 14–40. [CrossRef]
15. Leonard, T.C. Richard H. Thaler, Cass R. Sunstein, Nudge: Improving decisions about health, wealth, and happiness. *Const. Political Econ.* **2008**, *19*, 356. [CrossRef]
16. Kerr, N.L.; Tindale, R.S. Group performance and decision making. *Annu. Rev. Psychol.* **2004**, *55*, 623–655. [CrossRef]
17. Boix-Cots, D.; Pardo-Bosch, F.; Pujadas, P. A systematic review on multi-criteria group decision-making methods based on weights: Analysis and classification scheme. *Inf. Fusion* **2023**, *96*, 16–36. [CrossRef]
18. Tversky, A.; Kahneman, D. Judgment under Uncertainty: Heuristics and Biases: Biases in judgments reveal some heuristics of thinking under uncertainty. *Science* **1974**, *185*, 1124–1131. [CrossRef]
19. Tversky, A.; Kahneman, D. Advances in prospect theory: Cumulative representation of uncertainty. *J. Risk Uncertain.* **1992**, *5*, 297–323. [CrossRef]
20. Peterson, J.C.; Bourgin, D.D.; Agrawal, M.; Reichman, D.; Griffiths, T.L. Using large-scale experiments and machine learning to discover theories of human decision-making. *Science* **2021**, *372*, 1209–1214. [CrossRef]
21. Hardisty, D.J.; Thompson, K.F.; Krantz, D.H.; Weber, E.U. How to measure time preferences: An experimental comparison of three methods. *Judgm. Decis. Mak.* **2013**, *8*, 236–249. [CrossRef]
22. Tversky, A.; Shafir, E. Choice under conflict: The dynamics of deferred decision. *Psychol. Sci.* **1992**, *3*, 358–361. [CrossRef]
23. Zhu, Q.; Chen, Y.; Wang, H.; Zeng, Z.; Liu, H. A knowledge-enhanced framework for imitative transportation trajectory generation. In Proceedings of the 2022 IEEE International Conference on Data Mining (ICDM), Orlando, FL, USA, 30 November–3 December 2022; IEEE: New York, NY, USA, 2022; pp. 823–832.
24. Zhang, X.; Li, Y.; Zhou, X.; Zhang, Z.; Luo, J. Trajgail: Trajectory generative adversarial imitation learning for long-term decision analysis. In Proceedings of the 2020 IEEE International Conference on Data Mining (ICDM), Sorrento, Italy, 17–20 November 2020; IEEE: New York, NY, USA, 2020; pp. 801–810.
25. Abu-Nasser, B. Medical expert systems survey. *Int. J. Eng. Inf. Syst. (IJEAIS)* **2017**, *1*, 218–224.
26. Behzadian, M.; Otaghsara, S.K.; Yazdani, M.; Ignatius, J. A state-of the-art survey of TOPSIS applications. *Expert Syst. Appl.* **2012**, *39*, 13051–13069. [CrossRef]
27. Oh, J.; Yang, J.; Lee, S. Managing uncertainty to improve decision-making in NPD portfolio management with a fuzzy expert system. *Expert Syst. Appl.* **2012**, *39*, 9868–9885. [CrossRef]
28. Campbell, M. Knowledge discovery in deep blue. *Commun. ACM* **1999**, *42*, 65–67. [CrossRef]
29. Silver, D.; Hubert, T.; Schrittwieser, J.; Antonoglou, I.; Lai, M.; Guez, A.; Lanctot, M.; Sifre, L.; Kumaran, D.; Graepel, T. Mastering chess and shogi by self-play with a general reinforcement learning algorithm. *arXiv* **2017**, arXiv:1712.01815.
30. Petrović, V.M. Artificial Intelligence and Virtual Worlds—Toward Human-Level AI Agents. *IEEE Access* **2018**, *6*, 39976–39988. [CrossRef]
31. Pfeifer, R.; Bongard, J. *How the Body Shapes the Way We Think: A New View of Intelligence*; MIT Press: Cambridge, UK, 2007.
32. Lenat, D.B.; Guha, R.V.; Pittman, K.; Pratt, D.; Shepherd, M. Cyc: Toward programs with common sense. *Commun. ACM* **1990**, *33*, 30–49. [CrossRef]
33. Olive, J.; Christianson, C.; McCary, J. *Handbook of Natural Language Processing and Machine Translation: DARPA Global Autonomous Language Exploitation*; Springer Science & Business Media: Berlin/Heidelberg, Germany, 2011.
34. Harnad, S. The symbol grounding problem. *Phys. Nonlinear Phenom.* **1990**, *42*, 335–346. [CrossRef]
35. Shanahan, M. Perception as abduction: Turning sensor data into meaningful representation. *Cogn. Sci.* **2005**, *29*, 103–134. [CrossRef]
36. Dinur, A.R. Common and un-common sense in managerial decision making under task uncertainty. *Manag. Decis.* **2011**, *49*, 694–709. [CrossRef]
37. Garnelo, M.; Arulkumaran, K.; Shanahan, M. Towards deep symbolic reinforcement learning. *arXiv* **2016**, arXiv:1609.05518.
38. Leventhal, H.; Phillips, L.A.; Burns, E. The Common-Sense Model of Self-Regulation (CSM): A dynamic framework for understanding illness self-management. *J. Behav. Med.* **2016**, *39*, 935–946. [CrossRef]
39. Abbeel, P.; Ng, A.Y. Apprenticeship learning via inverse reinforcement learning. In Proceedings of the Twenty-First International Conference on MACHINE Learning, Banff, AB, Canada, 4–8 July 2004; p. 1.
40. Bain, M.; Sammut, C. A Framework for Behavioural Cloning. In Proceedings of the Machine Intelligence 15, Oxford, UK, 17 July 1995; pp. 103–129.
41. Ng, A.Y.; Russell, S. Algorithms for inverse reinforcement learning. In Proceedings of the ICML, San Francisco, CA, USA, 29 June–2 July 2000; Volume 1, p. 2.

42. Hussein, A.; Gaber, M.M.; Elyan, E.; Jayne, C. Imitation learning: A survey of learning methods. *ACM Comput. Surv. (CSUR)* **2017**, *50*, 1–35. [CrossRef]
43. Osa, T.; Pajarinen, J.; Neumann, G.; Bagnell, J.A.; Abbeel, P.; Peters, J. An algorithmic perspective on imitation learning. *Found. Trends Robot.* **2018**, *7*, 1–179. [CrossRef]
44. Pomerol, J.C.; Adam, F. Understanding Human Decision Making—A Fundamental Step Towards Effective Intelligent Decision Support. In *Intelligent Decision Making: An AI-Based Approach*; Springer: Berlin/Heidelberg, Germany, 2008; pp. 3–40.
45. Arora, P.; Varshney, S. Analysis of k-means and k-medoids algorithm for big data. *Procedia Comput. Sci.* **2016**, *78*, 507–512. [CrossRef]
46. The Gene Ontology Consortium; Aleksander, S.A.; Balhoff, J.; Carbon, S.; Cherry, J.M.; Drabkin, H.J.; Ebert, D.; Feuermann, M.; Gaudet, P.; Harris, N.L.; et al. The gene ontology knowledgebase in 2023. *Genetics* **2023**, *224*, iyad031.

Disclaimer/Publisher's Note: The statements, opinions and data contained in all publications are solely those of the individual author(s) and contributor(s) and not of MDPI and/or the editor(s). MDPI and/or the editor(s) disclaim responsibility for any injury to people or property resulting from any ideas, methods, instructions or products referred to in the content.

Article

Combining Balancing Dataset and SentenceTransformers to Improve Short Answer Grading Performance

Maresha Caroline Wijanto * and Hwan-Seung Yong

Department of Artificial Intelligence and Software, Ewha Womans University, Seoul 03760, Republic of Korea; hsyong@ewha.ac.kr
* Correspondence: mareshacw@ewhain.net

Abstract: Short-answer questions can encourage students to express their understanding. However, these answers can vary widely, leading to subjective assessments. Automatic short answer grading (ASAG) has become an important field of research. Recent studies have demonstrated a good performance using computationally expensive models. Additionally, available datasets are often unbalanced in terms of quantity. This research attempts to combine a simpler SentenceTransformers model with a balanced dataset, using prompt engineering in GPT to generate new sentences. Our recommended model also tries to fine-tune several hyperparameters to achieve optimal results. The research results show that the relatively small-sized all-distilroberta-v1 model can achieve a Pearson correlation value of 0.9586. The RMSE, F1-score, and accuracy score also provide better performances. This model is combined with the fine-tuning of hyperparameters, such as the use of gradient checkpointing, the split-size ratio for testing and training data, and the pre-processing steps. The best result is obtained when the new generated dataset from the GPT data augmentation is implemented. The newly generated dataset from GPT data augmentation achieves a cosine similarity score of 0.8 for the correct category. When applied to other datasets, our proposed method also shows an improved performance. Therefore, we conclude that a relatively small-sized model combined with the fine-tuning of the appropriate hyperparameters and a balanced dataset can provide performance results that surpass other models that require larger resources and are computationally expensive.

Keywords: short answer question; automatic short answer grading (ASAG); SentenceTransformers; dataset balancing; GPT; fine-tuning

Citation: Wijanto, M.C.; Yong, H.-S. Combining Balancing Dataset and SentenceTransformers to Improve Short Answer Grading Performance. *Appl. Sci.* **2024**, *14*, 4532. https://doi.org/10.3390/app14114532

Academic Editor: Andrea Prati

Received: 23 April 2024
Revised: 18 May 2024
Accepted: 23 May 2024
Published: 25 May 2024

Copyright: © 2024 by the authors. Licensee MDPI, Basel, Switzerland. This article is an open access article distributed under the terms and conditions of the Creative Commons Attribution (CC BY) license (https://creativecommons.org/licenses/by/4.0/).

1. Introduction

As a result of the COVID-19 pandemic that emerged at the end of 2019, E-learning or web-based distance learning platforms have become a viable alternative to facilitate the learning process [1]. Within this learning process, knowledge assessment plays a pivotal role in ensuring effective teaching [2]. Open-ended questions have been identified as a valuable method for determining students' level of knowledge and encouraging them to express their thoughts, perspectives, and experiences in their own words [1]. By inviting open-ended responses, teachers gain a more accurate and comprehensive insight into how students grasp domain-specific knowledge [3]. However, manually scoring these responses can introduce inconsistencies, as scoring may vary among markers or from one student to another [2]. Additionally, expecting a single definitive response to an open-ended question proves challenging for teachers, due to variations in students' vocabulary and writing structures [4]. This can lead to subjective judgements about the answers and compromise the objectivity of the assessment process [5].

According to Burrows et al. in Ref. [6], short answers have the following characteristics: the answer should not inferred just from the question's words (requiring external knowledge); the answer should be given in natural language; the length of the answer typically spans from one phrase to one paragraph; the content of the answer is relevant to the subject domain; and the answer is structured as closed-ended yet is not rigidly defined. However, some short-answer questions require students to express their subjective viewpoints within a defined context. Hence, short-answer questions also refer to semi-open-ended questions [7].

The grading system for short answers poses inherent challenges compared to automated multiple-choice grading systems. It is essential to thoroughly examine the nuances and variations in these answers to ensure accurate assessment [8]. The advancements in natural language processing (NLP) and machine learning applications have spurred interest among educators in creating exams comprising open-ended questions that can be automatically evaluated for a large number of students [5].

Automatic short answer grading (ASAG) is an emerging field of research, reflecting the educational sector's increasing adoption of technology to aid students and professionals. ASAG systems hold potential as valuable resources for educators, facilitating the enhanced integration of open-ended questions and providing more objective assessments for both formative and summative evaluations [9]. ASAG functions by analyzing students' answers in relation to a given question and the desired answer, as illustrated in Figure 1.

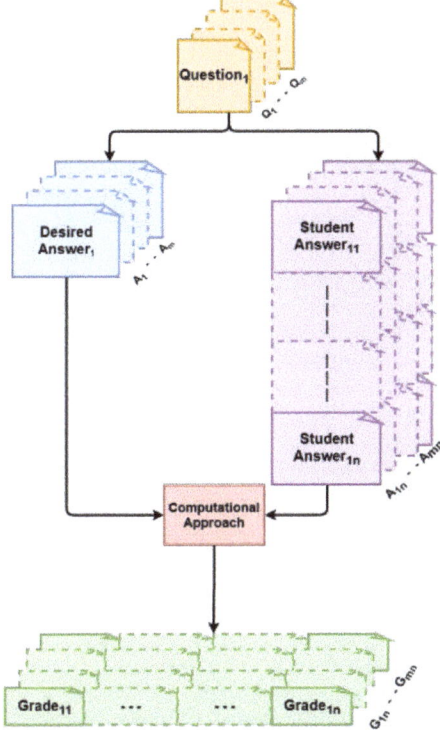

Figure 1. ASAG pipeline [10].

The recent advancements in natural language processing (NLP) and deep learning have introduced promising methodologies and frameworks capable of addressing various tasks. Across numerous NLP tasks, including ASAG, language models (LMs) have demonstrated considerable success. In modern approaches, LMs are trained using neural networks. Initial neural models were based on recurrent neural networks (RNNs), like long short-term memory networks (LSTMs and BiLSTm) [3]. The development of large language models like BERT (Bidirectional Encoder Representations from Transformers), based on the transformer architecture, and the increasing adoption of transfer learning have been instrumental in constructing custom ASAG systems [11].

Transformer employs self-attention for natural language processing, enabling the parallel computation of input and output vectors and addressing the sequential processing limitations of recurrent neural network (RNN), convolutional neural network (CNN), and long short-term memory (LSTM) approaches [12]. However, this self-attention mechanism can be computationally expensive. In the literature, ASAG studies often measure success by high correlations and a minimal loss value on standard benchmark tests using widely accessible datasets [5]. Many researchers indicate that the performance of ASAG systems is closely tied to the volume of training data available [13].

Reimers et al. [14], argued that, while the BERT and RoBERTa language models have a new state-of-the-art performance in sentence-pair regression tasks like semantic textual similarity, enabling the input of all the sentences into the network, the computational overhead is considerable. Though ASAG is crucial, implementing these expensive models may pose challenges. In this research, we explore some other Sentence-Bidirectional Encoder Representations (SBERT) models as mentioned in [15], and then propose a simpler model by fine-tuning certain hyperparameters to optimize the ASAG performance.

The available datasets exhibit an imbalance in the distribution of data across different labels. For instance, The SciEnts Bank (SEB) dataset [16] has a correct answer ratio of 39.9%, with label 4 representing the correct answer, while labels 0–3 are considered incorrect answers. And in contrast, the Mohler dataset from the University of North Texas predominantly contains correct answers, with approximately 78% coming from labels 4–5 only. As mentioned in Ref. [17], achieving a balanced dataset is crucial for developing optimal models, although this is challenging in practice. Augmentation, a method within oversampling, involves enhancing an existing dataset by adding supplementary data. Various methods used for augmenting data in NLP include random deletion, synonym replacement, random swap, and back translation [13,17]. In this research, we augment the dataset by utilizing GPT to paraphrase the answer, serving as an additional strategy of synonyms replacement. Despite GPT's inability to directly augment data, prompts can be utilized to generate synonyms and antonyms of words.

This paper is organized as follows: Section 2 reviews all works related to ASAG. Section 3 presents the proposed methods, including the datasets, evaluation metrics, and experiment setup used in this research. Section 4 presents a proof-of-concept implementation of the system and discussions of the experimental results. Section 5 concludes with all of the achievements from these experiments.

2. Related Works

This section starts by introducing the BERT network model, the baseline work in the area of ASAG, and describing how dataset augmentation influences the performance of ASAG.

In some various NLP tasks, BERT set a new state-of-the-art performance. BERT is a pre-trained transformer network [18]. Reimers and Gurevych recommended Sentence-BERT (SBERT), which uses Siamese BERT-Networks [14] to overcome some deficiencies in BERT. SBERT is a modification of the pre-trained BERT network that uses Siamese and triplet network structures to derive semantically meaningful sentence embeddings that can be compared using cosine similarity.

The SentenceTransformers framework provides various pre-trained models for NLP tasks. The model size and performance are different to each other. All these models have been evaluated for performance sentence embeddings and performance semantic search [14,19]. Table 1 shows some existing models related to sentence classification and question-answering.

Table 1. SentenceTransformers model size.

Model	Model Size
paraphrase-albert-small-v2 (AS)	43 MB
all-MiniLM-L6-v2 (M6)	80 MB
bert-base-uncased (BB)	110 MB
all-MiniLM-L12-v2 (M12)	120 MB
multi-qa-distilbert-cos-v1 (MD)	250 MB
all-distilroberta-v1 (DR)	290 MB
stsb-distilbert-base (SD)	330 MB
multi-qa-mpnet-base-dot-v1 (MM)	420 MB
quora-distilbert-base (QD)	500 MB
Sentence-T5-large (T5)	640 MB
all-roberta-large (RL)	1360 MB
Sentence-T5-XL (TX)	2370 MB

Several studies with varied models have achieved a good performance. Alreheli et al. proposed automatic short answer grading using paragraph vectors and transfer learning embedding [12]. In this work, they utilized the Texas dataset by Mohler to evaluate the models. The input for the ASAG model is the vector that represents the student answer (SA), along with the vector that represents the reference answer (RA). In each experiment, the vectors are inferred using two models; the paragraph vector (PV) model and the transfer learning model. Then, the similarity between SA and RA is measured using the cosine similarity. After that, the computed cosine similarity is used as a feature for a regression model to predict a particular answer score. They evaluate the models by comparing the actual score provided in the dataset, along with the predicted score using two evaluation metrics, the Pearson correlation coefficient and root mean square error (RMSE). For the PV vectors, it achieved 0.401 for the Pearson correlation and 0.893 for the RMSE. For the transfer learning, they applied the Roberta-large and Scibert models. The best accuracy achieved by fine-tuning the Roberta-large model on the domain-specific corpus was 0.620 for the Pearson correlation and 0.777 for the RMSE. This superiority is reasonable, because transformers can learn the context of the words from both directions. On the contrary, the pre-trained paragraph vectors perform better than the trained paragraph vectors on a domain-specific corpus. This indicates that paragraph vectors increase the model's generalizability.

The second approach for improving the performance of ASAG is to use transfer learning and augmentation described by the authors of [13]. They fine-tuned three-sentence transformer models on the SPRAG (Short Programming Related Answer Grading Dataset) corpus with five different augmentation techniques: viz., random deletion, synonym replacement, random swap, back translation, and NLPAug. The SPRAG corpus contains student responses involving keywords and special symbols. The dataset size is 4039 records, and it is a binary classification problem. They experimented with four different data sizes (25%, 50%, 75%, and 100%) with the augmented data to determine the impact of training data on the fine-tuned sentence transformer model. An SBERT architecture with a pretrained language model (PLM) was used for training. The experimentation used the stsb-distilbert-base, paraphrase-albertsmall-v2, and quora-distilbert-base pre-trained sentence transformer models. This paper provides an exhaustive analysis of fine-tuning pretrained sentence transformer models with varying sizes of data by applying text augmentation techniques. They found that applying random swap and synonym replacement techniques together while fine-tuning gave a significant improvement, with a 4.91% increase in accuracy (84.21%) and a 3.36% increase in the F1-score (88.11%).

The third approach, which achieved the best result so far, is integrating transformer-based embeddings and a BI-LSTM network [20]. The proposed model uses pretrained "transformer" models, specifically T5, in conjunction with a BI-LSTM architecture which is effective in processing sequential data by considering the past and future context. This research evaluated several pre-processing techniques and different hyperparameters to identify the most efficient architecture. Experiments were conducted using a standard benchmark dataset named the North Texas Dataset. This research achieved a state-of-the-art correlation value of 92.5%.

A recent study published in 2024 proposed paraphrase generation and supervised learning for improving ASAG performance [21]. First, they provided a sequence-to-sequence deep learning model that targets generating plausible paraphrased reference answers conditioned on the provided reference answer. Secondly, they proposed a supervised grading model based on sentence-embedding features. The grading model enriches features to improve accuracy, considering multiple reference answers. Experiments are conducted both in Arabic and English. They show that the paraphrase generator produces accurate paraphrases. Using multiple reference answers, the proposed grading model achieves a root mean square error of 0.6955, a Pearson correlation of 88.92% for the Arabic dataset, an RMSE of 0.779, and a Pearson correlation of 73.5% for the English dataset. While fine-tuning pre-trained transformers on the English dataset provided a state-of-the-art performance (RMSE: 0.762), our approach yields comparable results.

Data augmentation has become important in ASAG because more alternative answers can help accommodate the diversity of student answers. Howeve, generating these manually is difficult and needs significant effort. Some suggested methods used for augmenting data in NLPs include random deletion, synonym replacement, random swap, and back translation [13,17]. And in recent years, paraphrase generation become one of the effective strategies in data augmentation. Okur et al. in Ref. [22] used BART and GPT-2 as the paraphrasing model. With the development of GPT, we also think that GPT can be one of the good strategies to generate paraphrasing, especially with the existence of GPT-3.5 or GPT-4.

3. Materials and Methodology

As shown in Figure 2, our proposed method includes processing a dataset, training and fine-tuning a model, and evaluating the model. In this section, we will show the dataset, the pre-processing step for the data, the model implementation for this experiments the use of evaluation metrics, and the overall experimental setup.

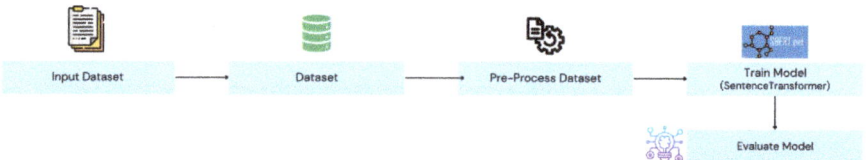

Figure 2. General architecture of proposed method.

For this experiment, we fine-tune the model by hyperparameter optimization. The details of this implementation will be discussed in the next section.

3.1. Dataset

The Mohler dataset comprises questions and answers in an introductory course in computer science provided by Texas University [23]. The goal of the dataset is to evaluate the model in grading the students' answers by comparing them with the evaluator's desired answer. It constitutes 2273 answers from 10 assignments and 2 examinations, collected from 31 students for 80 different questions.

Each answer in the assignment is graded from 0 (not correct) to 5 (totally correct) by two evaluators, who are specialized in the computer science major. The average of the two evaluators' scores is considered as the standard score of each answer. Each answer is graded from 0 to 5, in which grade 0 refers to (wrong), grade 5 refers to (correct), and grades 1 to 4 to partially correct answers. We used the average grade following the original research in this work. We show an example of the dataset in Table 2.

Table 2. Examples of questions and answers for the dataset.

Questions	Desired Answer	Student Answer	Score Avg
What is the role of a prototype program in problem solving?	To simulate the behavior of portions of the desired software product.	A prototype program simulates the behaviors of portions of the desired software product to allow for error checking.	4
		To simulate portions of the desired final product with a quick and easy program that does a small specific job. It is a way to help see what the problem is and how you may solve it in the final project.	5

Figure 3 shows the distribution of each grade label in the Mohler dataset. It can be seen that the grade label classification is not balanced, especially for grade label 0 and grade label 1. Since the amount of data affects the results, these kinds of data also make the performance less good.

Figure 3. Mohler dataset grade label distribution information.

Bonthu et al. [13] and Ouahrani et al.'s [21] research results show that data augmentation improves the ASAG performance, even if only slightly different. This is the basis for augmenting data so that the dataset become more balanced.

3.2. Data Pre-Processing

Before starting the analysis of the responses, we initially applied some pre-processing steps to remove irrelevant characters (e.g., numbers, punctuation) and turn the text into lowercase. After that, we have conducted only a tokenization. The same as Gaddipati et. al. in [10], we did not use any other checker. Since these transfer learning models are trained on a huge vocabulary, it is plausible to assume that they can understand the misspelled words to an extent. The versatility of transfer learning models to assign an embedding to the new words also assisted in disregarding the spelling mistakes. Other experiments also applied the removal of stopwords and lemmatization, to check whether the result is better or not.

3.3. Automatic Grading

Based on previous research, several strategies have been implemented to obtain a good ASAG performance. However, the best results from the research pf Gooma et al. [20] used the T5 model to achieve a correlation value of 92.5%. The T5 model has a large model size, as mentioned in Table 1, and is also computationally expensive. This research tries to find the right combination of models and hyperparameters to get better results with lower computational costs. Figure 4 depicts the recommended ASAG process.

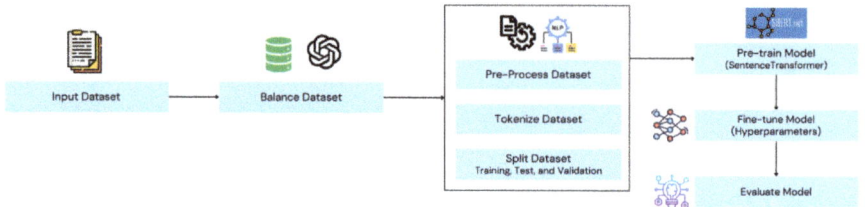

Figure 4. Modification of our proposed method.

This research will utilize eight SentenceTransformers models which have a relatively small size and fine-tune the models using some hyperparameters. Based on Table 1, we recommend several SentenceTransformers models, including paraphrase-albert-small-v2, all-MiniLM-L6-v2, bert-base-uncased, all-MiniLM-L12-v2, multi-qa-distilbert-cos-v1, all-distilroberta-v1, stsb-distilbert-base, and multi-qa-mpnet-base-dot-v1. All of these models are less than 500 MB in size.

In this study, we will fine-tune each model by exploring different combinations of hyperparameters. The parameters used include the size of the training–test data split, the number of epochs, learning rate, pre-processing steps, batch size, and the utilization of gradient checkpointing. Gradient checkpointing serves as a technique aimed at mitigating the memory requirements during deep neural network training, at the cost of having a small increase in computation time [24]. The system will be evaluated using various evaluation metrics, such as accuracy, F1-score, and Pearson correlation. All details about the experimental setup will be explained further in the next sub-section.

3.4. Dataset Balancing

To address the problem of data imbalance, we propose a data augmentation strategy utilizing GPT. The GPT methods used in this research are GPT-3.5 (model gpt-3.5-turbo-1106) and GPT-4 (model gpt-4). The method used is prompt engineering, using GPT to generate new sentences for each class. We implement the prompt engineering in GPT model based on the concept of each grade label-specific characteristics:

- Label 0: generate new opposite sentences of desired answer in dataset.
- Labels 1–4: generate new sentences by paraphrasing the existing student answer. The amount of data depends on the existing amount of data and the maximum amount of data in other labels.
- Label 5: generate new sentences by paraphrasing the existing desired answer.

By constructing appropriate prompts tailored to the paraphrasing task, we leverage the advanced natural language processing capabilities of GPT-3.5 and GPT-4 to generate linguistically diverse and context-specific rephrasings of student answers.

After generating new sentences, we check the quality of the new sentences using METEOR (Metric for Evaluation of Translation with Explicit Ordering) and cosine similarity. Figure 5 shows the augmentation process using GPT; the main idea is obtaining new sentences by paraphrasing the existing sentences.

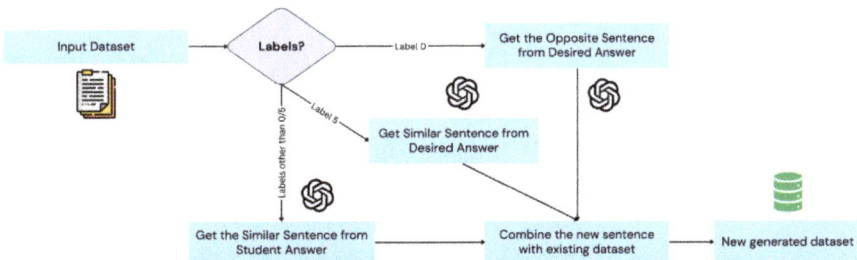

Figure 5. Proposed data augmentation process.

3.5. Evaluation Metrics

All processes need to be evaluated, both the new generated datasets and the short answer grading systems. The evaluation metrics that are commonly used to measure system performance can be seen in the following points.

1. Accuracy

Proportion of correctly graded answers. Accuracy is a straightforward metric; it is essential to consider its limitations, especially in scenarios with imbalanced datasets. In imbalanced datasets, where one class significantly outweighs the other, accuracy can be misleading. To address the limitations of accuracy, we use other evaluation metrics such as precision, recall, and F1-score [25].

2. F1-Score

The F1-score is a weighted comparison of average precision and recall. The formula for the F1-score is as follows:

$$F1 = \frac{2 \times Precision \times Recall}{Precision + Recall}, \quad (1)$$

The F1-score pays attention to the model's ability to handle imbalanced data classes. In addition, by using the F1-score as one of the evaluation metrics, we can compare the resulting model with other published models.

3. Pearson Correlation Score

Pearson's correlation is the most commonly used method in statistics to evaluate the strength and presence of a linear relationship between predicted and manual grades. Pearson's correlation coefficient is the covariance of the two variables divided by the product of their standard deviations [20].

4. Root Mean Square Error (RMSE)

Root mean square error or root mean square deviation is one of the most commonly used measures for evaluating the quality of predictions. It shows how far predictions fall from the measured true values using Euclidean distance. The use of RMSE is very common, and it is considered an excellent general-purpose error metric for numerical predictions. RMSE is calculated as follows:

$$RMSE = \sqrt{MSE}, \quad (2)$$

Meanwhile, mean square error (MSE) measures the square of differences between predictions and target values and computes the mean of them. MSE is calculated as follows:

$$MSE = \frac{1}{N} \sum_{i=1}^{N} (y_i - \hat{y}_i)^2 \quad (3)$$

5. METEOR

METEOR stands for Metric for Evaluation of Translation with Explicit Ordering; it is known for its higher correlation with human judgment, especially at the sentence level. These metrics always take a value between 0 and 1. This value indicates how similar the predicted text is to the reference texts, with values closer to 1 representing more-similar texts. METEOR's ability to measure the quality of the generated answer is based on unigram precision and recall. It significantly improves the correlation with human judgments. METEOR computes the similarity score of two texts by using a combination of unigram precision, unigram recall, and some additional measures like stemming and synonymy matching [21].

The cosine similarity technique is used for measuring the similarity between two vectors. The way it works is by measuring the cosine of the angle between two documents which are expressed in vectors. Haskova et al. stated that the angle between vectors determines whether they are pointing in the same or different directions. If vectors are pointing in the same direction, it means that the documents are similar; the closer they are expressed on the axis, the more similar they are. Vice versa, the farther they are expressed on the axis, the less similar they are [26].

3.6. Experimental Setup

As we ran many SBERT models and it required a more powerful graphical processor unit (GPU), we used Google Colab's T4 GPU with a high RAM (around 52 GB). Our study was based on the Hugging Face model based on Transformers for PyTorch 1.11.0 and TensorFlow 2.0. We also employed the OpenAI API, leveraging prompt-engineering techniques with GPT-3.5 and GPT-4, to paraphrase sentences for the data augmentation process.

Initially, the general scenario that was formed was to compare the implementation of each model by fine-tuning the various hyperparameters mentioned previously. We tried with epoch values of 8, 10, 12, or 16, then batch sizes of 8, 16, or 32, and learning rate values of 5×10^{-5} or 5×10^{-6}. We experimented with various scenarios that were combinations of these hyperparameters with the original dataset and the models mentioned. Based on these initial experiments, the best result was obtained from using epoch = 12, batch size = 16, and learning rate = 5×10^{-5}. So, the next experiment would use these three fixed parameters and the other parameters combined. These included whether or not to apply removing stopwords, whether to apply gradient checkpoints or not, and dataset-splitting sizes.

The next scenario was related to the data augmentation process. Initially, we used the GPT-3.5 model with 2 different temperature values. "Temperature" refers to a parameter that influences the randomness degree of the generated text. A low temperature (close to 0) leads to more deterministic outputs, where the model tends to choose words with higher probabilities, resulting in a more conservative and repetitive text. A high temperature (with a maximum value = 2) increases randomness in the generated text, causing the model to sample from less predictable words, resulting in a more diverse but potentially less coherent text. The results for these new generated sentences were evaluated using METEOR and cosine similarity, giving better results for temperature = 0.7. Next, with the parameter value temperature = 0.7, we also tried generating a new sentence with GPT-4. So, we will compare the results of generated new sentences between GPT-3.5 and GPT-4.

In the next scenario, we will compare the performance of the grading system using the dataset before augmentation and after the augmentation process. In addition, we also have a scenario for conducting experiments using larger SentenceTransformers models. Then the results will be compared between our recommended model and the larger model. The evaluation metrics used are as mentioned in the previous section. We will also check the running time of each model.

4. Result and Discussion

In this session, we will explain and discuss the experiment results based on the experimental setup described previously.

4.1. Initial Answer-Grading Process

As mentioned before, we will conduct experiments using the original dataset and some fixed parameters include epoch = 12, batch size = 16, and learning rate = 5×10^{-5}. The results shown in Table 3 include the best combination of hyperparameters from the overall results. We split the dataset into an 80–20 ratio or 70–30 ratio for training and testing data, respectively. The pre-processing step used for this experiment only removed special characters and changed the sentences into lowercase.

Table 3. Model performance for original dataset.

Model Name	Gradient Checkpointing	Split Size	RMSE	F1-Score	Accuracy	Pearson Correlation
bert-base-uncased			1.61902	0.19927	0.22287	0.21246
stsb-distilbert-base			1.04117	0.65803	0.66862	0.63649
all-distilroberta-v1			1.06047	0.66809	0.64516	0.66427
all-MiniLM-L6-v2	Yes	70–30	0.99158	0.60616	0.64367	0.64367
multi-qa-mpnet-base-dot-v1			0.89391	0.70164	0.70614	0.45196
multi-qa-distilbert-cos-v1			1.08879	0.63096	0.64474	0.47758
paraphrase-albert-small-v2			1.31466	0.58836	0.58333	0.43599
all-MiniLM-L12-v2			0.99299	0.63995	0.65351	0.43184
bert-base-uncased			1.91571	0.42417	0.40351	0.25566
stsb-distilbert-base			1.31803	0.67083	0.67105	0.59558
all-distilroberta-v1			1.24106	0.73221	0.72807	0.65142
all-MiniLM-L6-v2	No	80–20	1.19878	0.58192	0.56579	0.64576
multi-qa-mpnet-base-dot-v1			1.22907	0.68833	0.68859	0.76269
multi-qa-distilbert-cos-v1			1.4047	0.67267	0.67105	0.71517
paraphrase-albert-small-v2			1.79938	0.50837	0.49123	0.62883
all-MiniLM-L12-v2			1.15167	0.63541	0.64474	0.71271

Based on the results in Table 3, the all-distilroberta-v1 and mul-ti-qa-mpnet-base-dot-v1 models show promising results, although their performance across all evaluation metrics is still not better than the existing research. The best RMSE value obtained is nearly 0.9, whereas even the smallest value in the previous research reached 0.77. Moreover, the F1-score, accuracy and Pearson correlation values are only around 0.7, while the previous research has achieved more than that. So, we conducted additional experiments using a new dataset that has undergone a data augmentation process with GPT.

4.2. Dataset Balancing by Data Augmentation with GPT

Based on the previously explained scenario, the process of balancing the dataset with data augmentation will use a parameter value of temperature = 0.7 and will also utilize both the GPT-3.5 and GPT-4 models. For grade label = 0, the prompt text used will be "Please make a completely different sentence from this following sentence: '{answer}' so it counts as an opposite sentence" to get the opposite or antonyms. Meanwhile, for other grade labels, the prompt text used will be "Please paraphrase the following sentence '{answer}'" to get similar sentences or synonyms. Figure 6 shows the new dataset, with additional data generated by the GPT-3.5 model. Subsequently, the results of this process will be referred to as BalMohler-3.5, which means Balanced Mohler dataset with GPT-3.5.

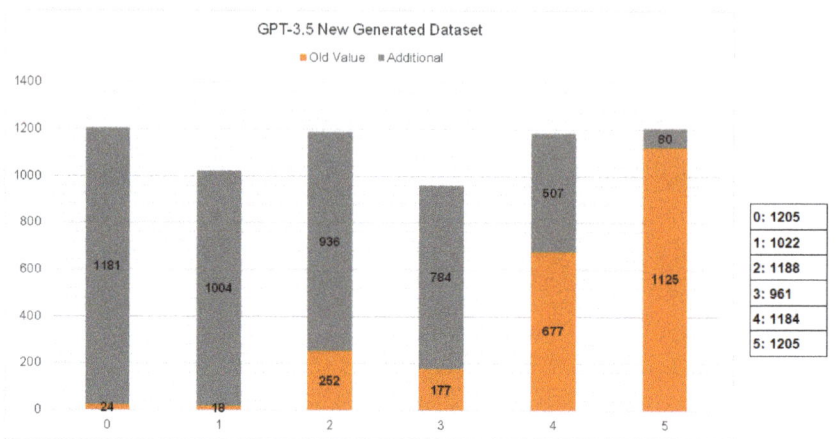

Figure 6. New generated dataset using GPT-3.5.

Figure 7 depicts the new dataset with added data resulting from data augmentation using GPT-4. Compared to Figure 3, this new dataset has more evenly distributed numbers. The standard deviation for the original Mohler dataset is 400, while the standard deviation for the new dataset is around 98 for both models. This dataset will be referred to as BalMohler-4.

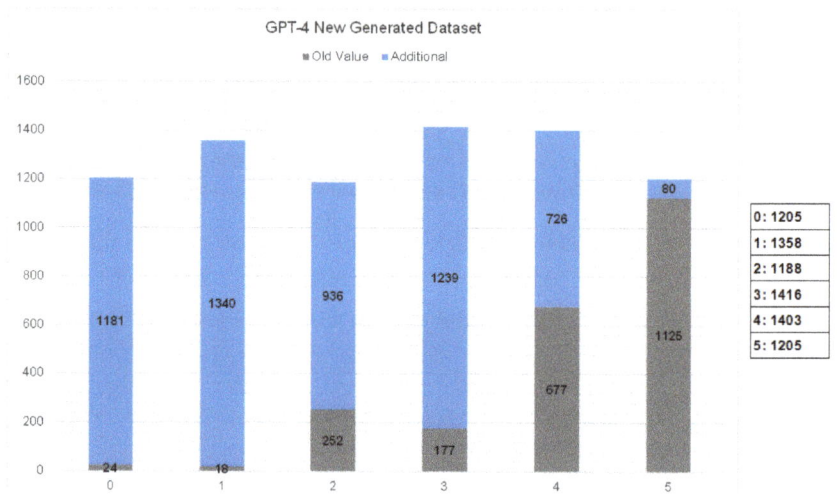

Figure 7. New generated dataset using GPT-4.

In addition to the distribution of data, we will also evaluate the results of these new generated sentences based on METEOR and cosine similarity scores. To facilitate evaluation, we will categorize grade labels into two categories: the correct category for grade labels 2–5 and the false category for grade labels 0–1. The grade labels 0–1 are considered as the false category because they represent answers that are far from the correct answer. Table 4 shows the evaluation results of data augmentation from both models.

Table 4. Data augmentation result evaluation.

Augmentation Model	Grade Category	Cosine Similarity	METEOR
GPT-3.5	False	0.4369	0.59722
GPT-3.5	Correct	0.8040	
GPT-4	False	0.3645	0.59309
GPT-4	Correct	0.8080	

These new generated sentences will be considered as student answers and will be compared with the desired answers for evaluation. So, a smaller value in the cosine similarity score indicates better results for the false category. Meanwhile, for the correct category, a larger value indicates better performance. The new sentences of the GPT-4 model are better, although there is only a small increase. Meanwhile, the METEOR score reflects the overall text quality, where a higher score indicates better quality. As seen in Table 4, the results of the new generated sentences with the GPT-3.5 model show a better performance, although the difference is not significant compared to the results from the GPT-4 model. Therefore, for future research, we will continue to evaluate the ASAG process using both of these new generated datasets.

4.3. Answer-Grading Process after Balancing Dataset

These experiments will use a new dataset with the additional data through data augmentation with the GPT-3.5 and GPT-4 models. We will conduct various experiments with existing model combinations, while fine-tuning the hyperparameters. From the several experiments that have been carried out, we will present the best results among them. The details of the scenarios conducted can be seen in Table 5. Exp 1 and Exp 5 represent the best scenarios for experiments using the original Mohler dataset, for which the results can be seen in Table 3. Removing stopwords is not included in the parameter combination because the experimental results are not significant, so it was not involved.

Table 5. Detail experiment scenario.

Experiment	Pre-Processing	Gradient Checkpointing	Split Size
Exp 1	SpCh LC	Yes	70–30
Exp 2	SpCh LC	No	70–30
Exp 3	SpCh LC	Yes	80–20
Exp 4	SpCh LC RS	Yes	80–20
Exp 5	SpCh LC	No	80–20

Each scenario is executed with a fixed parameters set based on the results of previous experiments. The results of this experiment will be evaluated using the evaluation metrics mentioned earlier, namely RMSE, F1-score, accuracy, and Pearson correlation. These experiments include the SentenceTranformers model mentioned in Table 1.

Figure 8 displays the RMSE values from our eight recommended models when implemented on three datasets: Original Mohler, BalMohler-3.5, and BalMohler-4, represented as Ori, Aug-3.5, and Aug-4, respectively, for each subsequent figure. Generally, better results are obtained from using BalMohler-3.5 and implementing the all-distilroberta-v1 model. To shorten the model names, what is shown in the figure is the abbreviation for each model based on Table 1.

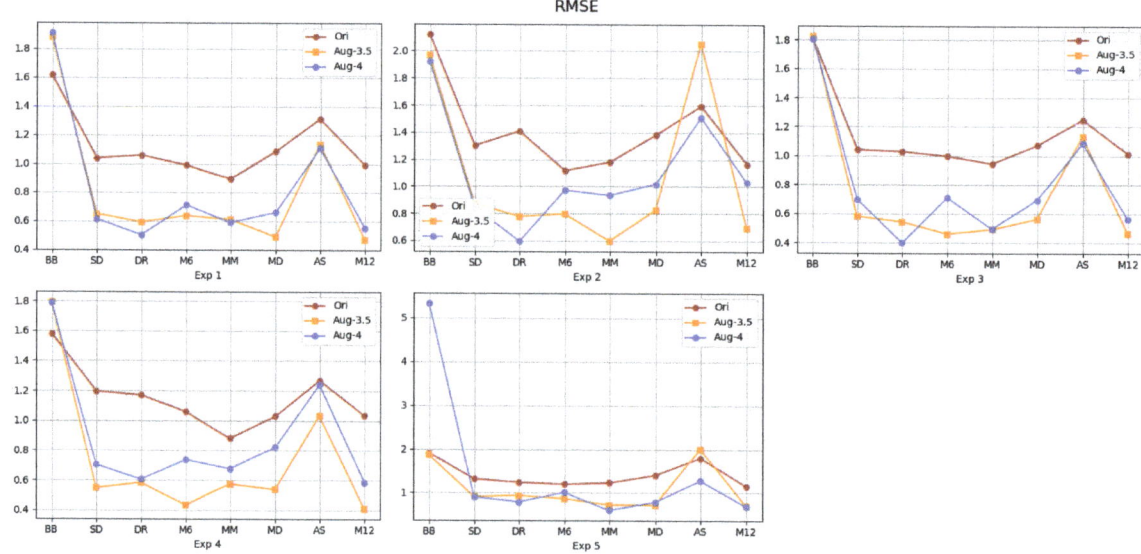

Figure 8. Comparing RMSE for all models and dataset.

The best RMSE score of 0.39913 was obtained from the implementation of Exp 3, using the all-distilroberta-v1 model with a new balanced dataset from data augmentation with the GPT-4 model. However, in general, using BalMohler-3.5 results in a smaller average RMSE score. Meanwhile, the worst RMSE scores were obtained from implementing Exp 5 on the bert-base-uncased model with BalMohler-4, with a value of 5.33808.

Figure 9 displays the experimental results based on the F1-score score obtained. Overall, balancing the dataset by implementing data augmentation improves the performance of the grading system.

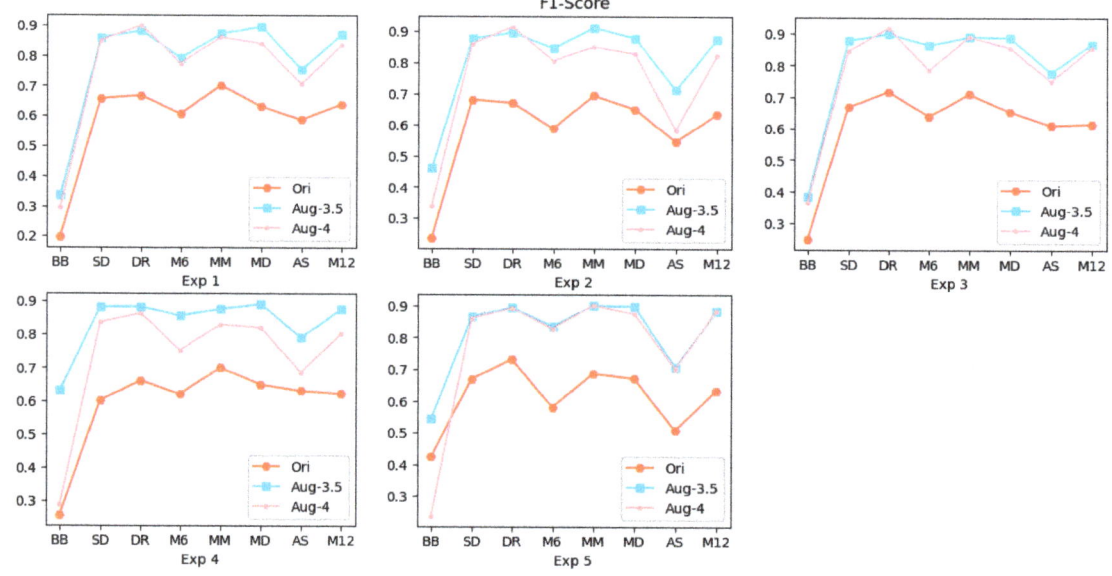

Figure 9. Comparing F1-score for all models and datasets.

The best F1-score result of 0.91886 was obtained from Exp 3 with BalMohler-4 and the implementation of the all-distilroberta-v1 model. Similar to the RMSE results, on average, the F1-score results from using BalMohler-3.5 are better than BalMohler-4. The worst result, 0.2391, comes from Exp 5, using the BalMohler-4 dataset and the bert-base-uncased model. This worst value is close to the worst F1-score value with the Original Mohler dataset (0.2379).

Figure 10 displays the performance evaluation in terms of accuracy. Similar to the other evaluation metrics, the resulting balanced dataset from implementing GPT data augmentation generally improves the performance of the grading system.

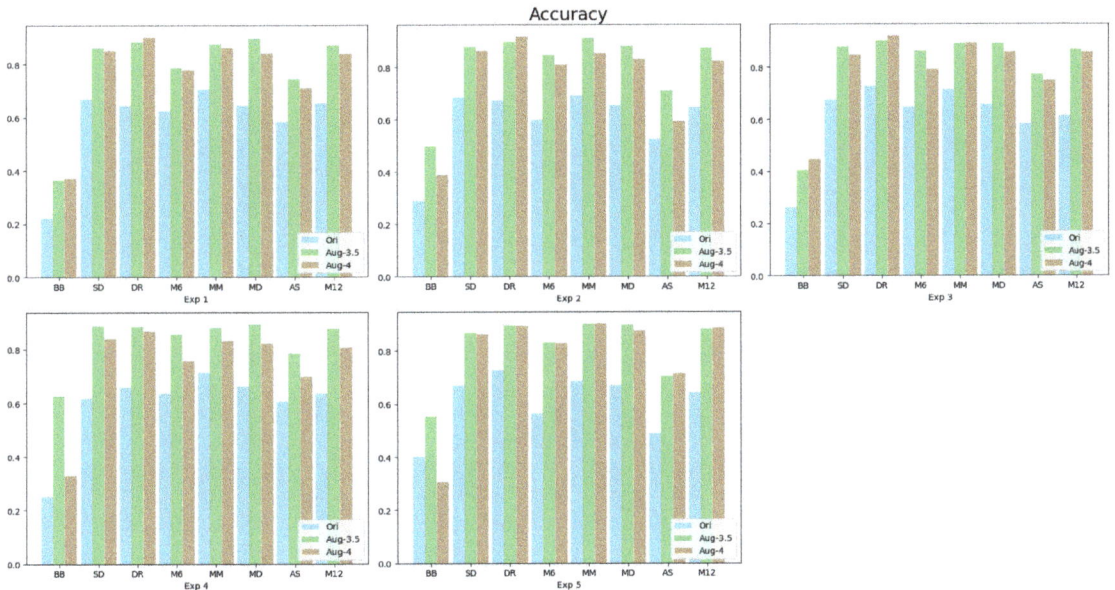

Figure 10. Comparing accuracy for all models and dataset.

The best accuracy result of 0.91969 was obtained from Exp 3 with the implementation of the BalMohler-4 dataset and the all-distilroberta-v1 model. On average, the accuracy value of BalMohler-3.5 implementation is also better than the implementation of BalMohler-4. The lowest accuracy score, 0.30488, comes from Exp 5, with the implementation of the BalMohler-4 dataset and the bert-base-uncased model.

Figure 11 displays the final evaluation results based on the Pearson correlation value. It is also clear that the balanced dataset from the implementation of GPT data augmentation improves performance results.

Slightly different from the previous evaluation, the best result was obtained from Exp 3 and the all-distilroberta-v1 model but with the implementation of BalMohler-3.5. The highest Pearson correlation score is 0.95855. The lowest score, 0.28748, was obtained from Exp 4 with the implementation of the BalMohler-4 dataset and the bert-base-uncased model.

Based on the experimental results so far, on average, the best implemented model is the all-distilroberta-v1 model, along with the use of BalMohler-3.5. The all-distilroberta-v1 model has a size of around 290 MB and can achieve good results for all the evaluation metrics conducted. As mentioned in Section 2, there have been other studies that have succeeded in achieving satisfactory evaluation metrics values, but those studies typically use larger models than those we recommend. For instance, Alreheli and Alghamdi used the all-roberta-large model [12], which is nearly 400% larger than our recommended model, and Gomaa used T5-XL [20] which is almost eight times larger than our recommended model.

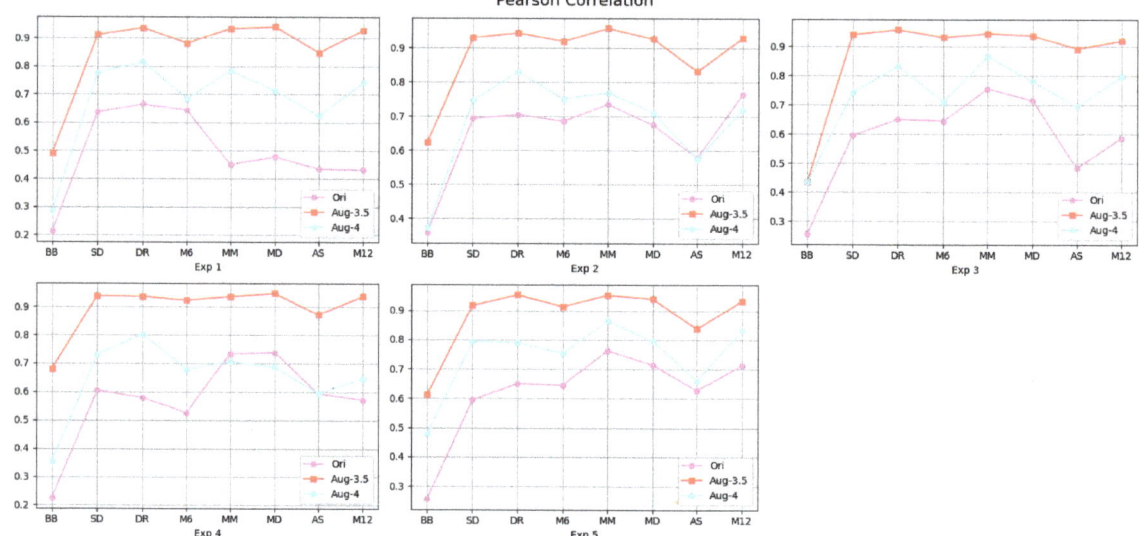

Figure 11. Comparing Pearson correlation for all models and datasets.

We also conducted additional experiments using larger models to observe their performance. Table 6 displays our overall best experimental results, as well as the results from previous research. Based on this summary, it can also be seen that data augmentation implementation to balance the dataset helps improve the performance for smaller-sized SentenceTransformers models. A smaller size means a smaller number of parameters and also results in a faster processing time [14,27].

Table 6. Comparison of all experimental results with previous research.

Model Name	Dataset	Pre-Processing	Gradient Checkpointing	Split Size	Batch Size	RMSE	F1-Score	Accuracy	Elation
all-distilroberta-v1 multi-qa-mpnet-base-dot-v1	Mohler	SpCh + LC	No	80–20 +	16	1.2411 1.2291	0.7322 0.6883	0.7281 0.6886	0.6514 0.7627
all-distilroberta-v1 multi-qa-mpnet-base-dot-v1 multi-qa-mpnet-base-dot-v1	BalMohler-3.5	SpCh + LC	Yes No	80–20 # 70–30 *	16	0.5449 0.4925 0.6012	0.8998 0.8907 0.9139	0.9009 0.8912 0.9145	0.9586 0.9454 0.9583
all-distilroberta-v1 multi-qa-mpnet-base-dot-v1	BalMohler-4	SpCh + LC	Yes	80–20 #	16	0.3991 0.4954	0.9189 0.8917	0.9197 0.8929	0.8329 0.8657
RoBERTa-large	BalMohler-4 BalMohler-3.5	SpCh + LC	Yes	80–20 #	16	0.4021 0.5096	0.9374 0.9328	0.9377 0.9334	0.9612 0.9579
RoBERTa-large [12]	Mohler	SpCh + RS	-	80–20	16	0.777	-	-	0.620
T5-XL [20]	Mohler	LC	-	80–20	3	0.109	-	-	0.928
paraphrase-albert-small-v2 [13]	SPRAG + Aug	SpCh	-	85–15	16	-	0.8811	0.8421	-
Ridge-LR [21]	Mohler + Aug-E	SpCh + RS	-	70–30	-	0.779	-	-	0.735
Ridge-LR [21]	Mohler + Aug-A	SpCh + RS	-	70–30	-	0.6955	-	-	0.8892

* Exp 2, # Exp 3, + Exp 5.

Our best results from the experiment are marked with green in the respective column. Results labeled in blue indicate additional experiments using larger models. The results obtained for F1-score, accuracy, and Pearson correlation are indeed better than those from experiments with smaller models, but the difference in performance is not significant compared to the average running time. In Figure 12 below, it can be seen that the average running time of the all-roberta-large model is many times longer than the other models. Given the disparity in performance, it is not proportional to the computational cost required. Balancing the dataset is also recommended in research by Bonthu et al. [13] and Ouahrani et al. [21]. There is an increase in performance when using the augmented data, which is also consistent with our experimental results.

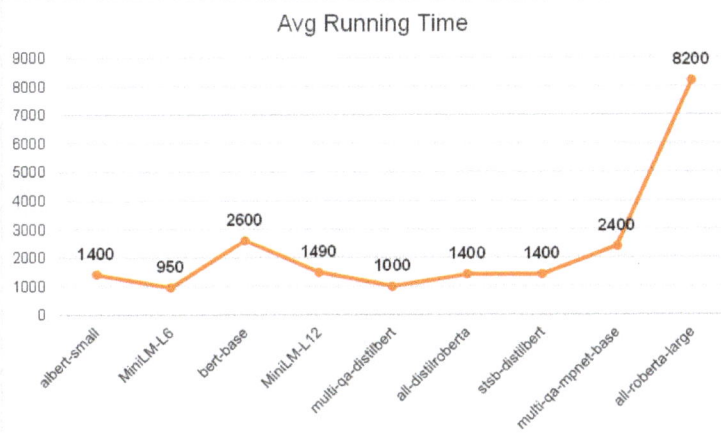

Figure 12. Average running time of each experiment.

The RMSE scores obtained from our experiments are indeed larger than those from Gomaa's research [20], but that does not have much impact because of other better evaluation metrics values. Moreover, the use of the T5-XL model would definitely require even larger and more expensive resources. The comparative F1-score and accuracy values were only obtained from research by Bonthu et al. [13], and our best experimental results on average also exceed those results. Bonthu et al. used the relatively small paraphrase-albert-small-v2 model, which is only about 40 MB in size. However, our recommended model can potentially perform better, due to differences in the datasets. Bonthu et al. used the SPRAG dataset, which is a binary classification problem [13], while the Mohler dataset consists of grade labels ranging from 0 to 5. With a more complex dataset, even though it requires a larger model, the results can still compete with simpler models. Note also that the paraphrase-albert-small-v2 model has an average running time that is not significantly different from the all-distilroberta-v1 model, which is about seven times larger in size. The Pearson correlation scores from our recommended model can also achieve better results than the existing research. When the same dataset and the same hyperparameter fine-tuning process are implemented on larger models, it indeed produces better results, but this performance improvement is not proportional to the larger and more expensive computational cost.

This research aims to find a simpler model with a proper fine-tuning process. The experiments have shown that relatively smaller-sized models with the proper fine-tuning can achieve a good performance. The data augmentation for balancing the dataset itself also contributes to a significant improvement in the performance results of this grading system. The all-distilroberta-v1 model, which is less than 300 MB in size, with the proper hyperparameter selection and combined with balancing the dataset, can compete with the results of larger and more complex models.

4.4. Additional Experiments

We conducted additional experiments to determine whether our proposed method also improves the performance of the ASAG system on other datasets. We utilized the SemEval-2013 dataset, a benchmark dataset from the SemEval-2013 Shared Task 7 [28]. Specifically, we used the two-way SciEnts Bank subset, which includes two grade labels: "correct" as grade label 1 and "incorrect" as grade label 0. This dataset consists of questions, desired answers, student answers, and two-way grade labels in the science domain.

We used both the original dataset and an augmented version. The initial dataset contained 4925 rows, with 2944 rows (60%) labeled as grade 0 and 1981 rows (40%) labeled as grade 1. We applied the same augmentation process to this dataset. Using prompt engineering in GPT models, we generated additional answers for grade label 0 by using antonyms of the desired answer and synonyms of words from the student answers to create new answers for grade label 1. Figure 13 illustrates the data distribution for both the original and the balanced datasets after augmentation with the GPT-3.5 and GPT-4.0 models. The blue bar represents data with a grade label of 0, while the orange bar represents data with a grade label of 1.

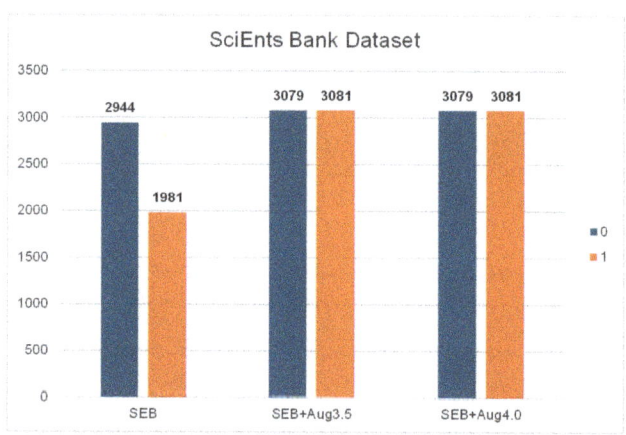

Figure 13. SciEnts Bank datasets data distribution.

We also evaluated the newly generated dataset using the METEOR score and cosine similarity score. The evaluation results are presented in Table 7, which shows that the dataset generated using the GPT-3.5 model performed better. These datasets were also implemented using the recommended models from the experiments in the previous section: the all-distilroberta-v1 and multi-qa-mpnet-base-dot-v1 models.

Table 7. SciEnts Bank data augmentation results evaluation.

Augmentation Model	Grade Category	Cosine Similarity	METEOR
GPT-3.5	False	0.4411	
GPT-3.5	Correct	0.7121	0.5921
GPT-4	False	0.4429	
GPT-4	Correct	0.7059	0.5498

From the five scenarios mentioned in Table 5, we only present the best two results for this additional experiment, using the same combination of hyperparameters. The results are displayed in Table 8.

Our best results from these additional experiments are also highlighted in green in the respective column. These results indicate that the augmentation process successfully increased system performance. Moreover, the results labeled in blue indicate experiments using larger models. The larger models achieved better scores in F1-score, accuracy, and Pearson correlation but, as with the previous dataset, the performance improvement was not significant compared to the average running time. When compared with previous research using the same dataset, our proposed method also improves performance quite well.

Table 8. Comparison of additional experimental results with previous research.

Model Name	Dataset	Pre-Processing	Gradient Checkpointing	Split Size	Batch Size	RMSE	F1-Score	Accuracy	Pearson Correlation	Average Runtime
all-distilroberta-v1	SEB SEB + Aug3.5 SEB + Aug4	SpCh + LC	Yes	80–20 #	16	0.8857 0.8205 0.7429	0.8186 0.8556 0.8375	0.8236 0.8556 0.8377	0.6399 0.7103 0.6738	1066 s 1406 s 1390 s
multi-qa-mpnet-base-dot-v1	SEB SEB + Aug3.5 SEB + Aug4	SpCh + LC	Yes	80–20 #	16	0.7818 0.7329 0.7956	0.8301 0.8507 0.8392	0.8296 0.8507 0.8393	0.6559 0.6999 0.6765	1806 s 2285 s 2285 s
RoBERTa-large	SEB SEB + Aug3.5	SpCh + LC	Yes	80–20 #	16	0.8648 1.0093	0.8722 0.8657	0.8722 0.8658	0.7411 0.7319	6381 s 6769 s
all-distilroberta-v1	SEB SEB + Aug3.5 SEB + Aug4	SpCh + LC	No	80–20 +	16	1.1348 1.1527 1.1986	0.8358 0.8313 0.8211	0.8357 0.8312 0.8215	0.6668 0.6582 0.6404	909 s 1114 s 1121 s
multi-qa-mpnet-base-dot-v1	SEB SEB + Aug3.5 SEB + Aug4	SpCh + LC	No	80–20 +	16	0.8226 0.8161 0.9179	0.8518 0.8639 0.8521	0.8519 0.8636 0.8523	0.6928 0.7265 0.7058	1898 s 2409 s 2381 s
Graph Convolutional Networks [29]	SEB SEB + BT	-	-	90–10 90–10	32 32	- -	0.705 0.725	0.710 0.732	- -	- -
XLNET base [3]	SEB	-	-	90–10	16	-	0.693	0.702	-	-

Exp 3, + Exp 5, BT: back translation.

When compared with the Mohler dataset, the improvement in experiments with the SciEnts Bank dataset was not very significant. Based on further observations, we found that the Mohler dataset had an average of 20 words per row of data, while the SciEnts Bank dataset had an average of only 12 words per row of data. This difference in word count may contribute to the less significant performance improvement, as fewer words were considered in the evaluation.

5. Conclusions

In this study, we proposed a simpler SentenceTransformers model combined with balancing the dataset and fine-tuning the hyperparameters of the model to handle an automatic short answer grading system. Our recommended SentenceTransformers model has a relatively small size, resulting in manageable resource requirements. We also balanced the dataset using GPT data augmentation, employing prompt engineering in the GPT model to generate new sentences based on existing student answers or desired answers. Additionally, we also fine-tuned the model by combining appropriate hyperparameters to achieve optimal grading performance. From the experiments conducted, the new balanced dataset significantly improved the performance of the grading system, as observed through RMSE, F1-score, accuracy, and Pearson correlation metrics.

The newly generated answer data from GPT also display satisfactory results, with a cosine similarity score reaching 0.8 for the correct category and 0.3 for the false category. This data augmentation aims to create a more balanced distribution of grade labels in the dataset. The implementation of this new balanced dataset also resulted in a significant performance improvement. The best result we obtained reached a Pearson correlation value of 0.9586 from the implementation of the all-distilroberta-v1 model. This model has a relatively small size and underwent a fine-tuning of hyperparameters, as well as the utilization of the new balanced dataset. Key hyperparameters include (1) the use of gradient checkpointing to reduce memory consumption; (2) the split-size ratio for the training and testing datasets, with 80% for training and 20% for testing; (3) a pre-processing step involving the removal of special characters and converting text to lowercase. Other parameters remained fixed across all the experiments, as mentioned earlier. Furthermore, the RMSE, F1-score, and accuracy score also consistently achieved better results compared to the previous research. This has also been demonstrated by additional experimental results. Although the performance increase for the new dataset is not very significant, there is still an improvement. This difference can be attributed to the varying characteristics of the datasets themselves.

In terms of future works, there are several areas that might be explored further. Currently, the dataset used consists only of English-language data. Hence, future research could explore the implementation of models trained on datasets from other languages. GPT has proven to be able to generate new sentences to enhance performance. Therefore, it is also possible to leverage GPT for language translation when using datasets other than in English.

Author Contributions: Conceptualization, M.C.W. and H.-S.Y.; methodology, M.C.W.; software, M.C.W.; validation, H.-S.Y.; formal analysis, M.C.W.; investigation, M.C.W.; resources, M.C.W.; data curation, M.C.W.; writing—original draft preparation, M.C.W.; writing—review and editing, M.C.W. and H.-S.Y.; visualization, M.C.W.; supervision, H.-S.Y. All authors have read and agreed to the published version of the manuscript.

Funding: This research is partially supported by a Korea Agency for Infrastructure Technology Advancement (KAIA) grant funded by the Ministry of Land, Infrastructure, and Transport (Grant RS-2022-00143782).

Institutional Review Board Statement: Not applicable.

Informed Consent Statement: Not applicable.

Data Availability Statement: The original contributions presented in the study are included in the article, further inquiries can be directed to the corresponding author.

Conflicts of Interest: The authors declare no conflicts of interest.

References

1. Wijanto, M.C.; Karnalim, O.; Tan, R. Work in progress: High School Students' Perspective on Assessment Question Types during Online Learning—Preliminary Study for Automated Assessments of Open-ended Questions. In Proceedings of the EDUNINE 2022—6th IEEE World Engineering Education Conference: Rethinking Engineering Education after COVID-19: A Path to the New Normal, Santos, Brazil, 13–16 March 2022; Institute of Electrical and Electronics Engineers Inc.: Piscataway, NJ, USA, 2022. [CrossRef]
2. Süzen, N.; Gorban, A.N.; Levesley, J.; Mirkes, E.M. Automatic short answer grading and feedback using text mining methods. *Procedia Comput. Sci.* **2020**, *169*, 726–743. [CrossRef]
3. Ghavidel, H.; Zouaq, A.; Desmarais, M. Using BERT and XLNET for the Automatic Short Answer Grading Task. In Proceedings of the CSEDU 2020—Proceedings of the 12th International Conference on Computer Supported Education, Prague, Czech Republic, 2–4 May 2020; SciTePress: Setúbal, Portugal, 2020; pp. 58–67. [CrossRef]
4. Bagaria, V.; Badve, M.; Beldar, M.; Ghane, S. An Intelligent System for Evaluation of Descriptive Answers. In Proceedings of the 3rd International Conference on Intelligent Sustainable Systems, ICISS 2020, Thoothukudi, India, 3–5 December 2020; Institute of Electrical and Electronics Engineers Inc.: Piscataway, NJ, USA, 2020; pp. 19–24. [CrossRef]
5. Tulu, C.N.; Ozkaya, O.; Orhan, U. Automatic Short Answer Grading with SemSpace Sense Vectors and MaLSTM. *IEEE Access* **2021**, *9*, 19270–19280. [CrossRef]
6. Burrows, S.; Gurevych, I.; Stein, B. The Eras and Trends of Automatic Short Answer Grading. *Int. J. Artif. Intell. Educ.* **2014**, *25*, 60–117. [CrossRef]
7. Zhang, L.; Huang, Y.; Yang, X.; Yu, S.; Zhuang, F. An automatic short-answer grading model for semi-open-ended questions. *Interact. Learn. Environ.* **2019**, *30*, 177–190. [CrossRef]
8. Wijaya, M.C. Automatic Short Answer Grading System in Indonesian Language Using BERT Machine Learning. *Rev. d'Itelligence Artif.* **2021**, *35*, 503–509. [CrossRef]
9. Condor, A.; Litster, M.; Pardos, Z. Automatic short answer grading with SBERT on out-of-sample questions. In Proceedings of the 14th International Conference on Educational Data Mining (EDM 2021), Online, 29 June–2 July 2021; International Educational Data Mining Society: Paris, France, 2021; pp. 345–352.
10. Gaddipati, S.K.; Nair, D.; Plöger, P.G. Comparative Evaluation of Pretrained Transfer Learning Models on Automatic Short Answer Grading. *arXiv* **2020**, arXiv:2009.01303.
11. Poulton, A.; Eliens, S. Explaining transformer-based models for automatic short answer grading. In Proceedings of the ACM International Conference Proceeding Series, Association for Computing Machinery, Busan, Republic of Korea, 15–17 September 2021; pp. 110–116. [CrossRef]
12. Alreheli, A.S.; Alghamdi, H.S. Automatic Short Answer Grading Using Paragraph Vectors and Transfer Learning Embeddings. *J. King Abdulaziz Univ. Comput. Inf. Technol. Sci.* **2022**, *11*, 25–31. [CrossRef]
13. Bonthu, S.; Sree, S.R.; Prasad, M.K. Improving the performance of automatic short answer grading using transfer learning and augmentation. *Eng. Appl. Artif. Intell.* **2023**, *123*, 106292. [CrossRef]
14. Reimers, N.; Gurevych, I. Sentence-BERT: Sentence Embeddings using Siamese BERT-Networks. *arXiv* **2019**, arXiv:1908.10084.

15. Ndukwe, I.G.; Amadi, C.E.; Nkomo, L.M.; Daniel, B.K. Automatic Grading System Using Sentence-BERT Network. In *Lecture Notes in Computer Science (Including Subseries Lecture Notes in Artificial Intelligence and Lecture Notes in Bioinformatics)*; Springer: Cham, Switzerland, 2020; pp. 224–227. [CrossRef]
16. Dzikovska, M.O.; Nielsen, R.D.; Brew, C. Towards Effective Tutorial Feedback for Explanation Questions: A Dataset and Baselines. In Proceedings of the 2012 Conference of the North American Chapter of the Association for Computational Linguistics: Human Language Technologies, Montreal, BC, Canada, 3–8 June 2012; pp. 200–210.
17. Wiratmo, A.; Nopember, I.T.S.; Fatichah, C. Indonesian Short Essay Scoring Using Transfer Learning Dependency Tree LSTM. *Int. J. Intell. Eng. Syst.* **2020**, *13*, 278–285. [CrossRef]
18. Devlin, J.; Chang, M.-W.; Lee, K.; Toutanova, K. BERT: Pre-training of Deep Bidirectional Transformers for Language Understanding. *arXiv* **2018**, arXiv:1810.04805.
19. Reimers, N. Sentence Transformers Documentation. Available online: https://www.sbert.net/docs/pretrained_models.html (accessed on 18 January 2024).
20. Gomaa, W.H.; Nagib, A.E.; Saeed, M.M.; Algarni, A.; Nabil, E. Empowering Short Answer Grading: Integrating Transformer-Based Embeddings and BI-LSTM Network. *Big Data Cogn. Comput.* **2023**, *7*, 122. [CrossRef]
21. Ouahrani, L.; Bennouar, D. Paraphrase Generation and Supervised Learning for Improved Automatic Short Answer Grading. *Int. J. Artif. Intell. Educ.* **2024**, 1–44. [CrossRef]
22. Okur, E.; Sahay, S.; Nachman, L. Data Augmentation with Paraphrase Generation and Entity Extraction for Multimodal Dialogue System. In Proceedings of the 13th Conference on Language Resources and Evaluation (LREC 2022), Marseille, France, 21–23 June 2022; European Language Resources Association (ELRA): Paris, France, 2022; pp. 4114–4125.
23. Mohler, M.; Bunescu, R.; Mihalcea, R. Learning to Grade Short Answer Questions using Semantic Similarity Measures and Dependency Graph Alignments. In Proceedings of the 49th Annual Meeting of the Association for Computational Linguistics, Portland, OR, USA, 19–24 June 2011; Association for Computational Linguistics: Stroudsburg, PA, USA, 2011; pp. 752–762.
24. Chen, T.; Xu, B.; Zhang, C.; Guestrin, C. Training Deep Nets with Sublinear Memory Cost. *arXiv* **2016**, arXiv:1604.06174.
25. Goodfellow, I.; Bengio, Y.; Courville, A. *Deep Learning*; The MIT Press: Cambridge, MA, USA, 2016.
26. Januzaj, Y.; Luma, A. Cosine Similarity—A Computing Approach to Match Similarity between Higher Education Programs and Job Market Demands Based on Maximum Number of Common Words. *Int. J. Emerg. Technol. Learn.* **2022**, *17*, 258–268. [CrossRef]
27. Sanh, V.; Debut, L.; Chaumond, J.; Wolf, T. DistilBERT, a distilled version of BERT: Smaller, faster, cheaper and lighter. In Proceedings of the 5th EMC2—Energy Efficient Machine Learning and Cognitive Computing, Vancouver, BC, Canada, 13 December 2019.
28. Dzikovska, M.O.; Nielsen, R.; Brew, C.; Leacock, C.; Giampiccolo, D.; Bentivogli, L.; Clark, P.; Dagan, I.; Dang, H.T. SemEval-2013 Task 7: The Joint Student Response Analysis and 8th Recognizing Textual Entailment Challenge. In Proceedings of the Second Joint Conference on Lexical and Computational Semantics (*SEM), Volume 2: Seventh International Workshop on Semantic Evaluation (SemEval 2013), Atlanta, GA, USA, 14–15 June 2013; Association for Computational Linguistics: Stroudsburg, PA, USA, 2013; pp. 263–274.
29. Tan, H.; Wang, C.; Duan, Q.; Lu, Y.; Zhang, H.; Li, R. Automatic short answer grading by encoding student responses via a graph convolutional network. *Interact. Learn. Environ.* **2020**, *31*, 1636–1650. [CrossRef]

Disclaimer/Publisher's Note: The statements, opinions and data contained in all publications are solely those of the individual author(s) and contributor(s) and not of MDPI and/or the editor(s). MDPI and/or the editor(s) disclaim responsibility for any injury to people or property resulting from any ideas, methods, instructions or products referred to in the content.

Article

Text Mining and Multi-Attribute Decision-Making-Based Course Improvement in Massive Open Online Courses

Pei Yang, Ying Liu *, Yuyan Luo *, Zhong Wang and Xiaoli Cai

College of Management Science, Chengdu University of Technology, Chengdu 610059, China; yangpei@cdut.edu.cn (P.Y.); wangzhong2012@cdut.edu.cn (Z.W.); caixiaoli@cdut.edu.cn (X.C.)
* Correspondence: liuying3@stu.cdut.edu.cn (Y.L.); luoyuyan13@mail.cdut.edu.cn (Y.L.)

Abstract: As the leading platform of online education, MOOCs provide learners with rich course resources, but course designers are still faced with the challenge of how to accurately improve the quality of courses. Current research mainly focuses on learners' emotional feedback on different course attributes, neglecting non-emotional content as well as the costs required to improve these attributes. This limitation makes it difficult for course designers to fully grasp the real needs of learners and to accurately locate the key issues in the course. To overcome the above challenges, this study proposes an MOOC improvement method based on text mining and multi-attribute decision-making. Firstly, we utilize word vectors and clustering techniques to extract course attributes that learners focus on from their comments. Secondly, with the help of some deep learning methods based on BERT, we conduct a sentiment analysis on these comments to reveal learners' emotional tendencies and non-emotional content towards course attributes. Finally, we adopt the multi-attribute decision-making method TOPSIS to comprehensively consider the emotional score, attention, non-emotional content, and improvement costs of the attributes, providing course designers with a priority ranking for attribute improvement. We applied this method to two typical MOOC programming courses—C language and Java language. The experimental findings demonstrate that our approach effectively identifies course attributes from reviews, assesses learners' satisfaction, attention, and cost of improvement, and ultimately generates a prioritized list of course attributes for improvement. This study provides a new approach for improving the quality of online courses and contributes to the sustainable development of online course quality.

Keywords: MOOC; text mining; multi-attribute decision-making; course improvement; deep learning

Citation: Yang, P.; Liu, Y.; Luo, Y.; Wang, Z.; Cai, X. Text Mining and Multi-Attribute Decision-Making-Based Course Improvement in Massive Open Online Courses. *Appl. Sci.* **2024**, *14*, 3654. https://doi.org/10.3390/app14093654

Academic Editors: Rafal Rzepka, Pawel Dybala and Michal Ptaszynski

Received: 15 March 2024
Revised: 20 April 2024
Accepted: 23 April 2024
Published: 25 April 2024

Copyright: © 2024 by the authors. Licensee MDPI, Basel, Switzerland. This article is an open access article distributed under the terms and conditions of the Creative Commons Attribution (CC BY) license (https://creativecommons.org/licenses/by/4.0/).

1. Introduction

Massive Open Online Courses (MOOCs), a significant innovation in teaching technology, offer a diverse array of high-quality open online courses globally [1], effectively overcoming numerous limitations of traditional offline learning in terms of cost, space, and background [2,3]. Simultaneously, the openness of MOOCs presents a novel opportunity for higher education institutions to create a richer and enhanced learning experience for learners through strengthened collaboration in knowledge sharing [4]. The emergence of MOOCs has not only facilitated the global sharing of educational resources but has also breathed new life into educational equity and popularization efforts [5]. MOOCs are now at the forefront of education [6]. Since the inception of the "Year of MOOCs" in 2012, the utilization of MOOCs has been steadily increasing worldwide. This trend has been further accelerated by the COVID-19 pandemic, which prompted the transition of numerous offline courses to online formats [7]. Consequently, people have become increasingly reliant on MOOCs, with millions of new users signing up on their platforms [8]. By 26 January 2024, China had launched over 76,800 MOOCs, catering to a staggering 1.277 billion learners within the country. MOOCs are now at the forefront of education [9].

However, despite the booming development of MOOCs, the rapid surge in the number of courses has given rise to numerous challenges, including a high dropout rate [10–12]

and inconsistent quality [13], thereby hampering the sustainable progress of MOOCs. Since learner satisfaction plays a pivotal role in extending the duration of product usage [14], it has become imperative to pinpoint the issues prevailing in MOOCs and furnish targeted recommendations to course designers, aiming to enhance learner satisfaction.

MOOCs enable learners to share their views and perceptions of courses through posted reviews. These reviews encompass a diverse array of learner needs, expectations, and suggestions [15], serving as a rich resource for enhancing course quality. By delving deeply into these comments, we can gain insights into the learners' satisfaction and dissatisfaction levels towards various course attributes [16], enabling us to provide targeted improvement directions to course designers [17], such as the research of Geng et al. [18] and Liu et al. [19].

Text mining is the process of extracting valuable information from text data [20]. By analyzing the satisfaction and dissatisfaction expressed by reviewers, valuable insights can be gained into consumer requirements, which in turn can guide product improvements [21–23]. For the product attributes that people are less satisfied with, it may be the product attributes that need to be improved [24,25]. These studies usually first mine the attributes of the product or service from the reviews, as well as the sentiment related to the attribute, and then determine the satisfaction and attention to the attribute. Ultimately, the decision on which attributes require improvement is made based on a combination of satisfaction levels and attention received from customers.

Text mining can also be employed to identify attributes in MOOCs that learners are dissatisfied with, enabling instructors to subsequently target these attributes for improvement.

However, there is a relative lack of research on MOOC improvement through text mining, and previous studies rarely considered the non-emotional content and improvement cost of product attributes. Indeed, online reviews encompass both emotional and non-emotional content, both of which are equally valuable for analysis [26]. One of the primary reasons that individuals write online reviews on e-commerce platforms is to alleviate consumer uncertainty during the shopping process [27]. Therefore, we can infer that the non-emotional reviews penned by learners on MOOCs serve to assist those reading the reviews in better comprehending the course content. Therefore, we can infer that the non-emotional reviews written by learners on MOOCs are meant to help the readers of the reviews understand the course content better. In turn, the attributes contained in non-emotional reviews may be important to future learners, in a reviewer's opinion.

When determining which product attributes require improvement, the associated cost is a critical consideration. In the realm of product development, cost often serves as a pivotal factor in prioritizing attribute enhancements [28]. Similarly, the same principle applies to course improvement. If learners express dissatisfaction with a particular course attribute but the cost of enhancing that attribute is substantial, it may not necessarily receive a high priority for improvement. Alternatively, even if learners express satisfaction with a particular attribute, it may still be prioritized for improvement if the cost of enhancing it is relatively low.

Hence, a comprehensive approach is necessary for course improvement, considering learner satisfaction, non-emotional content, attention, and the cost of enhancing various attributes. This makes it a multi-attribute decision-making (MADM) problem to identify course attributes that need to be improved. MADM refers to the problem of making decisions for multiple alternatives with multiple evaluation criteria [29].

To address the aforementioned problems, we aim to introduce a novel approach that integrates text mining and multi-attribute decision-making (MADM) for the improvement of MOOCs. This framework will enable us to extract course attributes that learners prioritize, along with their emotional and non-emotional responses towards these attributes. Subsequently, it will provide recommendations for course enhancement.

The remainder of this paper is organized as follows: In Section 2, we introduce related work. Section 3 describes the research approach. Section 4 presents the details of our experiments. Section 5 presents our conclusions and suggestions for future research.

2. Related Work

In this section, we review works that are pertinent to our topic. Specifically, we consider recent studies focusing on MOOC review text mining, as well as those that explore product or service improvement through review text mining.

2.1. MOOC Review Text Mining

As the accumulation of learner reviews continues to grow on MOOC platforms and text mining technology becomes more sophisticated, there has been a surge in research exploring the analysis of MOOC review texts.

Kastrati et al. [30] proposed a weakly supervised aspect-based sentiment analysis method for MOOC reviews, which can effectively reduce the dependence of the model on the number of training samples. Mrhar et al. [31] proposed a sentiment analysis method that combines Bayesian neural networks, convolutional neural networks (CNNs), and long short-term memory (LSTM). This method demonstrates promising results in the sentiment analysis of MOOC comments. Wang et al. [32] employed the ALBERT-BiLSTM model for sentiment analysis of MOOC reviews, revealing superior performance compared to existing methodologies. To enable data-driven design automation, Dina et al. [33] proposed a text-mining-based approach that automatically extracts feature sentiment pairs from MOOC reviews. In order to improve the performance of sentiment analysis, Liu et al. [34] combined BERT and part-of-speech information in MOOC reviews of sentiment. Liu et al. [35] proposed an MOOC-BERT model to automatically identify learners' cognition from online reviews, which has been verified to be superior to representative deep learning models in recognition and cross-curriculum.

Lundqvist et al. [36] categorized learners in MOOCs into beginner, experienced, and unknown categories and conducted a comparative analysis of the attitude differences among these learner groups in MOOCs. To gain insights into learners' preferences and concerns, Geng et al. [18] employed a combination of machine learning techniques and statistical analyses to mine reviews from NetEase, and they found learners pay more attention to the teaching and platform rather than the course content. Hew et al. [37] identified the influencing factors of learner satisfaction through text mining techniques. Nie et al. [38] introduced a comprehensive scoring methodology for MOOCs. This approach initially mined individuals' satisfaction levels with various course aspects from MOOC reviews, subsequently employing the Analytic Hierarchy Process (AHP) to generate an overall course rating. Chen et al. [39] used a structural topic model in MOOC reviews and found that the topics vary depending on the emotions or categories of the courses involved. Li et al. [6] undertook a comprehensive investigation to identify the key factors that contribute to the success of MOOCs. They categorized MOOCs into two distinct types and conducted a detailed analysis to uncover the influencing factors associated with each type. Gomez et al. [9] performed an extensive analysis of a large corpus of review text data, uncovering that numeric ratings alone provide limited guidance to learners in the course selection process. However, they identified that text mining can be effectively employed to aid course selection, as MOOC reviews contain abundant valuable information. Nilashi et al. [40] introduced a multi-stage text-mining-driven approach aimed at uncovering the influencing factors of learner satisfaction in MOOCs. Furthermore, to ensure the reliability and practical implications of their proposed method, they validated its efficacy through a survey methodology. Based on online reviews, Wang et al. [41] discovered that not all online review topics exert an influence on overall learner satisfaction.

It is evident that recent research on MOOC text mining has primarily focused on methodological studies, with a strong emphasis on extracting valuable content from MOOC reviews. Notably, there has been a significant focus on exploring the influencing factors of learner satisfaction. However, a notable gap exists in the literature regarding the provision of course improvement suggestions based on these mined influencing factors. This study aims to bridge this gap and provide practical recommendations for enhancing the quality of MOOCs and learner satisfaction.

2.2. Review Text-Mining-Based Product or Service Improvement

Online reviews encompass consumers' varying degrees of satisfaction and dissatisfaction towards various products or services. Consequently, numerous studies have been devoted to the text mining of online reviews, aiming to facilitate product enhancement and the creation of innovative offerings.

Zhou et al. [24] utilized Latent Dirichlet Allocation (LDA) [42] to extract product attributes from Amazon reviews. Subsequently, they employed a rule-based sentiment analysis approach to categorize the sentiment associated with these attributes. Finally, they leveraged the Kano model to identify the attributes that required improvement. Their approach is a typical approach of text-mining-based product improvement, such as Lee et al. [25]. Liu et al. [43] enhanced the efficacy of the traditional Kano model for product improvement by integrating it with the fuzzy Analytic Hierarchy Process (AHP) and Quality Function Deployment (QFD). This integrated approach provided a comprehensive framework for prioritizing and optimizing product attributes, leading to more targeted and effective product enhancements. Chen et al. [44] undertook a comprehensive text mining analysis of high-speed rail reviews, successfully extracting six passenger demands. Utilizing large-scale group decision-making techniques, they aggregated the satisfaction levels of 100 passengers across these six demands, finally deriving the six demands' degrees of satisfaction and rankings. Chen et al. [22] took into account both explicit and implicit features during the extraction of product attributes. They employed neural network training to determine the relative importance of each feature, subsequently mapping these attributes to the Kano model. Previous studies employing text mining and the Kano model have not considered the consumers' affective needs. In order to address this issue, Jin et al. [45] adopted the Kansei-integrated Kano method to obtain the priority of product feature improvement. This method incorporates affective design and is more comprehensive than the traditional Kano model. Goldberg et al. [46] used the attribute mapping framework introduced by MacMillan and McGrath, successfully extracting attributes from reviews to offer innovation insights. Their approach to mining attributes from online reviews surpasses previous methods in terms of depth and precision. Huang et al. [47] combined the Latent Dirichlet Allocation (LDA) topic model and Quality Function Deployment (QFD) to mine customer requirements and competition information from online reviews for new product development. Ji et al. [48] proposed a large-scale group decision-making method to integrate the satisfaction of individual consumers on product attributes to form group satisfaction. Joung and Kim [49] utilized text mining techniques to cluster consumers based on their preferences. Initially, they mined the product attributes that consumers prioritize from online reviews. Subsequently, they employed neural networks to determine the importance of these attributes and clustered consumers accordingly, leveraging the attribute importance weights. Zhang and Song [50] combined text mining and large group decision-making methods to make product improvement decisions. Through a comparative analysis with intuitive and expert decision-making approaches, they demonstrated the reliability and effectiveness of their proposed method, surpassing both comparison methods. While numerous studies have primarily centered on enhancing products or services through text mining, often neglecting non-emotional content and the associated costs of attribute improvement, they primarily focus on consumer attention and satisfaction.

2.3. Review the Ranking Method of Multi-Attribute Decision-Making

Whether it be a purchase decision in daily life or a task scheduling decision at work, decision-making forms a crucial aspect of people's daily lives and professional endeavors [51]. Some decisions can be relatively straightforward, with minor consequences if a wrong choice is made. However, other decisions can be highly intricate, such that even a slight error can lead to significant repercussions, necessitating a profound and cautious approach. Generally speaking, a decision problem in real life typically involves numerous criteria or attributes that must be taken into account concurrently to arrive at a well-informed decision. The exploration of such challenges is frequently labeled as multi-criterion decision-making or multi-attribute decision-making, which entails selecting the most suitable option from a finite set of alternatives.

The literature introduces some of the most renowned MADM methods, including PROMETHEE, TODIM, VIKOR, and TOPSIS, which are designed to tackle ranking problems effectively [52–55]. Qin et al. [56] initially extracted product attributes, weight values, and emotional tendencies from online reviews. Subsequently, they employed the Random Multi-criteria Acceptance Analysis (SMAA)-PROMETHEE method, to derive product ranking outcomes. Zhang et al. [57] proposed an innovative product selection model that integrates sentiment analysis with the intuitionistic fuzzy TODIM method. This model aims to assist potential customers in ranking alternative products based on consumers' opinions regarding product performance. Liang et al. [51] introduced a quantitative approach for hotel selection leveraging online reviews. Furthermore, they innovatively developed the DL-VIKOR method, which ranks hotels based on customer satisfaction scores and the weights of extracted attributes. Nilashi et al. [58] conducted a cluster analysis using self-organizing mapping (SOM) to categorize hotel features. Subsequently, they employed the similarity to Ideal Solution Prioritization technique (TOPSIS) to rank these features. Additionally, neural fuzzy technology was utilized to reveal customer satisfaction levels, providing a comprehensive understanding of hotel performance. Li et al. [55] introduced a novel approach for product ranking that integrates the mining of online reviews with an interval-valued intuitionistic fuzzy technique (TOPSIS). This method aims to assist consumers in selecting products that align with their individual preferences.

No single multi-attribute decision analysis method (MADM) can be unilaterally designated as the best or worst, as each has its own unique strengths and limitations [59]. Each MADM method possesses distinct advantages and limitations, and its effectiveness depends heavily on how it is tailored to specific outcomes and objectives within the planning process. Consequently, the selection of an appropriate MADM model should be guided by specific scenarios and requirements rather than solely relying on general evaluations. TOPSIS, originally developed by Hwang and Yoon in 1981, is a straightforward ranking method, both conceptually and in terms of its application [60]. TOPSIS offers three notable benefits: it is comprehensive; it requires minimal data; and it produces intuitive and easily comprehensible results [55]. Given its excellent performance in our investigation [61], the method combining the interval intuitionistic fuzzy set with TOPSIS was chosen as the ranking method for this study.

After considering the above studies, we first use text mining techniques to extract the attributes that learners pay attention to, the corresponding emotional tendency and attention degree of the attributes from online reviews, and determine the satisfaction degree of the attributes. We then use a multi-attribute decision-making approach to prioritize attribute improvements.

3. Methods

Our objective is to leverage text mining techniques in MOOC reviews to identify priority areas for improvement in MOOC attributes. Our method consists of the following steps: (1) collection and preprocessing of data; (2) extraction of course attributes; (3) identification of emotional and non-emotional content linked to the course attributes; (4) ascertainment of the cost associated with attribute improvements; and (5) generation of a prioritized list of attribute improvements (see Figure 1).

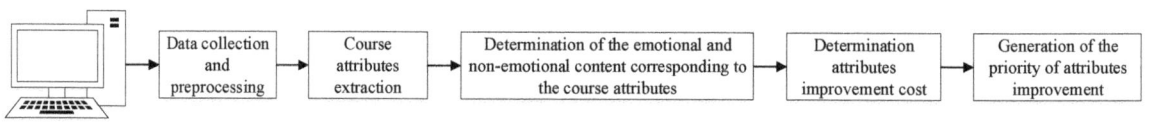

Figure 1. Overall steps of our method.

3.1. Data Collection and Preprocessing

We implemented a Python crawler to scrape online reviews on MOOC platforms. After that, we conducted the following preprocessing:

Step 1: Removing duplicate reviews. Removing duplicate reviews can help us improve the efficiency and quality of text mining.

Step 2: Text segmentation. As some reviews are long in length, they may contain multiple course keywords. We first used symbols (e.g., ".", "?", "!", "...") to segment longer reviews into short sentences, trying to make each sentence contain only one course keyword. Then we segmented the sentences into word lists to facilitate course attribute extraction and sentiment analysis.

Step 3: Removing stop words, including meaningless punctuation (e.g., !?#@). Removing stop words not only prevents subsequent steps from wasting time on non-informative words but also enhances the accuracy and efficiency of determining emotional and non-emotional content. As we intend to employ deep-learning-based text classification methods for the identification of emotional and non-emotional content, the presence of stop words can potentially compromise the performance of these models.

Step 4: Data labeling. Given that our sentiment analysis task relies on machine learning algorithms, labeled datasets are essential for training classifiers. Therefore, it is imperative to label the data, classifying them into four distinct categories: positive, negative, neutral, and non-emotional.

Step 5: Part-of-speech tagging. To extract alternative attribute words for courses, which are typically nouns, we must first determine the part-of-speech of each word in the course reviews.

Among them, for word segmentation and part-of-speech tagging, we utilized the jieba word segmentation tool, a highly renowned Chinese word segmentation utility in Python.

3.2. Course Attributes Extraction

Previous studies mainly used the LDA topic model or word embedding to extract and classify product attributes, where word embedding methods do not require manual interpretation of each topic [49]. Therefore, we used the word embedding method to extract and classify course attributes. Our approach consisted of the following steps:

Step 1: Nouns were extracted from all the reviews for a course, that is, words that start with "n" in the POS tags. Course attributes are usually nouns, so we extract the nouns in course reviews as our alternative attribute words. The number of extracted nouns is usually large, and the main attributes of a course may be limited to a few, so we needed to find a way to cluster these nouns and then treat each cluster as an attribute. Clustering algorithms usually take input in vector form, so we needed to convert nouns into vector form.

Step 2: To convert the selected nouns into a vectorized format, we employed the FastText word embedding approach. Word embedding is a technique that maps words onto a vector space, allowing each word to be represented by a unique vector. Furthermore, words with similar meanings tend to have word vectors that are proximate in Euclidean space. This characteristic makes word embedding an ideal approach for converting words into vectors and subsequently clustering them. A representative word embedding method is FastText [62], which considers the morphology of words, that is, the n-grams of word characters. For example, the 2-g of the word "where" includes "wh", "he", "er", and "re". FastText for words that do not occur during training can be represented by the sum of its character n-grams, which allows FastText to represent any word. The pre-trained Chinese FastText word vectors (https://dl.fbaipublicfiles.com/fasttext/vectors-crawl/cc.zh.300.bin.gz (accessed on 1 February 2024)) are used to represent the nouns in our work.

Step 3: For noun clustering, we employed a combined approach incorporating Affinity Propagation (AP) clustering [63] and manual intervention. AP is one of the most classical and widely used clustering methods that clusters the close words into the same cluster and the far words into different clusters by calculating the distance. A significant advantage of AP clustering is that it does not require the specification of the number of clusters upfront.

After AP clustering, we manually proofread the clustering results to make the meaning of words in each cluster as close as possible and delete meaningless words.

Step 4: The words in each cluster after clustering are induced to summarize the course attributes. By summarizing the nouns in each cluster into attributes, we can know the main attributes of a course that learners pay attention to.

Step 5: The nouns' frequency of different attributes was counted, as was the total word frequency of an attribute. This approach is based on the assumption that a higher frequency of an attribute word indicates greater learner attention and importance placed on that specific attribute.

3.3. Determination of the Emotional and Non-Emotional Content Corresponding to the Course Attributes

To identify the emotional and non-emotional content associated with the course attributes, we built a deep-learning-based text classification model. This model was trained to categorize review text into four distinct classes: positive, negative, neutral, and non-emotional. For ease of presentation, the classification of a review belonging to these four classes was collectively referred to as sentiment classification. We used a simple means to determine the course attribute and its corresponding sentiment class in a review: if a review contains a word belonging to an attribute whose sentiment tendency has been determined to be non-emotional, we consider the attribute to be non-emotional in this review. For example, in the sentence "The teacher's mandarin is good", obviously, this is a positive review, and if mandarin belongs to the attribute "expressive ability", we can assume that the emotional tendency of expressive ability in this sentence is positive (see Figure 2). We have highlighted the keywords that appear in the sentence in red in the Figure 2 and highlighted the corresponding attribute in the box. First and foremost, we carefully selected a specific number of comments from the vast pool of online feedback to be manually annotated. Positive sentiment comments are labeled as 1, negative sentiment as −1, neutral sentiment as 0, and comments lacking emotional expression are designated as 2. Subsequently, a subset of these annotated comments was utilized to train a deep learning model, while the remaining comments served as a validation set to assess the model's accuracy. After rigorous testing, the model exhibiting the highest predictive accuracy was chosen to analyze the emotional tendencies of the remaining unlabeled comments.

The pivotal aspect lies in developing a deep-learning-based sentiment classification model that can effectively categorize and analyze the sentiment expressed in reviews. The reason why we choose deep learning for sentiment classification is that in recent years, deep learning has gradually become the mainstream model for sentiment classification, and with the help of pre-trained language models such as BERT [64], the accuracy of sentiment classification has also been greatly improved. Deep learning is a machine learning technique rooted in artificial neural networks, requiring the integration of various deep learning modules to construct a complete model. Our deep-learning-based sentiment classification model is structured into three distinct modules: the embedding layer, the intermediate layer, and the output layer.

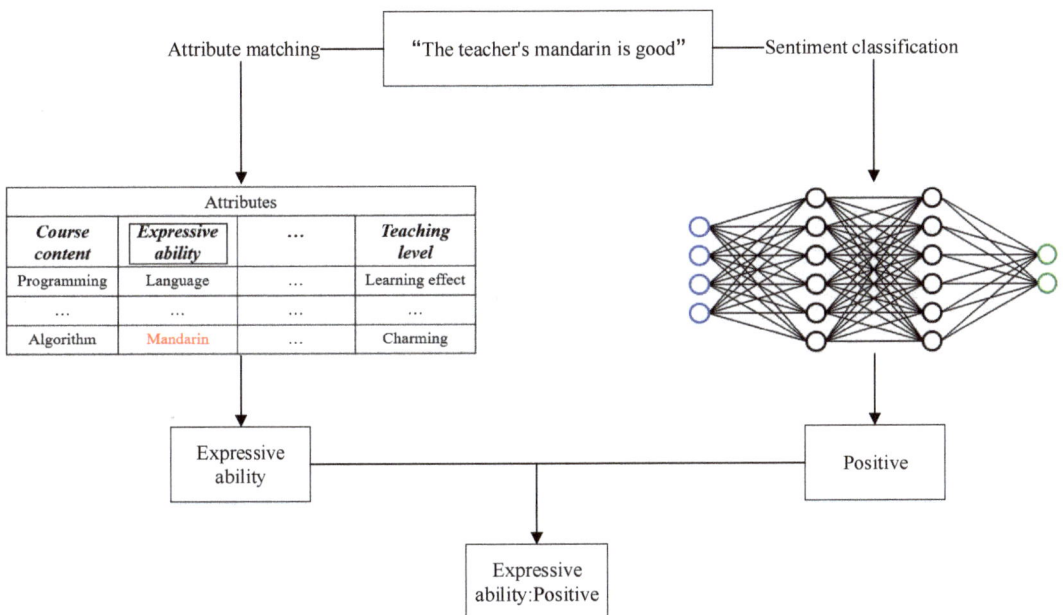

Figure 2. The way to determine the course attribute and its corresponding sentiment class in a review.

3.3.1. Embedding Layer

As previously mentioned in Section 3.2, word embeddings serve as a mechanism to transform words into a vectorized format. The embedding layer specifically handles the vectorization of words within the deep learning model. We utilized the BERT model directly for word embedding due to its exceptional performance in downstream tasks, particularly sentiment classification. In this paper, we directly used Chinese BERT (https://huggingface.co/bert-base-chinese (accessed on 2 February 2024)) pre-trained by Hugging Face to help us with word embedding. Hugging Face (https://huggingface.co/ (accessed on 1 February 2024)) is an artificial intelligence community, providing people with a large number of artificial intelligence code and application programming interface (API), so that people can implement artificial intelligence with low cost. The inputs and outputs in BERT-based embedding can be seen in Figure 3.

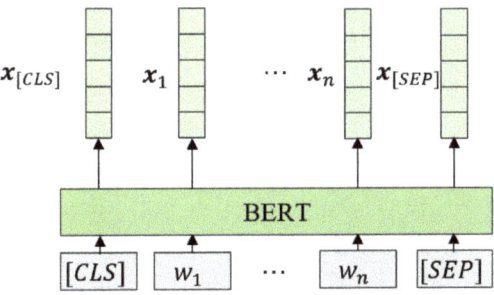

Figure 3. The input and output of BERT-based embedding.

The inputs in BERT embedding are words (for English) or characters (for Chinese) of a sentence, which are [CLS], $w_1, w_2, \ldots, w_n$, [SEP], where $w_1, w_2, \ldots, w_n$ are words or characters of a sentence of length n (the number of English words or Chinese characters), "[CLS]" is a symbol specifically used for text classification, and "[SEP]" represents a sentence separator or terminator. When we input sentences, Hugging Face automatically adds these two symbols to the original sentences without us having to manually do so. The outputs are $x_{[CLS]}, x_1, \ldots, x_n,$ and $x_{[SEP]}$, which are the vector representation of all inputs.

3.3.2. Intermediate Layer

Compared with the embedding layer and output layer, which are relatively fixed, the intermediate layer changes more. In order to avoid using only one intermediate layer and resulting in poor performance of the trained classifier, we compared six intermediate layers: base BERT, feed-forward neural network (FFNN), CNN, LSTM, LSTM combined with attention mechanism (LSTM + attention), and LSTM combined with CNN (LSTM + CNN).

(1) Base BERT

One of the simplest uses of the BERT architecture is baseBERT, which uses BERT for word embeddings and then feeds the vector representation of "[CLS]" into a linear layer classifier (see Figure 4).

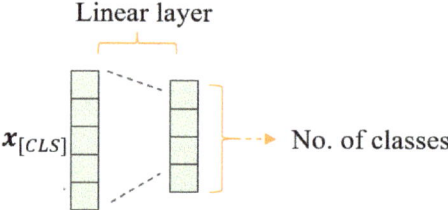

Figure 4. The structure of base BERT in our work.

(2) FFNN

FFNNs are the most primitive deep learning modes. A representative model is the deep average network (DAN) [65], as shown in Figure 5. Its main structure includes the embedded, average, linear, and output layers. In our work, we adopted the following structure depicted in Figure 5. The embedding size within our DAN was set to 100.

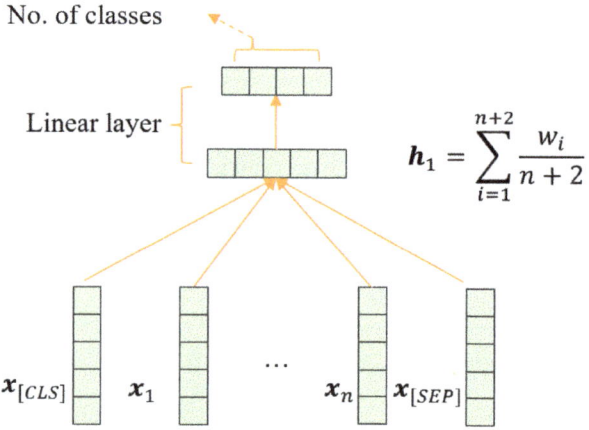

Figure 5. DAN in our work.

(3) CNN

CNNs (convolutional neural networks) extract features from the input text by applying convolutional filters, enabling the capture of local n-gram patterns. Max-pooling layers then select the most important features, and fully connected layers classify the text. In our work, we used the structure based on the work of Kim [66]. We employed 100 kernels of three different sizes (2, 3, and 4) for feature extraction, followed by max-pooling for pooling (see Figure 6).

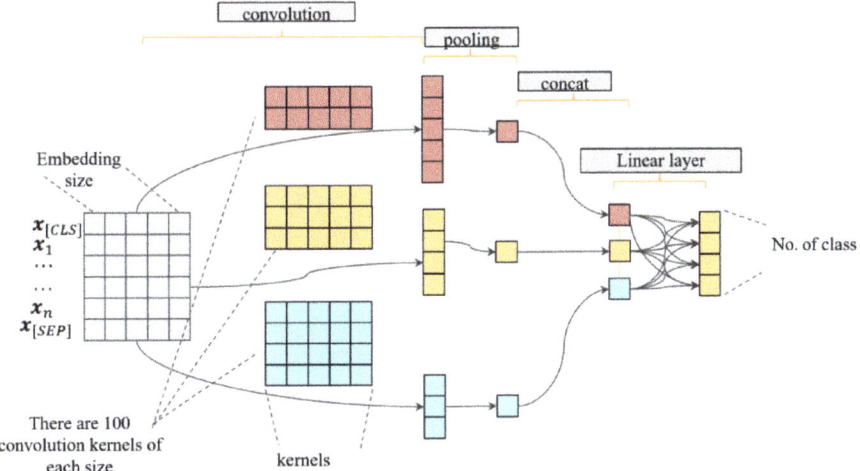

Figure 6. CNN in our work.

(4) LSTM

LSTM networks are a type of RNN designed to handle sequential data. In sentiment classification, LSTMs process text one token at a time, capturing contextual information through their memory cells. The final hidden state is then fed into a classification layer to predict the label. We used Bidirectional LSTM (BiLSTM) in our work (see Figure 7). The hidden layer size of the LSTM is 100 and the output size of the first linear layer is 128.

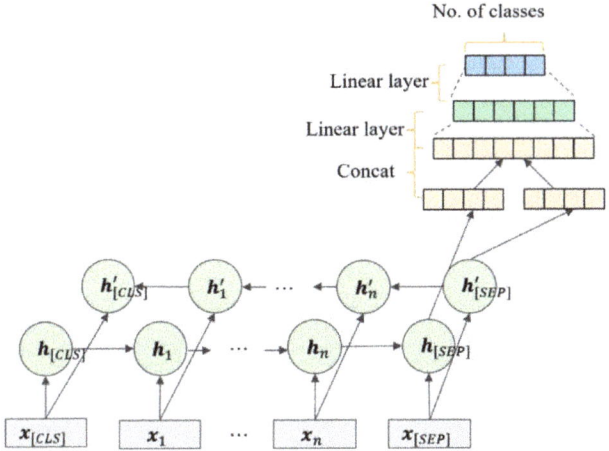

Figure 7. LSTM in our work.

(5) LSTM+attention

The ordinary LSTM-based text classification method only uses the output of the last time step as the input of the classifier, while the LSTM+attention-based method aims to weight the output of all time steps to obtain the input of the classifier (see Figure 8), which can consider more semantic information.

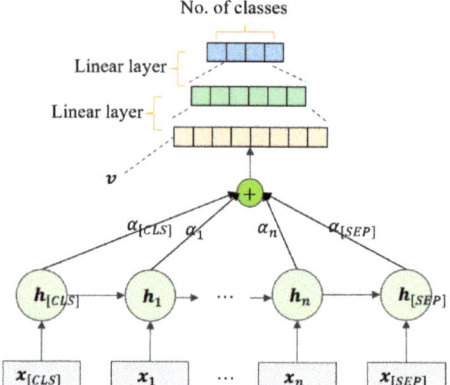

Figure 8. LSTM+attention in our work.

We used the calculation of Yang et al. [67] to achieve α_t as follows:

$$u_t = \tanh(W_s h_t + b_s) \tag{1}$$

$$\alpha_t = \frac{\exp(u_t^T u_s)}{\sum_t \exp(u_t^T u_s)}, \tag{2}$$

Then we can obtain the weighted sentence representation v as follows:

$$v = \sum_t \alpha_t h_t, \tag{3}$$

where W_s, b_s, u_s are parameters, α_t is the weight of the output h_t, and v is the sentence representation after the weighted sum.

(6) LSTM+CNN

The LSTM+CNN structure is able to combine the advantages of LSTM and CNN, where LSTM is able to capture long-term dependencies and CNN is able to extract local features. We added residual structure to help convergence. The model we used can be seen in Figure 9.

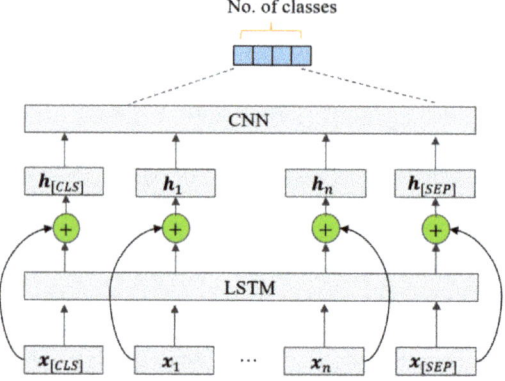

Figure 9. LSTM+CNN in our work.

3.3.3. Output Layer

The output layer is unified as softmax layer for most deep-learning-based sentiment classification. The softmax layer can be calculated as follows:

$$\begin{aligned}
\text{softmax}(z_{i1}, z_{i2}, \ldots, z_{ic}, \ldots, z_{iM}) &= \left(\frac{e^{z_{i1}}}{\sum_{i=1}^{n} e^{z_{i1}}}, \frac{e^{z_{i2}}}{\sum_{i=1}^{n} e^{z_{i2}}}, \ldots, \frac{e^{z_{iM}}}{\sum_{i=1}^{n} e^{z_{iM}}} \right) \\
&= (\mathrm{P}(cls_1|s_i), \mathrm{P}(cls_2|s_i), \ldots, \mathrm{P}(cls_M|s_i))
\end{aligned} \quad (4)$$

By utilizing our proposed sentiment classification method, we can extract the attributes mentioned in each review along with their corresponding sentiment. Subsequently, we calculate the overall count of reviews that involve distinct attributes and various emotional responses, as depicted in Figure 10.

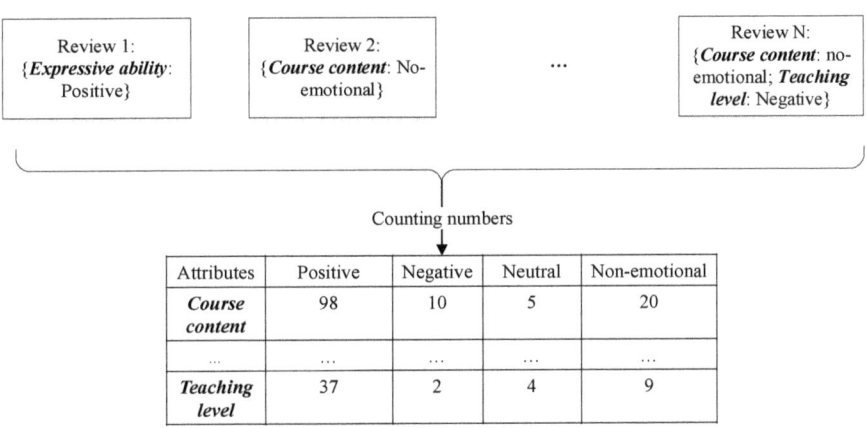

Figure 10. An illustration of the total number of reviews involving different attributes and different emotions.

3.4. Determination Attributes Improvement Cost

For different course attributes of a course, it is difficult to obtain precise attribute improvement costs, but we can ask experts to judge the difficulty of different attributes to be improved and rank the difficulty. Given the difficulty rankings, we can use the rank sum weight method (RSW) to estimate the cost and obtain the proportion of each attribute improvement cost in the total cost. The RSW method is a method to determine attribute weights in MADM, which is able to generate quantitative weights of attributes based on the attribute importance ranking given by experts [68]. The proportion of each attribute improvement cost in the total cost can be calculated as follows:

$$cost_i = \frac{n - r_i + 1}{\sum_{i=1}^{n}(n - r_i + 1)} \quad (5)$$

In Formula (5), r_i represents the improvement difficulty ranking of the ith attribute; the more difficult it is, the higher the ranking. n represents the total number of attributes. It is easy to see that in Formula (5), for all attributes, the denominator is a constant value, while the value of the numerator varies according to the ranking of the ith attribute; the higher the ranking, the larger the numerator, and then the larger the proportion of the estimated cost.

3.5. Generation of the Priority of Attributes Improvement

After the previous steps, we obtained six evaluation criteria for each course attribute: the number of positive reviews (c_1), the number of neutral reviews (c_2), the number of negative reviews (c_3), the number of non-emotional reviews (c_4), the attribute frequency (c_5), and the improvement cost (c_6) (see Table 1).

Table 1. Evaluation matrix for all course attributes.

Attributes	c_1	c_2	c_3	c_4	c_5	c_6
a_1	x_{11}	x_{12}	x_{13}	x_{14}	x_{15}	x_{16}
...	...	...	...	...	...	...
a_n	x_{n1}	x_{n2}	x_{n3}	x_{n4}	x_{n5}	x_{n6}

In Table 2, $c_1,\ldots,c_6$ represent evaluation criteria, $a_1,\ldots,a_n$ represent course attributes, and x_{ij} represents the value of the jth evaluation criterion of the ith attribute.

Table 2. An example of attributes and attribute words.

Attributes	Attribute Words
Teacher's expressive ability	Language, speaking speed, voice, ...
Teaching level	Thinking, clear thinking, enlightening, ...
Teaching method	Method, format, teaching, ...
Exercise	Question, example problem, mistake, ...
Learning threshold	Getting started for beginners, beginners, advanced level, ...
Course organization	Teaching content, basics, key points of the course, ...
Supporting materials	Teaching material, data, courseware, ...
Platform	Video, subtitle, clarity, ...
Experience of learning	Effect, feeling, classic, ...
Professional knowledge	C language, code, algorithm, ...

Given the varying number of attribute words across different attributes, it is unfair to simply tally the total number of reviews expressing various sentiments for all the attribute words within a single attribute, especially for those with fewer attribute words. We normalize the values of c_1, c_2, c_3, and c_4 using the number of reviews involving the attribute with positive, negative, neutral and no sentiment divided by the total number of reviews involving the attribute. For c_5, we take the average of the number of reviews involving each word in an attribute.

So, we use the following formulas to calculate x_{i1}, x_{i2}, x_{i3}, x_{i4}, and x_{i5}:

$$x_{i1} = \frac{\sum_{w \in S_{a_i}} N_w^{pos}}{N_{a_i}} \quad (6)$$

$$x_{i2} = \frac{\sum_{w \in S_{a_i}} N_w^{neu}}{N_{a_i}} \quad (7)$$

$$x_{i3} = \frac{\sum_{w \in S_{a_i}} N_w^{neg}}{N_{a_i}} \quad (8)$$

$$x_{i4} = \frac{\sum_{w \in S_{a_i}} N_w^{non}}{N_{a_i}} \quad (9)$$

$$x_{i5} = \frac{\sum_{w \in S_{a_i}} N_w}{|S_{a_i}|} \quad (10)$$

x_{i6} can by calculated by Formula (5). In Formulas (6)–(10), w represents the index of word, S_{a_i} represents the set of attribute words for attribute a_i, $|S_{a_i}|$ represents the cardinality of set S_{a_i}, N_w represents the number of reviews involving wth word, N_{a_i} represents the number of reviews involving any words in S_{a_i} and N_w^{pos}, N_w^{neg}, N_w^{neu}, and N_w^{non} represent the number of positive, neutral, negative, and non-emotional reviews involving wth word, respectively.

Our task is to decide the priority of course attribute improvement. c_1 represents the degree of learners' satisfaction. Obviously, the higher the degree of satisfaction, the lower the priority for improvement. Therefore, it is a cost criterion. The neutral reviews indicate that learners are not very satisfied with the mentioned attributes; that is, they still have a certain degree of dissatisfaction, and thus, c_2 is the benefit criterion. Similarly, c_3 is a benefit criterion. Although reviews without emotional content do not express satisfaction or dissatisfaction, if certain attributes are mentioned by learners in reviews, it means that learners may pay more attention to these attributes. Thus, c_4 the is benefit criterion. c_5 represents the total attention and is therefore a benefit criterion. The higher the improvement cost, the more difficult the improvement. Obviously, c_6 is the cost criterion.

Next, we employed the multi-attribute decision-making (MADM) approach to assess the attributes. Specifically, we utilized the Technique for Order Preference by Similarity to Ideal Solution (TOPSIS) method [69] to determine the priority scores for attribute improvement. TOPSIS determines a score for each alternative by calculating the distance between the true alternative and two ideal alternatives: the positive ideal alternative (a dummy alternative constructed with the best value of each attribute of all alternatives) and the negative ideal alternative (a dummy alternative constructed with the worst value of each attribute of all alternatives). The calculation process for TOPSIS has been detailed in our previous work [70].

The TOPSIS method requires the weights of various indicators to be determined beforehand. Therefore, we adopted the entropy weight method (EWM) to determine the weights of each indicator. Since the entropy weight method is an objective weight calculation method that determines the weight of an indicator based on the amount of information, it can reduce the deviation caused by subjective factors when determining the weight of indicators and make the results more practical. It has been widely used [71]. The calculation steps are as follows:

Step 1: Data standardization.

To eliminate the influence of varying scales and magnitudes, the original data must be normalized. The normalization formula is as follows:

$$z_{ij} = \frac{x_{ij} - \min(x_j)}{\max(x_j) - \min(x_j)} \tag{11}$$

where x_{ij} is the original value of the jth evaluation criterion of the ith alternative, z_{ij} is the normalized value, and $\min(x_j)$ and $\max(x_j)$ are the minimum and maximum values of the jth evaluation criterion, respectively.

Step 2: Calculate the entropy value of each criterion.

Compute the entropy value for each criterion. The entropy value reflects the degree of dispersion in the criterion data. The formula is as follows:

$$e_j = -\frac{1}{\ln n} \sum_{i=1}^{n} p_{ij} \ln p_{ij} \tag{12}$$

where e_j is the entropy value of the jth evaluation criterion, n is the number of alternatives, p_{ij} is the proportion of the jth evaluation criterion value for the ith alternative to the sum of that criterion, which can be calculated by $p_{ij} = \frac{z_{ij}}{\sum_{i=1}^{n} z_{ij}}$, and ln is the natural logarithm.

Step 3: Calculate the dispersion coefficient of each criterion.

The dispersion coefficient is the reciprocal of the entropy value and reflects the ability of a criterion to distinguish between evaluation objects. The formula is as follows:

$$d_j = 1 - e_j \tag{13}$$

Step 4: Calculate the weight of each criterion.

Determine the weight for each criterion based on its dispersion coefficient. The weights are proportional to the dispersion coefficients. The formula is as follows:

$$w_j = \frac{d_j}{\sum_{j=1}^{m} d_j} \quad (14)$$

where w_j is the weight of the jth evaluation criterion, and m is the number of criteria.

4. Results

4.1. Data Collection and Preprocessing

As we said in Section 3.1, we used a Python crawler to crawl the review data from the Chinese University MOOC (https://www.icourse163.org/ (accessed on 1 February 2024)). As the largest online course platform in China, the Chinese University MOOC boasts a significant volume of reviews. We gathered reviews for two courses: C language and Java language, spanning from January 2018 to November 2022. The numbers of reviews for the two courses are 12,560 and 3980, respectively.

We then segmented the reviews by splitting the reviews with multiple sentences into a single sentence and filtering out those with sentences less than five characters in length. It should be noted that we added some turning words for sentence segmentation, such as "but" and "however", because the text before and after these words often have different sentiments. Although some sentences were removed because they were less than five characters, some long reviews were split into multiple sentences, resulting in 11,327 and 3865 sentences, respectively.

Subsequently, we conducted part-of-speech (POS) tagging on both datasets, focusing on extracting nouns and their respective document frequencies. Document frequency refers to the count of review sentences in which a particular term appears. We then filter out words with too few document frequencies. Since the number of reviews in the two courses is different, it is not fair to filter words with the same criterion. Hence, for the C language course, we eliminated nouns with a document frequency of less than five, while for the Java language course, we excluded those with a document frequency of less than two. We manually weeded out nouns or noun phrases unrelated to the course content, such as "thank you", "not interesting", and "a waste of time". Consequently, we arrived at 340 alternative attribute words for the C language course and 350 for the Java language course.

4.2. Course Attributes Extraction

As previously mentioned in Section 3.2, we implemented loading FastText word vectors and AP clustering in Python. We then filtered and reorganized the clustered results, resulting in 10 attributes for both courses. The original results are in Chinese; for the convenience of readers, we use the translation results of the original words as attribute words and show a part of the attribute words for each attribute (see Table 2). The complete results in Chinese are shown on github (https://github.com/yangpeiailong/Text-Mining-and-Multi-Attribute-Decision-Making-based-Course-Improvement-in-MOOCs/tree/main/mooc/data/1attributes (accessed on 10 February 2024)).

4.3. Determination of the Emotional and Non-Emotional Content Corresponding to the Course Attributes

As stated in Section 3.3, we intend to utilize deep learning for sentiment analysis. However, deep learning models require labeled samples for training. Therefore, we manually annotated a subset of samples selected from our collected dataset. To prevent a decline in training accuracy due to imbalanced data, we aim to maintain a similar number of samples across all classes during the labeling process. Consequently, apart from the reviews collected for the two courses, we also gathered feedback from other similar courses. Despite this, neutral reviews remain relatively scarce. Therefore, we can only ensure that the number of other classes of data is similar and the number of neutral reviews is relatively small. The training sample consisted of 572 positive, 125 neutral, 537 negative, and 531 emotionless comments, amounting to a total of 1765 comments.

Subsequently, we trained six deep-learning-based classifiers using these samples and evaluated their performance. The settings of our deep learning models are as follows:

(1) Our deep learning models have been implemented using Pytorch 2.2.0 with a GTX 1080Ti graphics card and 32 GB of RAM.
(2) The loss functions were the cross-entropy loss functions, and the gradient descent algorithm was adaptive motion estimation or stochastic gradient descent, with learning rates ranging from 1×10^{-1} to 1×10^{-5}.
(3) The number of epochs is 40.
(4) The numbers of training and test batches were 32 and 100, respectively.
(5) The embedding word vector size of the BERT-based models was 768.
(6) The pre-trained BERT model utilized was the BERTModel, in conjunction with the BERTTokenizer tools from the Hugging Face transformer toolkit. This toolkit provides a Pytorch interface, enabling training and utilization with various intermediate structures, all implemented using Pytorch.
(7) We use accuracy to evaluate the performance of all six models, which can be calculated as follows:

$$accuracy = \frac{N_T}{N} \quad (15)$$

where N_T represents the number of samples that are correctly classified, and N represents the number of samples. To calculate the accuracy, we employed five-fold cross-validation, a technique that evaluates machine learning algorithms by dividing the available data into k equal-sized subsets. The model is trained k times, with k^{-1} subsets used for training and one subset reserved for validation. This iterative process ensures that each subset is utilized for validation once during the k iterations, thereby providing an unbiased assessment of the model's performance. The source code is available at github (https://github.com/yangpeiailong/Text-Mining-and-Multi-Attribute-Decision-Making-based-Course-Improvement-in-MOOCs/tree/main/mooc/codes (accessed on 10 February 2024)). The accuracy of the six classifiers is shown in Figure 11.

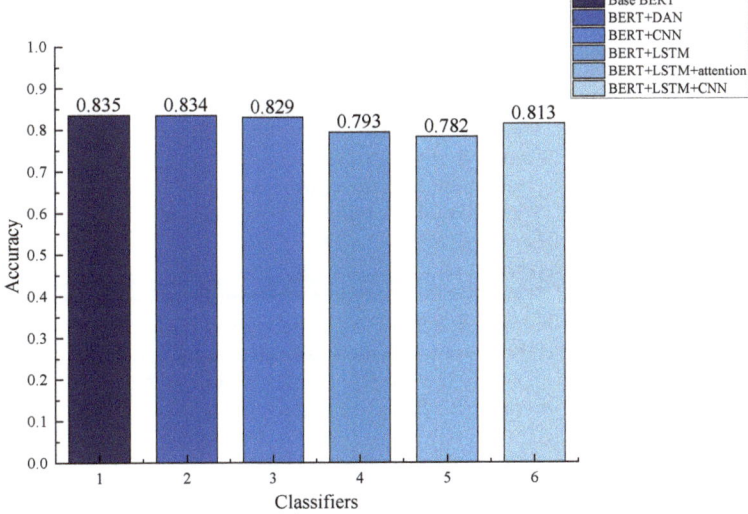

Figure 11. The accuracy of the six classifiers.

It is evident that base BERT exhibits superior performance among the six classifiers, making it the preferred choice for data labeling. Following the labeling process, we tallied the number of reviews expressing different sentiments for each attribute. These results are presented in Appendix A.

Appendix A reveals that across all attributes, the number of positive reviews consistently dominates, indicating a general satisfaction with the attributes of these two courses. Nevertheless, the extent of satisfaction exhibits considerable variation among different attributes. Notably, attributes such as "Teaching Level" and "Teaching Method" enjoy high satisfaction levels, with the combined count of neutral and negative reviews comprising less than one-third of the total. In contrast, attributes such as "Exercise" exhibit lower satisfaction levels, with neutral and negative reviews comprising nearly half of the total feedback. This underscores the varying degrees of satisfaction across different attributes.

4.4. Determination Attributes Improvement Cost

To minimize the risk of biased judgments resulting from reliance on a sole expert's opinion, we disbursed the extracted attributes and corresponding attribute words to six experts for independent review. This approach guaranteed that each expert, leveraging their unique professional expertise, would base their judgments solely on their thorough comprehension of the attributes. Consequently, a more balanced and comprehensive evaluation of the cost of attribute improvement was achieved.

Since both courses belong to the programming category and share identical attributes, it is reasonable to presume that the cost of enhancing these attributes would be comparable. Therefore, in order to streamline the ranking process, we did not ask experts to individually rank each attribute for each course. Instead, we requested a unified ranking for the attributes across both courses. The resulting rankings are presented in Table 3.

Table 3. The ranking results of improvement cost of all attributes.

Attributes	Expert 1	Expert 2	Expert 3	Expert 4	Expert 5	Expert 6
Teacher's expressive ability	7	9	10	10	10	10
Teaching level	2	1	2	1	1	1
Teaching method	8	5	6	5	5	6
Exercise	9	6	7	9	7	7
Learning threshold	3	3	5	6	4	4
Course organization	5	3	1	2	3	3
Supporting materials	10	8	8	8	8	8
Platform	6	8	8	7	9	9
Experience of learning	4	6	4	4	5	5
Professional knowledge	1	2	3	3	2	2

Table 3 reveals that while there exist minor disparities in experts' assessments regarding the cost of attribute improvement, the overall trend remains consistent. All experts agree that enhancing attributes such as "Teaching level", "Course organization", and "Professional knowledge" involves a relatively high cost. Conversely, attributes like "Teacher's expressive ability", "Exercise", and "Supporting materials" are considered easier to improve.

This finding resonates strongly with our intuitive understanding. The enhancement of attributes such as teachers' professional knowledge, teaching level, and course organization is deeply ingrained in their extensive learning and experience. Consequently, these aspects are inherently challenging to improve in the short term. In contrast, other attributes, which are not intrinsically tied to teachers, can be relatively easier to improve upon.

The calculation of Formula (5) requires a set of rankings, but our six experts have provided us with a total of six sets of rankings, which makes it impossible for us to directly bring the results of Table 3 into Formula (5) for calculation. Nonetheless, considering that there is little difference in the judgments of the experts, we directly calculate the averages of the rankings across six experts for each attribute and then bring it into Equation (5) to calculate the improvement cost.

4.5. Generation of the Priority of Attributes Improvement

Utilizing Equations (6)–(10) and the preceding results, we constructed the decision matrices for both courses, as presented in Tables 4 and 5.

Table 4. The decision matrix of C language.

Attributes	c_1	c_2	c_3	c_4	c_5	c_6
Teacher's expressive ability	0.701	0.044	0.067	0.188	41.200	0.040
Teaching level	0.868	0.036	0.032	0.064	167.167	0.162
Teaching method	0.775	0.053	0.067	0.105	26.682	0.093
Exercise	0.468	0.115	0.363	0.054	32.313	0.068
Learning threshold	0.829	0.037	0.044	0.090	69.148	0.119
Course organization	0.754	0.046	0.121	0.079	53.631	0.139
Supporting materials	0.641	0.086	0.206	0.066	24.000	0.056
Platform	0.620	0.068	0.193	0.118	45.500	0.063
Experience of learning	0.715	0.059	0.141	0.085	21.017	0.111
Professional knowledge	0.604	0.051	0.133	0.211	54.333	0.149

Table 5. The decision matrix of Java language.

Attributes	c_1	c_2	c_3	c_4	c_5	c_6
Teacher's expressive ability	0.753	0.087	0.060	0.100	19.500	0.040
Teaching level	0.849	0.040	0.034	0.076	57.452	0.162
Teaching method	0.765	0.074	0.078	0.083	10.700	0.093
Exercise	0.352	0.125	0.460	0.063	10.789	0.068
Learning threshold	0.849	0.043	0.043	0.065	26.500	0.119
Course organization	0.735	0.059	0.111	0.095	23.237	0.139
Supporting materials	0.744	0.041	0.140	0.074	10.333	0.056
Platform	0.654	0.067	0.202	0.077	12.667	0.063
Experience of learning	0.746	0.057	0.122	0.075	5.598	0.111
Professional knowledge	0.582	0.085	0.116	0.217	10.714	0.149

Based on the results in Tables 4 and 5, we can use Formulas (11)–(14) to calculate the criteria weights. Because our evaluation criteria include benefit and cost criteria, for cost criteria, we transform them into benefit criteria by calculating their reciprocals. The results are shown in Table 6.

Table 6. The criteria weights of both courses.

Attributes	c_1	c_2	c_3	c_4	c_5	c_6
C language	0.178	0.173	0.156	0.169	0.156	0.169
Java language	0.177	0.175	0.149	0.172	0.156	0.170

Then, we can use TOPSIS to calculate the improvement priority score and ranking of each attribute of the two courses. The results are shown in Table 7.

Table 7. The decision matrix of Java language.

Attributes	C Language		Java Language	
	Scores	Rankings	Scores	Rankings
Teacher's expressive ability	0.384	6	0.344	4
Teaching level	0.412	3	0.395	2
Teaching method	0.224	8	0.221	8
Exercise	0.518	1	0.527	1
Learning threshold	0.224	9	0.220	9
Course organization	0.216	10	0.244	7
Supporting materials	0.392	5	0.270	6
Platform	0.413	2	0.331	5
Experience of learning	0.244	7	0.179	10
Professional knowledge	0.392	4	0.392	3

5. Discussion

5.1. Single Criterion Analysis

We first analyzed the results in Tables 4 and 5, which reflect the performance of the attributes of the two courses on the six criteria.

We first analyzed the three criteria c_1, c_2, and c_3, which are the proportion of positive, neutral, and negative reviews among all learners' comments on an attribute and directly reflect learners' satisfaction with the attribute. For an attribute, the higher c_1 is, or the lower c_2 and c_3 are, the higher the learner's satisfaction is; otherwise, the lower the satisfaction is. It is not difficult to find that for the two programming courses, there is little difference in the attribute satisfaction of learners for between the two courses. The two attributes that learners are most satisfied with are "Teaching level" and "Learning threshold". The majority of the learners believed that the teachers in both courses possess a commendable level of expertise, and the courses were tailored to cater to beginners. The attributes that the learners felt performed well are "Teachers' expressive ability", "Teaching methods", "Course organization" and "Learning experience". It means that the teachers of these two courses have good expression ability, can basically tell the course content clearly, and adopt appropriate teaching methods. The schedule of the course is reasonable, and the key and difficult points are prominent. The learner's listening experience was good. For "Supporting material", Java language performs well, while C language performs moderately. For C, it may be necessary to provide more high-quality supporting material. The satisfaction of the "Platform" and "Professional knowledge" of the two courses is general, which means that learners are dissatisfied with the services provided by the MOOC platform to a certain extent, and the amount and depth of professional knowledge provided by the courses may need to be improved. Finally, learners expressed dissatisfaction with the "Exercise" aspect of both courses, indicating that there is ample room for improvement in terms of exercise difficulty and quantity.

We then analyzed c_4. It is the proportion of non-emotional reviews for an attribute. As previously mentioned in Section 1, although non-emotional reviews do not directly reflect satisfaction or dissatisfaction, the attributes that reviewers mentioned in non-emotional reviews are the attributes that they thought might be of importance to future learners. It is evident that "Professional knowledge" is a relatively crucial attribute. A possible reason for this could be that in technical courses, learners tend to prioritize the quantity and quality of professional knowledge imparted by the instructor. Additionally, the expression of professional knowledge in reviews tends to utilize more non-emotional descriptions. This does not imply that other attributes are less significant; however, they may be more easily expressed through emotional reviews.

For c_5, it is the average of the number of reviews involving each word in an attribute. A higher value of c_5 means that people pay more attention to the attribute word in this attribute and, accordingly, pay more attention to this attribute. Learners of both courses pay significant attention to attributes such as "Teaching level", "Learning threshold", "Course organization", and "Teacher's expressive ability". These four attributes directly influence the learning effectiveness of students. Hence, this finding aligns with our intuitive understanding. "Professional knowledge" and "Platform" are of much interest in C language but not so much in Java. This divergence might be attributed to the widespread perception of C as a programming language closer to hardware, distinguished by its low-level and intricate syntax and structure. Consequently, learners often require a deeper level of proficiency and educational guidance to comprehend and utilize it effectively. On the contrary, Java's more refined and straightforward syntax may not necessitate extensive expertise, particularly for novice learners. The remaining attributes are of lesser concern to learners. The potential reason is that these attributes may have a minimal impact on the learning outcomes and may not hold significant value for future learners, as perceived by those who have already completed the learning process.

c_6 has already been discussed in Section 4.4, so we will not add anything further.

5.2. Comprehensive Analysis

We then analyzed the results in Table 7, which reflect the comprehensive performance of the attributes of the two courses.

We began our analysis with the top-ranked attributes: "Exercise", "Teaching level", "Professional knowledge", and "Platform". Among these, "Exercise" receives the lowest satisfaction rating and is relatively inexpensive to improve, thus carrying the highest priority. For "Teaching level", although it has the highest satisfaction and improvement cost, it still has a high improvement priority because it is the attribute that learners pay the most attention to. For "Professional knowledge", despite its low attention and high improvement cost, it still has a high improvement priority because of its low satisfaction. For "Platform", although it does not have high attention, it has low improvement cost and low satisfaction, so it has high priority.

Next, we analyzed the attributes that are in the middle ranking: "Teacher's expressive ability" and "Supporting materials". For "Teacher's expressive ability" and "Supporting materials", although their improvements are less costly, they have medium satisfaction, medium attention, and therefore medium improvement priorities.

Finally, we delved into the lower-ranked attributes: "Course organization", "Experience of learning", "Teaching method", and "Learning threshold". "Course organization" and "Experience of learning" carry low improvement priority due to their medium satisfaction ratings, low attention levels, and high costs of improvement. For "Teaching method", since they have high satisfaction, low attention, and a medium cost of improvement, they have a low priority for improvement. For "Learning threshold", despite its high attention, it has low priority for improvement due to its high satisfaction and high cost of improvement.

Following the aforementioned analysis, our method is capable of comprehensively taking into account various facets of improving course attributes, thereby encouraging a more comprehensive approach when contemplating improvements to MOOCs.

5.3. Implications and Future Research

The limitations of this study and potential future research avenues primarily encompass the following aspects: Firstly, a sentiment analysis might not be sufficiently precise. The sentence-level sentiment analysis method we employed assumes that the sentiment expressed in a sentence corresponds to the attributes mentioned within it. However, this approach can be limited when a sentence contains multiple attributes and varying emotions. Future research should focus on employing more fine-grained sentiment analysis methods, such as aspect-level sentiment analysis, to enhance the precision of sentiment analysis. Second, the way we determine costs is not necessarily accurate. As our attribute improvement costs are determined through the voting of six experts, this sample size may not be sufficient and introduces a degree of subjectivity, potentially leading to inaccuracies in cost judgments. Therefore, a comprehensive and thorough investigation into the costs associated with course improvements is crucial. Thirdly, we have employed only one method for multi-attribute decision-making, namely TOPSIS. However, there may be other, more suitable methods available. It is imperative to explore and discuss whether other multiple-attribute decision-making techniques might outperform TOPSIS in certain scenarios. Fourthly, our experiments were limited to only two courses: C language and Java language. Consequently, the experimental results obtained may not be universally applicable. To validate the generalizability of our method, it is necessary to apply it to courses across diverse fields.

6. Conclusions

Based on the learner's perspective, this paper presents a method that leverages text mining and multi-attribute decision-making frameworks to enhance the attributes of MOOCs, taking into account both non-emotional reviews and improvement costs. Initially, the method collected online reviews from MOOC platforms and extracted pertinent course attributes from these reviews. Subsequently, the review text was categorized into four sentiment groups—positive, neutral, negative, and non-emotional—utilizing deep learning techniques. To ascertain the improvement costs associated with each attribute, an expert voting mechanism was utilized. Finally, a decision matrix was constructed, and the TOPSIS method was employed to prioritize the attributes for improvement. To validate the effectiveness of our proposed method, it was applied to two programming courses: C language and Java language. The experimental results demonstrate the utility of our approach. The main contributions of this study are as follows:

(1) Regarding data selection, this study employed learners' online comments as the analytical dataset to establish a text mining model specifically tailored for MOOC online comments. The outcomes offer a real-time and visual representation of learners' preferences and needs.

(2) Given the limited resources available, realizing their rational utilization holds significant importance in enhancing the overall quality of courses. Previous studies have primarily focused on determining the priority of attribute improvements solely based on learners' satisfaction levels, neglecting the crucial aspect of cost. Nevertheless, cost plays a pivotal role in the improvement process of course attributes. In scenarios where the cost of enhancing a specific attribute is prohibitively high, it becomes imperative to deliberate on whether to pursue such improvements despite their low satisfaction rating. Consequently, this study builds upon previous research, incorporating both cost considerations and non-emotional reviews, thereby enabling course managers to undertake a more comprehensive evaluation when aiming to enhance courses.

(3) This study introduces a novel approach that integrates text mining and multi-attribute decision-making frameworks to effectively enhance MOOC attributes from the learners' perspective. By incorporating non-emotional reviews and considering improvement costs, our method offers a comprehensive and practical solution to course managers. The utilization of deep learning techniques in sentiment analysis ensures accurate categorization of reviews, while the expert voting mechanism provides a reliable estimation of improvement costs. The TOPSIS method then facilitates the prioritization of attributes for improvement, enabling course managers to make informed decisions. This approach not only addresses the challenges associated with managing overwhelming information but also ensures that limited resources are allocated efficiently.

(4) Overall, at the theoretical level, this study contributes to enriching relevant research on MOOC improvement by adopting a learner-centric perspective. It also serves as a valuable reference for similar studies. From a managerial perspective, the utilization of text mining and multi-attribute decision-making techniques enables a more precise analysis of curriculum attributes, thus addressing the challenge of managing overwhelming information faced by course managers. Furthermore, it offers course managers a more comprehensive framework to prioritize the improvement of course attributes across six dimensions: positive emotion, neutral emotion, negative emotion, no emotion, attention, and cost.

Author Contributions: Methodology, P.Y.; writing—original draft, P.Y.; visualization, P.Y.; Writing—review and editing, Y.L. (Ying Liu) and Z.W.; investigation, Y.L. (Ying Liu) and X.C.; formal analysis, Y.L. (Yuyan Luo), Z.W. and X.C.; supervision, Y.L. (Yuyan Luo). All authors have read and agreed to the published version of the manuscript.

Funding: This research was funded by the Higher Education Scientific Research Planning Project in 2022 of Chinese Association of Higher Education (Grant No. 22CX0406), the Ideological and Political Education Research Project in 2023 of University Ideological and Political Work Team Training and Research Center (Grant No. CJSFZ23-33), and the Key Research Base of Humanities and Social Sciences of Sichuan Province-Sichuan Education Informatization Application and Development Research Center Project (Project No. JYXX20-004). Major Projects on Talent Cultivation and Teaching

Reform in Higher Education in Sichuan Province in 2023 (JG2023-49). The General Project of Graduate Education and Teaching Reform in Chengdu University of Technology (Grant No. 2023YJG203).

Institutional Review Board Statement: Not applicable.

Informed Consent Statement: Not applicable.

Data Availability Statement: The data presented in this study are available on request from the corresponding authors.

Conflicts of Interest: The authors declare no conflicts of interest.

Appendix A

Table A1. The number of reviews in different sentiments for each attribute for C language.

Attributes	Positive	Neutral	Negative	Non-Emotional
Teacher's expressive ability	284	18	27	76
Teaching level	4278	177	159	313
Teaching method	428	29	37	58
Exercise	224	55	174	26
Learning threshold	1382	62	74	150
Course organization	2199	135	353	230
Supporting materials	193	26	62	20
Platform	273	30	85	52
Experience of learning	851	70	168	101
Professional knowledge	611	52	135	214

Table A2. The number of reviews in different sentiments for each attribute for Java language.

Attributes	Positive	Neutral	Negative	Non-Emotional
Teacher's expressive ability	113	13	9	15
Teaching level	1682	80	68	150
Teaching method	156	15	16	17
Exercise	62	22	81	11
Learning threshold	558	28	28	43
Course organization	823	66	124	106
Supporting materials	90	5	17	9
Platform	68	7	21	8
Experience of learning	329	25	54	33
Professional knowledge	185	27	37	69

References

1. Ahmadi, S.; Nourmohamadzadeh, Z.; Amiri, B. A hybrid DEMATEL and social network analysis model to identify factors affecting learners' satisfaction with MOOCs. *Heliyon* **2023**, *9*, e17894. [CrossRef]
2. Huang, H.; Jew, L.; Qi, D.D. Take a MOOC and then drop: A systematic review of MOOC engagement pattern and dropout factor. *Heliyon* **2023**, *9*, e15220. [CrossRef]
3. Zhang, C.H.; Su, W.H.; Chen, S.C.; Zeng, S.Z.; Liao, H.C. A Combined Weighting Based Large Scale Group Decision Making Framework for MOOC Group Recommendation. *Group Decis. Negot.* **2023**, *32*, 537–567. [CrossRef]
4. Ossiannilsson, E.; Altınay, Z.; Altınay, F. Towards fostering quality in open online education through OER and MOOC practices. In *Open Education: From OERs to MOOCs*; Springer: Berlin, Heidelberg, Germany, 2017; pp. 189–204.
5. Guerrero, M.; Heaton, S.; Urbano, D. Building universities' intrapreneurial capabilities in the digital era: The role and impacts of Massive Open Online Courses (MOOCs). *Technovation* **2021**, *99*, 102139. [CrossRef]
6. Li, L.; Johnson, J.; Aarhus, W.; Shah, D. Key factors in MOOC pedagogy based on NLP sentiment analysis of learner reviews: What makes a hit. *Comput. Educ.* **2022**, *176*, 104354. [CrossRef]
7. Su, B.H.; Peng, J. Sentiment Analysis of Comment Texts on Online Courses Based on Hierarchical Attention Mechanism. *Appl. Sci.* **2023**, *13*, 4204. [CrossRef]
8. Castillo, N.M.; Lee, J.; Zahra, F.T.; Wagner, D.A. MOOCS for development: Trends, challenges, and opportunities. *Int. Technol. Int. Dev.* **2015**, *11*, 35–42.
9. Gomez, M.J.; Calderón, M.; Sánchez, V.; Clemente, F.J.G.; Ruipérez-Valiente, J.A. Large scale analysis of open MOOC reviews to support learners' course selection. *Expert Syst. Appl.* **2022**, *210*, 118400. [CrossRef]

10. Talebi, K.; Torabi, Z.; Daneshpour, N. Ensemble models based on CNN and LSTM for dropout prediction in MOOC. *Expert Syst. Appl.* **2024**, *235*, 121187. [CrossRef]
11. Li, S.; Du, J.L.; Yu, S. Diversified resource access paths in MOOCs: Insights from network analysis. *Comput. Educ.* **2023**, *204*, 104869. [CrossRef]
12. Pandey, M.; Litoriya, R.; Pandey, P. Scrutinizing student dropout issues in MOOCs using an intuitionistic fuzzy decision support system. *J. Intell. Fuzzy Syst.* **2023**, *44*, 4041–4058. [CrossRef]
13. Wang, Y. Where and what to improve? Design and application of a MOOC evaluation framework based on effective teaching practices. *Distance Educ.* **2023**, *44*, 458–475. [CrossRef]
14. Almufarreh, A.; Arshad, M. Exploratory Students' Behavior towards Massive Open Online Courses: A Structural Equation Modeling Approach. *Systems* **2023**, *11*, 223. [CrossRef]
15. Qi, C.; Liu, S.D. Evaluating On-Line Courses via Reviews Mining. *IEEE Access* **2021**, *9*, 35439–35451. [CrossRef]
16. Wei, X.X.; Taecharungroj, V. How to improve learning experience in MOOCs an analysis of online reviews of business courses on Coursera. *Int. J. Manag. Educ.* **2022**, *20*, 100675. [CrossRef]
17. Ahadi, A.; Singh, A.; Bower, M.; Garrett, M. Text mining in education—A bibliometrics-based systematic review. *Educ. Sci.* **2022**, *12*, 210. [CrossRef]
18. Geng, S.; Niu, B.; Feng, Y.; Huang, M. Understanding the focal points and sentiment of learners in MOOC reviews: A machine learning and SC-LIWC-based approach. *Br. J. Educ. Technol.* **2020**, *51*, 1785–1803. [CrossRef]
19. Liu, Q.; Ding, Y.; Qian, P.; Li, R.; Zhou, J. Analysis of the influencing factors of online classes satisfaction based on text mining Take MOOC platform art education online classes as an example. *J. Educ. Humanit. Soc. Sci.* **2022**, *2*, 249–256. [CrossRef]
20. Jung, Y.; Lee, B.G. Research trends in text mining: Semantic network and main path analysis of selected journals. *Expert Syst. Appl.* **2020**, *162*, 113851. [CrossRef]
21. Kou, G.; Yang, P.; Peng, Y.; Xiao, H.; Xiao, F.; Chen, Y.; Alsaadi, F.E. A cross-platform market structure analysis method using online product reviews. *Technol. Econ. Dev. Econ.* **2021**, *27*, 992–1018. [CrossRef]
22. Chen, K.; Jin, J.; Luo, J. Big consumer opinion data understanding for Kano categorization in new product development. *J. Ambient. Intell. Humaniz. Comput.* **2022**, *13*, 2269–2288. [CrossRef]
23. Shi, L.L.; Lin, J.; Liu, G.Q. Product feature extraction from Chinese online reviews: Application to product improvement. *Rairo-Oper. Res.* **2023**, *57*, 1125–1147. [CrossRef]
24. Zhou, F.; Ayoub, J.; Xu, Q.; Yang, X.J. A machine learning approach to customer needs analysis for product ecosystems. *J. Mech. Des.* **2020**, *142*, 011101. [CrossRef]
25. Lee, H.; Cha, M.S.; Kim, T. Text mining-based mapping for kano quality factor. *ICIC Express Letters. Part B, Applications: An International. J. Res. Surv.* **2021**, *12*, 185–191.
26. Guo, J.; Wang, X.; Wu, Y. Positive emotion bias: Role of emotional content from online customer reviews in purchase decisions. *J. Retail. Consum. Serv.* **2020**, *52*, 101891. [CrossRef]
27. Picazo-Vela, S.; Chou, S.Y.; Melcher, A.J.; Pearson, J.M. Why provide an online review? An extended theory of planned behavior and the role of Big-Five personality traits. *Comput. Hum. Behav.* **2010**, *26*, 685–696. [CrossRef]
28. Li, G.; Reimann, M.; Zhang, W. When remanufacturing meets product quality improvement: The impact of production cost. *Eur. J. Oper. Res.* **2018**, *271*, 913–925. [CrossRef]
29. Zavadskas, E.K.; Turskis, Z.; Kildienė, S. State of art surveys of overviews on MCDM/MADM methods. *Technol. Econ. Dev. Econ.* **2014**, *20*, 165–179. [CrossRef]
30. Kastrati, Z.; Imran, A.S.; Kurti, A. Weakly supervised framework for aspect-based sentiment analysis on students' reviews of MOOCs. *IEEE Access* **2020**, *8*, 106799–106810. [CrossRef]
31. Mrhar, K.; Benhiba, L.; Bourekkache, S.; Abik, M. A Bayesian CNN-LSTM model for sentiment analysis in massive open online courses MOOCs. *Int. J. Emerg. Technol. Learn.* **2021**, *16*, 216–232. [CrossRef]
32. Wang, C.; Huang, S.; Zhou, Y. Sentiment analysis of MOOC reviews via ALBERT-BiLSTM model. *EDP Sci.* **2021**, *336*, 05008. [CrossRef]
33. Dina, N.; Yunardi, R.; Firdaus, A. Utilizing text mining and feature-sentiment-pairs to support data-driven design automation massive open online course. *Int. J. Emerg. Technol. Learn.* **2021**, *16*, 134–151. [CrossRef]
34. Liu, W.; Lin, S.; Gao, B.; Huang, K.; Liu, W.; Huang, Z.; Feng, J.; Chen, X.; Huang, F. BERT-POS: Sentiment analysis of MOOC reviews based on BERT with part-of-speech information. In Proceedings of the 23rd International Conference on Artificial Intelligence in Education, Durham, UK, 27–31 July 2022; Springer International Publishing: Cham, Switzerland, 2022; pp. 371–374.
35. Liu, Z.; Kong, X.; Chen, H.; Liu, S.; Yang, Z.K. MOOC-BERT: Automatically Identifying Learner Cognitive Presence from MOOC Discussion Data. *IEEE Trans. Learn. Technol.* **2023**, *16*, 528–542. [CrossRef]
36. Lundqvist, K.; Liyanagunawardena, T.; Starkey, L. Evaluation of student feedback within a MOOC using sentiment analysis and target groups. *Int. Rev. Res. Open Distrib. Learn.* **2020**, *21*, 140–156. [CrossRef]
37. Hew, K.F.; Hu, X.; Qiao, C.; Tang, Y. What predicts student satisfaction with MOOCs: A gradient boosting trees supervised machine learning and sentiment analysis approach. *Comput. Educ.* **2020**, *145*, 103724. [CrossRef]
38. Nie, Y.; Luo, H.; Sun, D. Design and validation of a diagnostic MOOC evaluation method combining AHP and text mining algorithms. *Interact. Learn. Environ.* **2021**, *29*, 315–328. [CrossRef]

39. Chen, X.; Cheng, G.; Xie, H.; Chen, G.; Zou, D. Understanding MOOC reviews: Text mining using structural topic model. *Hum.-Cent. Intell. Syst.* **2021**, *1*, 55–56. [CrossRef]
40. Nilashi, M.; Abumalloh, R.A.; Zibarzani, M.; Samad, S.; Zogaan, W.A.; Ismail, M.Y.; Mohd, S.; Akib, N.A.M. What factors influence students satisfaction in massive open online courses? Findings from user-generated content using educational data mining. *Educ. Inf. Technol.* **2022**, *27*, 9401–9435. [CrossRef]
41. Wang, W.; Liu, H.W.; Wu, Y.J.; Goh, M. Disconfirmation effect on online reviews and learner satisfaction determinants in MOOCs. *Educ. Inf. Technol.* **2023**, *28*, 15497–15521. [CrossRef]
42. Blei, D.M.; Ng, A.Y.; Jordan, M.I. Latent dirichlet allocation. *J. Mach. Learn. Res.* **2003**, *3*, 993–1022.
43. Liu, C.; Jia, G.; Kong, J. Requirement-oriented engineering characteristic identification for a sustainable product–service system: A multi-method approach. *Sustainability* **2020**, *12*, 8880. [CrossRef]
44. Chen, Z.S.; Liu, X.L.; Chin, K.S.; Pedrycz, W.; Tsui, K.L.; Skibniewski, M.J. Online-review analysis based large-scale group decision-making for determining passenger demands and evaluating passenger satisfaction: Case study of high-speed rail system in China. *Inf. Fusion* **2021**, *69*, 22–39. [CrossRef]
45. Jin, J.; Jia, D.; Chen, K. Mining online reviews with a Kansei-integrated Kano model for innovative product design. *Int. J. Prod. Res.* **2022**, *60*, 6708–6727. [CrossRef]
46. Goldberg, D.M.; Abrahams, A.S. Sourcing product innovation intelligence from online reviews. *Decis. Support Syst.* **2022**, *157*, 113751. [CrossRef]
47. Huang, S.; Zhang, J.; Yang, C.; Gu, Q.; Li, M.; Wang, W. The interval grey QFD method for new product development: Integrate with LDA topic model to analyze online reviews. *Eng. Appl. Artif. Intell.* **2022**, *114*, 105213. [CrossRef]
48. Ji, F.; Cao, Q.; Li, H.; Fujita, H.; Liang, C.; Wu, J. An online reviews-driven large-scale group decision making approach for evaluating user satisfaction of sharing accommodation. *Expert Syst. Appl.* **2023**, *213*, 118875. [CrossRef]
49. Joung, J.; Kim, H. Interpretable machine learning-based approach for customer segmentation for new product development from online product reviews. *Int. J. Inf. Manag.* **2023**, *70*, 102641. [CrossRef]
50. Zhang, F.; Song, W. Product improvement in a big data environment: A novel method based on text mining and large group decision making. *Expert Syst. Appl.* **2024**, *245*, 123015. [CrossRef]
51. Liang, X.; Liu, P.D.; Wang, Z.H. Hotel selection utilizing online reviews: A novel decision support model based on sentiment analysis and dl-vikor method. *Technol. Econ. Dev. Econ.* **2019**, *25*, 1139–1161. [CrossRef]
52. Lenort, R.; Wicher, P.; Zapletal, F. On influencing factors for Sustainable Development goal prioritisation in the automotive industry. *J. Clean Prod.* **2023**, *387*, 135718. [CrossRef]
53. Cao, P.P.; Zheng, J.; Li, M.Y. Product Selection Considering Multiple Consumers' Expectations and Online Reviews: A Method Based on Intuitionistic Fuzzy Soft Sets and TODIM. *Mathematics* **2023**, *11*, 3767. [CrossRef]
54. Vyas, V.; Uma, V.; Ravi, K. Aspect-based approach to measure performance of financial services using voice of customer. *J. King Saud Univ.-Comput. Inf. Sci.* **2022**, *34*, 2262–2270. [CrossRef]
55. Li, K.; Chen, C.Y.; Zhang, Z.L. Mining online reviews for ranking products: A novel method based on multiple classifiers and interval-valued intuitionistic fuzzy TOPSIS. *Appl. Soft. Comput.* **2023**, *139*, 110237. [CrossRef]
56. Qin, J.D.; Zeng, M.Z. An integrated method for product ranking through online reviews based on evidential reasoning theory and stochastic dominance. *Inf. Sci.* **2022**, *612*, 37–61. [CrossRef]
57. Zhang, Z.Y.; Guo, J.; Zhang, H.R.; Zhou, L.X.; Wang, M.J. Product selection based on sentiment analysis of online reviews: An intuitionistic fuzzy TODIM method. *Complex Intell. Syst.* **2022**, *8*, 3349–3362. [CrossRef]
58. Nilashi, M.; Mardani, A.; Liao, H.C.; Ahmadi, H.; Manaf, A.A.; Almukadi, W. A Hybrid Method with TOPSIS and Machine Learning Techniques for Sustainable Development of Green Hotels Considering Online Reviews. *Sustainability* **2019**, *11*, 6013. [CrossRef]
59. Zayat, W.; Kilic, H.S.; Yalcin, A.S.; Zaim, S.; Delen, D. Application of MADM methods in Industry 4.0: A literature review. *Comput. Ind. Eng.* **2023**, *177*, 109075. [CrossRef]
60. Behzadian, M.; Otaghsara, S.K.; Yazdani, M.; Ignatius, J. A state-of the-art survey of TOPSIS applications. *Expert Syst. Appl.* **2012**, *39*, 13051–13069. [CrossRef]
61. Liu, Y.; Bi, J.W.; Fan, Z.P. A Method for Ranking Products through Online Reviews Based on Sentiment Classification and Interval-Valued Intuitionistic Fuzzy TOPSIS. *Int. J. Inf. Technol. Decis. Mak.* **2017**, *16*, 1497–1522. [CrossRef]
62. Bojanowski, P.; Grave, E.; Joulin, A.; Mikolov, T. Enriching word vectors with subword information. *Trans. Assoc. Comput. Linguist.* **2017**, *5*, 135–146. [CrossRef]
63. Frey, B.J.; Dueck, D. Clustering by passing messages between data points. *Science* **2007**, *315*, 972–976. [CrossRef] [PubMed]
64. Devlin, J.; Chang, M.W.; Lee, K.; Toutanova, K. Bert: Pre-training of deep bidirectional transformers for language understanding. *arXiv* **2018**, arXiv:1810.04805.
65. Iyyer, M.; Manjunatha, V.; Boyd-Graber, J.; Daumé, H. Deep Unordered Composition Rivals Syntactic Methods for Text Classification. In Proceedings of the 53rd Annual Meeting of the Association for Computational Linguistics and the 7th International Joint Conference on Natural Language Processing, Beijing, China, 26–31 July 2015; Association for Computational Linguistics: Toronto, ON, Canada, 2015; Volume 1: Long Papers, pp. 1681–1691.
66. Kim, Y. Convolutional neural networks for sentence classification. *arXiv* **2014**, arXiv:1408.5882.

67. Yang, Z.; Yang, D.; Dyer, C.; He, X.; Smola, A.; Hovy, E. Hierarchical attention networks for document classification. In Proceedings of the 2016 Conference of the North American Chapter of the Association for Computational Linguistics: Human Language Technologies, San Diego, CA, USA, 12–17 June 2016; pp. 1480–1489.
68. Stillwell, W.G.; Seaver, D.A.; Edwards, W. A comparison of weight approximation techniques in multiattribute utility decision making. *Organ. Behav. Hum. Perform.* **1981**, *28*, 62–77. [CrossRef]
69. Khan, M.N.; Gupta, N.; Matharu, M.; Khan, M.F. Sustainable E-Service Quality in Tourism: Drivers Evaluation Using AHP-TOPSIS Technique. *Sustainability* **2023**, *15*, 7534. [CrossRef]
70. Kou, G.; Yang, P.; Peng, Y.; Xiao, F.; Chen, Y.; Alsaadi, F.E. Evaluation of feature selection methods for text classification with small datasets using multiple criteria decision-making methods. *Appl. Soft Comput.* **2020**, *86*, 105836. [CrossRef]
71. Ji, J.; Wang, D.Y. Evaluation analysis and strategy selection in urban flood resilience based on EWM-TOPSIS method and graph model. *J. Clean Prod.* **2023**, *425*, 138955. [CrossRef]

Disclaimer/Publisher's Note: The statements, opinions and data contained in all publications are solely those of the individual author(s) and contributor(s) and not of MDPI and/or the editor(s). MDPI and/or the editor(s) disclaim responsibility for any injury to people or property resulting from any ideas, methods, instructions or products referred to in the content.

Article

Automatic Vulgar Word Extraction Method with Application to Vulgar Remark Detection in Chittagonian Dialect of Bangla

Tanjim Mahmud [1,2,*], Michal Ptaszynski [1,*] and Fumito Masui [1]

[1] Text Information Processing Laboratory, Kitami Institute of Technology, Kitami 090-8507, Japan; f-masui@mail.kitami-it.ac.jp

[2] Department of Computer Science and Engineering, Rangamati Science and Technology University, Rangamati 4500, Bangladesh

* Correspondence: tanjim_cse@yahoo.com (T.M.); michal@mail.kitami-it.ac.jp (M.P.)

Abstract: The proliferation of the internet, especially on social media platforms, has amplified the prevalence of cyberbullying and harassment. Addressing this issue involves harnessing natural language processing (NLP) and machine learning (ML) techniques for the automatic detection of harmful content. However, these methods encounter challenges when applied to low-resource languages like the Chittagonian dialect of Bangla. This study compares two approaches for identifying offensive language containing vulgar remarks in Chittagonian. The first relies on basic keyword matching, while the second employs machine learning and deep learning techniques. The keyword-matching approach involves scanning the text for vulgar words using a predefined lexicon. Despite its simplicity, this method establishes a strong foundation for more sophisticated ML and deep learning approaches. An issue with this approach is the need for constant updates to the lexicon. To address this, we propose an automatic method for extracting vulgar words from linguistic data, achieving near-human performance and ensuring adaptability to evolving vulgar language. Insights from the keyword-matching method inform the optimization of machine learning and deep learning-based techniques. These methods initially train models to identify vulgar context using patterns and linguistic features from labeled datasets. Our dataset, comprising social media posts, comments, and forum discussions from Facebook, is thoroughly detailed for future reference in similar studies. The results indicate that while keyword matching provides reasonable results, it struggles to capture nuanced variations and phrases in specific vulgar contexts, rendering it less robust for practical use. This contradicts the assumption that vulgarity solely relies on specific vulgar words. In contrast, methods based on deep learning and machine learning excel in identifying deeper linguistic patterns. Comparing SimpleRNN models using Word2Vec and fastText embeddings, which achieved accuracies ranging from 0.84 to 0.90, logistic regression (LR) demonstrated remarkable accuracy at 0.91. This highlights a common issue with neural network-based algorithms, namely, that they typically require larger datasets for adequate generalization and competitive performance compared to conventional approaches like LR.

Keywords: vulgar remark detection; vulgar term extraction; low-resource language; logistic regression; recurrent neural network

1. Introduction

Bangladesh has seen a remarkable increase in its use of the Internet over the past two decades. There were more than 125 million Internet users in Bangladesh as of November 2022, according to the Bangladesh Telecommunication Regulatory Commission (BTRC) [1]. Additionally, with the help of the implementation of the Digital Bangladesh initiative [2], the vast majority of people in Chittagong [3], Bangladesh's second-largest city, now have access to the Internet and actively use social media. According to a survey [4], the number of Facebook users in Bangladesh is the highest among social media (see

Figure 1). Moreover, with the benefit of Unicode being widely used on most communication devices, such as tablets or smartphones, speakers of underrepresented languages, such as Chittagonian, can express their thoughts in their native languages and dialects. Many people in Chittagong now use social media on a regular basis, regularly using platforms like Facebook [5], imo [6], various blogs, and WhatsApp [7]. These platforms offer a venue where people can express themselves freely and informally. However, the pervasiveness of social media has also resulted in unfavorable influences that are difficult to shake. Excessive use of social media has the potential to cause addiction [8], which as a result could cause young people to spend more time on these platforms than they spend with their family and friends [9]. Their general health and social interactions may suffer as a result of this addiction. Additionally, social media witnesses the growing problem of the increase in online abuse and cyberbullying, which can have a negative impact on a person's self-esteem and even violate their privacy [10]. The spread of misinformation and hatred online has also contributed to an uptick in violent crimes in society [11]. Receiving messages with vulgar language is a startling realization of this unwelcome and damaging phenomenon. The likelihood of encountering such vulgar remarks rises as social media use increases.

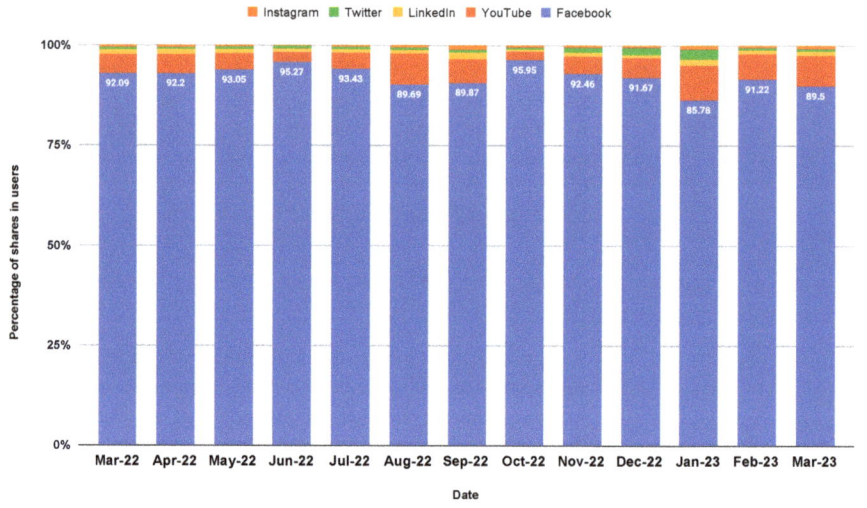

Figure 1. Yearly social media usage statistics in Bangladesh as of March 2023 [4].

Vulgarity or obscenity with regards to language refers to terms used to describe the use of vulgar language, such as swearing, taboo words, or offensive expressions [12,13]. Unfortunately, such language has become more and more common in contemporary culture [14], especially on social media sites like Twitter [15]. The majority of the time, however, vulgar language is used in the context of online harassment and negativity. While there are some instances where vulgar language may be used in a positive context to convey informality, express anger, or establish a sense of belonging with a particular group [16], these are usually rare in comparison to its use in negative contexts or in closed message groups. Therefore, detecting such use of language quickly and effectively is necessary to allow social media platforms to efficiently moderate their contents.

Therefore, the goal of this research was to find and evaluate vulgar remarks in the Chittagonian dialect of Bangla. Chittagonian is a member of the Indo-Aryan language family [17] and is spoken by between 13 and 16 million people, most of whom live in Bangladesh [18]. Although some linguists categorize it as a distinct language, Chittagonian is frequently used together with Bengali and has its own pronunciation, vocabulary, and grammar [19]. In this paper, we present a system for automatic detection of vulgar remarks in an effort to combat the growing problem of vulgar language online. To achieve this we apply various machine learning (ML) and deep learning (DL) algorithms, such as logistic regression (LR) or recurrent neural networks (RNN). We also use a variety of feature extraction techniques to expand the system's functionality. The performance of these ML and DL algorithms in detecting vulgar remarks was thoroughly investigated through rigorous experimentation, which is particularly important in a low-resource language scenario, such as Chittagonian, where linguistic resources are scarce.

The goal of this research is to advance the field of vulgar remark detection for the Chittagonian dialect by achieving the following key objectives:

1. Collect a comprehensive dataset of 2500 comments and posts exclusively from publicly accessible Facebook accounts in the Chittagonian dialect.
2. Ensure the dataset's quality and reliability through rigorous manual annotation and validation using established metrics like Cohen's Kappa statistics [20] and Krippendorff's alpha [21].
3. Develop a simple keyword-matching-based baseline method for vulgar remark detection using a hand-crafted vulgar word lexicon.
4. Create a method for automatically expanding the vulgar word lexicon to ensure future-proofing of the baseline method.
5. Implement various matching algorithms to detect sentences containing vulgar remarks, beginning with a simple method using a manually crafted vulgar word lexicon, an automatic method using simple TF-IDF statistics for vulgar term extraction with no additional filtering of non-vulgar words, as well as a more robust method applying additional filtering of words with a high probability of being non-vulgar.
6. Evaluate various ML- and DL-based approaches to identify vulgar remarks in Chittagonian social media posts, aiming for over 90% accuracy for practical applicability.
7. Conduct a thorough comparison between the keyword-matching baseline method and machine learning (ML) and deep learning (DL) models to achieve the highest possible performance in vulgar remark detection.

The subsequent sections of this paper are organized as follows: In Section 2, we present a comprehensive review of related works on vulgar word detection and related content filtering in the languages of Bangladesh. Section 3 elaborates on the dataset collection and preprocessing techniques specific to the Chittagonian dialect as well as outlines the proposed automatic vulgar word extraction method, detailing the NLP, ML, and DL techniques employed. In Section 4, we present the experimental results and performance evaluation of our approach. Finally, Section 5 summarizes the contributions and discusses potential future directions for expanding this research.

2. Literature Review

This section presents a thorough analysis of prior work on the identification and categorization of vulgarity. We have also included studies from closely related fields because vulgar language in user-generated content on social media frequently includes expressions of sexism, racism, hate speech, and other types of online abuse [12]. Table 1 shows research in Bengali on topics related to detecting vulgarity.

Traditionally, vulgar expression lexicons have been developed as a means of vulgarity detection [22]. These lexicon-based approaches need to be updated frequently to remain effective, however. In contrast, machine learning (ML) techniques provide a more dynamic approach by classifying new expressions as either vulgar or non-vulgar without relying on predetermined lexicons. Deep learning has made significant contributions to the field of signal and image processing [23], diagnosis [24], wind forecasting [25] and time series forecasting [26].

Beyond lexicon-based techniques, vulgarity detection has been the subject of several studies. Moreover, numerous linguistic and psychological studies [27] have been carried out to comprehend the pragmatic applications [13] and various vulgar language forms [28].

For machine learning-related studies, for example, Eshan et al. [29] ran an experiment in which they classified data obtained by scraping the Facebook pages of well-known celebrities using the traditional machine learning classifiers multinomial naive Bayes, random forest, and SVM (support vector machine). They gathered unigram, bigram, and trigram features and weighted them using *TF-IDF* vectorizers. On datasets of various sizes, containing 500, 1000, 1500, 2000, and 2500 samples. The results showed that when using unigram features, a sigmoid kernel had the worst accuracy performance, and SVM with a linear kernel had the best accuracy performance. However, MNB demonstrated the highest level of accuracy for bigram and trigram features. In conclusion, TfidfVectorizer features outperformed CountVectorizer features when combined with an SVM linear kernel.

Akhter et al. [30] suggested using user data and machine learning techniques to identify instances of cyberbullying in Bangla. They used a variety of classification algorithms, such as naive Bayes (NB), J48 decision trees, support vector machine (SVM), and k-nearest neighbors (KNN). A 10-fold cross-validation was used to assess how well each method performed. The results showed that SVM performed better than the other algorithms when it came to analyzing Bangla text, displaying the highest accuracy score of 0.9727.

Holgate et al. [16] introduced a dataset of 7800 tweets from users whose demographics were known. Each instance of vulgar language use was assigned to one of six different categories by the researchers. These classifications included instances of aggression, emotion, emphasis, group identity signaling, auxiliary usage, and non-vulgar situations. They sought to investigate the practical implications of vulgarity and its connections to societal problems through a thorough analysis of this dataset. Holgate et al. obtained a macro F1 score of 0.674 across the six different classes by thoroughly analyzing the data that were gathered.

Emon et al. [31] created a tool to find abusive Bengali text. They used various deep learning and machine learning-based algorithms to achieve this. A total of 4700 comments from websites like Facebook, YouTube, and Prothom Alo were collected in a dataset. These comments were carefully labeled into seven different categories. Emon et al. experimented with various algorithms to find the best one. The recurrent neural network (RNN) algorithm demonstrated the highest accuracy among the investigated methods, achieving a satisfying score of 0.82.

Awal et al. [32] demonstrated a naive Bayes system made to look for abusive comments. They gathered a dataset of 2665 English comments from YouTube in order to evaluate their system. They then translated these English remarks into Bengali utilizing two techniques: (i) Bengali translation directly; (ii) Bengali translation using dictionaries. Awal et al. evaluated the performance of their system after the translations. Their system impressively achieved the highest accuracy of 0.8057, demonstrating its potency in identifying abusive content in the context of the Bengali language.

Hussain et al. [33] suggested a method that makes use of a root-level algorithm and unigram string features to identify abusive Bangla comments. They gathered 300 comments for their dataset from a variety of websites, including Facebook pages, news websites, and YouTube. The dataset was split into three subsets, each of which contained 100, 200, and 300 comments. These subsets were used to test their system, which resulted in an average accuracy score of 0.689.

Das et al. [34] carried out a study on detecting hate speech in Bengali and Romanized Bengali. They extracted samples from Twitter in order to gather the necessary information, producing a dataset with 5071 samples in Bengali and Romanized Bengali. They used a variety of training models in their study, including XML-RoBERTa, MuRIL, m-BERT, and IndicBERT. Following testing, they discovered that XML-RoBERTa had the highest accuracy, at 0.796.

Sazzed [35] collected 7245 YouTube reviews manually and divided them into two categories: vulgar and non-vulgar. The purpose of this process was to produce two benchmark corpora for assessing vulgarity detection algorithms. Following the testing of several methods, the bidirectional long short-term memory (BiLSTM) model showed the most promising results, achieving the highest recall scores for identifying vulgar content in both datasets.

Jahan et al. [36] created a dataset by using online comment scraping tools to collect comments from public Facebook pages, such as news and celebrity pages. SVM, random Forest, and AdaBoost were the three machine learning techniques used to categorize the comments for the detection of abusive content. Their approach, which was based on the random forest classifier, outperformed other methods in terms of accuracy and precision, scoring 0.7214 and 0.8007, respectively. AdaBoost, on the other hand, demonstrated the best recall performance, earning a score of 0.8131.

Ishmam et al. [37] collected a dataset sourced from Facebook, categorized into six distinct classes. The dataset was enriched with linguistic and quantitative features, and the researchers employed a range of text preprocessing techniques, including punctuation removal, elimination of bad characters, handling hashtags, URLs, and mentions, as well as tokenization and stemming. They utilized neural networks, specifically GRUs (gated recurrent units), alongside other machine learning classifiers, to conduct classification tasks based on the historical, religious, cultural, social, and political contexts of the data.

Karim et al. [38] used a combination of machine learning classifiers and deep neural networks to detect hate speech in Bengali. They analyzed datasets containing comments from Facebook, YouTube, and newspaper websites using a variety of models, including logistic regression, SVM, CNN, and Bi-LSTM. The researchers divided hate speech into four distinct categories: political, religious, personal, and geopolitical. With F1 scores of 0.78 for political hate speech, 0.91 for personal hate speech, 0.89 for geopolitical hate speech, and 0.84 for religious hate speech detection in the Bengali language, their results showed satisfying performance.

Sazzed [39] created a transliterated corpus of 3000 comments from Bengali, 1500 of which were abusive and 1500 of which were not. As a starting point, they used a variety of supervised machine learning methods, such as deep learning-based bidirectional long short-term memory networks (BiLSTM), support vector machines (SVM), logistic regression (LR), and random forest (RF). The SVM classifier displayed the most encouraging results (with an F1 score of 0.827 ± 0.010) in accurately detecting abusive content.

User comments from publicly viewable Facebook posts made by athletes, officials, and celebrities were analyzed in a study by Ahmed et al. [40]. The researchers distinguished between Bengali-only comments and those written in English or a mix of English and other languages. Their research showed that 14,051 initial comments in total, or approximately 31.9% of them, were directed at male victims. However, a significant number of the 29,950 comments, or 68.1% of the total, were directed at female victims. The study also highlighted how comments were distributed according to the different types of victims. A total of 9375 comments were directed at individuals who are social influencers. Among these, 5.98% (equivalent to 2633 comments) were aimed at politicians, while 4.68% (or 2061 comments) were focused on athletes. Additionally, 6.78% (about 2981 comments) of the comments were centered around singers, and the majority, which is 61.25% (totaling 26,951 comments), were directed at actors.

For the classification of hate speech in the Bengali language, Romim et al. [41] used neural networks, including LSTM (long short-term memory) and BiLSTM (bidirectional LSTM). They used word embeddings that had already been trained using well-known algorithms such as FastText, Word2Vec, and Glove. The largest dataset of its kind to date, the extensive Bengali dataset they introduced for the research includes 30,000 user comments. The researchers thoroughly compared different deep learning models and word embedding combinations. The outcomes were encouraging as all of the deep learning models performed well in the classification of hate speech. However, the support vector machine (SVM) outperformed the others with an accuracy of 0.875.

Islam et. al. [42] used large amounts of data gathered from Facebook and YouTube to identify abusive comments. To produce the best results, they used a variety of machine learning algorithms, such as multinomial naive Bayes (MNB), multilayer perceptron (MLP), support vector machines (SVM), decision tree, random forest, and SVM with stochastic gradient descent-based optimization (SGD), ridge classifier, and k-nearest neighbors (k-

NN). They used a Bengali stemmer for preprocessing and random undersampling of the dominant class before processing the dataset. The outcomes demonstrated that, when applied to the entire dataset, SVM had the highest accuracy of 0.88.

In their study, Aurpa et al. [43] used transformer-based deep neural network models, like BERT [44] and ELECTRA [45], to categorize abusive comments on Facebook. For testing and training, they used a dataset with 44,001 Facebook comments. The test accuracy for their models, which was 0.85 for the BERT classifier and 0.8492 for the ELECTRA classifier, showed that they were successful in identifying offensive content on the social media platform.

Table 1. Research on vulgarity detection or related topics in Bengali (Facebook (F), YouTube (Y)).

Paper	Classifier	Highest Score	Language	Sample Size	Class and Ratio	Data Sources
[29]	Multinomial Naive Bayes, Random Forest, Support Vector Machines,	80% (Accuracy)	Bengali	2.5K	-	F
[30]	Support Vector Machines, Naive Bayes, Decision Tree, K-Nearest Neighbors	97% (Accuracy)	Bengali	2.4 K	Non-Bullying Bullying (10%)	F, T
[31]	Linear Support Vector Classification, Logistic Regression, Multinomial Naive Bayes, Random Forest Artificial Neural Network, RNN + LSTM	82.2% (Accuracy)	Bengali	4.7 K	Slang (19.57%), Religious, Politically, Positive, Neutral, violated (13.28%), Anti-feminism (0.87%), Hatred (13.15%), Personal attack (12.36%)	F, Y, News portal
[32]	Naive Bayes	80.57% (Accuracy)	Bengali	2.665 K	Non-Abusive, Abusive (45.55%)	Y
[33]	Root-Level approach	68.9% (Accuracy)	Bengali	300	Not Bullying, Bullying	F, Y News portal
[35]	Logistic Regression, Support Vector Machines, Stochastic Gradient Descent, Bidirectional LSTM	89.3% (F1 Score) 82.4% (F1 Score)	Bengali	7.245 K	Non Vulgar, Vulgar	Y
[36]	Support Vector Machines, RF, Adaboost	72.14% (Accuracy) 80% (Precision)	Bengali	2 K	Non Abusive, Abusive (78.41%)	F

Table 1. *Cont.*

Paper	Classifier	Highest Score	Language	Sample Size	Class and Ratio	Data Sources
[37]	Gated Recurrent Units, Support Vector Classification, LinearSVC, Random Forest, Naive Bayes	70.1% (Accuracy)	Bengali	5.126 K	Religious comment (14.9%), Hate speech (19.2%), Inciteful (10.77%), Communal hatred (15.67%), Religious hatred (15.68%), Political comment (23.43%)	F
[38]	Logistic Regression, Support Vector Machines, Convolutional Neural Network, BIdirectional LSTM, BERT, LSTM	78% 91% 89% 84% (F1 Score)	Bengali	8.087 K	Personal (43.44%), Religious (14.97%), Geopolitical (29.23%), Political (12.35%)	F
[39]	Logistic Regression, Support Vector Machines, Random Forest, Bidirectional LSTM	82.7% (F1 Score)	Bengali	3 K	Non abusive, Abusive (10%)	Y
[41]	Long Short-term Memory, Bidirectional LSTM	87.5% (Accuracy)	Bengali	30 K	Not Hate speech, Hate speech (33.33%)	F, Y
[42]	Multinomial Naive Bayes, Multilayer Perceptron, Support Vector Machines, Decision Tree, Random Fores, Stochastic Gradient Descent, K-Nearest Neighbors	88% (Accuracy)	Bengali	9.76 K	Non Abusive, abusive (50%)	F, Y
[43]	ELECTRA, Deep Neural Network, BERT	85% (Accuracy) (BERT), 84.92% (Accuracy) (ELECTRA)	Bengali	44.001 K	Troll (23.78%), Religious (17.22%), Sexual (20.29%), Not Bully (34.86%), Threat (3.85%)	F

Based on our comprehensive analysis of papers related to vulgarity detection and related topics like abusive and bullying detection, as well as detection in the low-resource language Bengali, several critical research gaps emerge. These gaps include the absence of a clear problem definition in some papers, the prevalence of small-sized datasets without a well-defined annotation process, and the lack of benchmarking efforts to assess dataset quality. Additionally, class imbalance in datasets remains an issue, and limited attention has been given to vulgarity detection in low-resource language Bengali, with only a single work [35] addressing this area. Many papers fail to specify the source of their datasets and conduct limited experiments. Field surveys are often superficial or nonexistent. Furthermore, none of the papers considered ethical considerations in data collection, such as preserving user privacy through dataset anonymization. Addressing these research gaps is essential for advancing the field of vulgarity detection and related areas, ensuring the development of more robust, ethical, and well-defined detection systems.

Although many of the above-mentioned studies focus on the detection of bullying or hate speech, which often contain vulgar remarks, the presence of vulgarities specifically in the Chittagonian dialect of Bangla has not previously been investigated. By concentrating on information taken from posts on social media that were written in the Chittagonian dialect, this study seeks to close this gap. It is the start of an effort to accurately identify and gauge the frequency of vulgar language used in these social media posts.

3. Proposed Methodology

The experimental procedures are depicted in Figure 2, and the methodology is explained as follows:

3.1. Data Collection

Due to the lack of a good quality dataset designed specifically for vulgar text detection in the Chittagonian dialect, gathering data posed one major challenge in this study. The ML models need to be trained and tested on a sizable dataset in order to produce trustworthy results for classifying vulgar remarks. The data collection procedure used for the vulgar word detection in the paper is described as follows (see also Figure 3):

1. The dataset used in this study was made up of text excerpts from social media sites and open comment sections. A dataset with a wide range of topics, writing styles, and user demographics was intended to be both diverse and representative.
2. Facebook comments were manually gathered from a variety of independent sources, such as the public profiles and pages of well-known people.
3. Random sampling was used to guarantee a balanced and representative dataset. Each data source's popularity, user activity, and content suitability had to be taken into account when selecting random text samples from it. The objective was to gather a significant amount of information while keeping a variety of vulgar words and their context.

Figure 4 shows six examples from the dataset.

3.2. Data Annotation Process

Annotating data is required for the creation of machine learning models, such as those used to detect vulgar remarks. To properly train the model, the data must be labeled or annotated with relevant information. Below, we provide an outline of the data annotation process for vulgar remark detection.

Data annotators: Three native speakers of the Chittagonian dialect were hired, two of whom had Bachelor of Science degrees in engineering and one of whom had a master's degree in linguistics.

Experts responsible for preparing annotation standard and guidelines: Two people work in NGO organization [46], where they work on, among others, Internet-based surveys about harassment. One of them was a male, and the other one was a female, both with higher education (master's degree, sociology and social work).

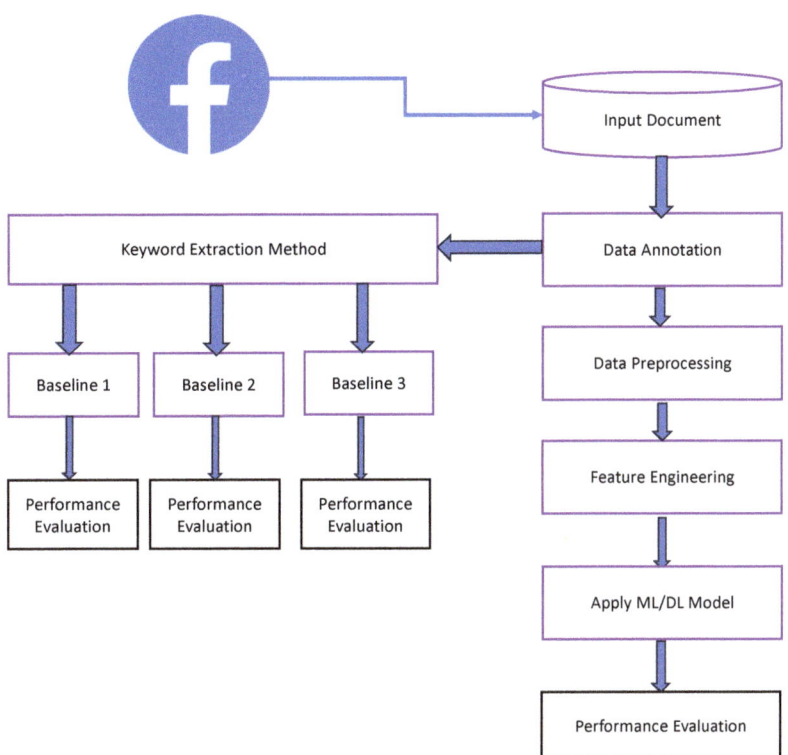

Figure 2. Outline of performed experiments.

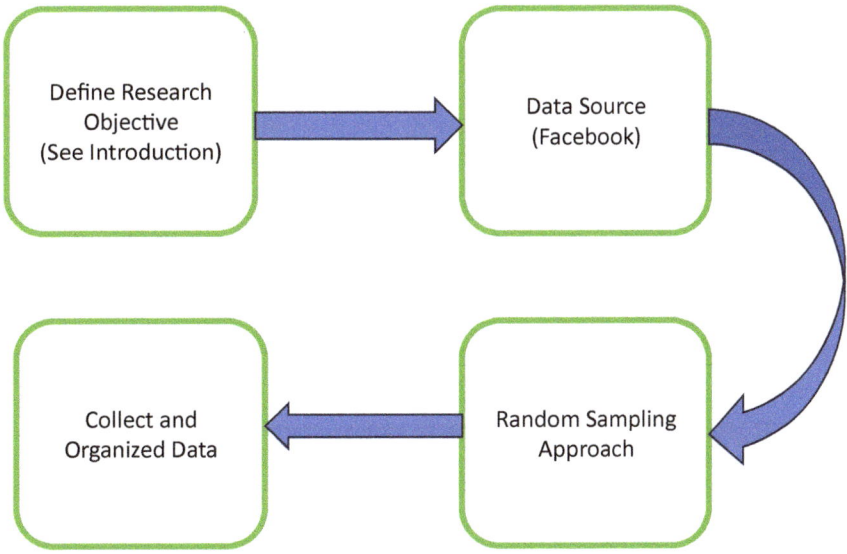

Figure 3. Steps that were used to collect and organize data.

Chittagonian Dialect	English Translation
অডা তুই শুয়োরের বাচ্চা ।	You are P*glet.
তোরা বেগ্গুন খানকির ফুয়া	You are all son of wh*res
বাংলাদেশত নতুন বেইশ্যাদেহা যার।	New pr*stitutes are appearing in Bangladesh.
মাগির দালাল শিমুল্লে	Shimulleh is a sl*t's broker
কুত্তার বাচ্চা	Son of a b*tch
সানার দুধ এক্কান ১০ কেজি	Sana's t*t is 10kg

Figure 4. Examples from the dataset.

3.2.1. Standard and Guidelines

To identify vulgarities, a dataset required annotations adhering to a predefined set of guidelines [47,48]. Therefore, prior to beginning the annotation process, clear and detailed annotation guidelines were developed. Using these guidelines, human annotators were taught how to identify and label vulgar remarks in the text. We provided definitions and usage examples of vulgar words and procedures for dealing with ambiguous cases of vulgar expressions in the guidelines.

The following are the essential standards and recommendations used in this study during the annotation process:

1. Definition of vulgar words: In this research, we defined vulgar words as unpleasant words such as *sl*t, motherf*cker, b*tch*, etc., from the Chittagong dialect of Bangla used to harass other people, institutions, groups, and society.
2. Severity scale: A number between 1–100 was assigned to each vulgar word from the Chittagong dialect by three language experts.
3. Annotator training: Three annotators were trained in the interpretation of vulgar words from the Chittagong dialect, so that the annotation process could be conducted properly. This includes training to maintain professional attitude towards the annotated text in all annotations. This includes avoiding any personal bias or judgment.
4. Consideration of context: Depending on the context, vulgar words can mean different things and offend people in different ways. The context of the message as well as any cultural or social elements that might affect how a vulgar word is perceived should be taken into account when annotating the text.
5. Evaluation of annotation integrity: All data annotations were evaluated for their integrity using inter-rater agreement measures like Cohen's Kappa [20] and Krippendorff's alpha [21].
6. Respect of privacy: Treat any personally identifying information in the annotated text in accordance with any applicable laws or policies and respect the individual's privacy.

3.2.2. Data Annotation Evaluation

The dataset comprised 2500 samples, with each sample being manually annotated following the process depicted in Figure 5. Initially, three annotators independently annotated each review, generating a total of 7500 judgments. In case of any disagreements among the annotators, a majority voting approach was employed to resolve them. As a result, the raw dataset included 1009 samples marked as vulgar and 1476 samples marked as non-vulgar. Additionally, 15 conflicting samples were identified during the annotation process, and after discussion with the annotators, these were excluded and discarded from the final dataset. An example of a conflict was, e.g., a sentence like মঙ্গল শোভাযাত্রা অমঙ্গল (English translation: *"Mangal shovajatra* (Mass procession) is inauspicious"). Three annotators gave three different judgments to this comment, i.e., the first judged this comment as non-vulgar, the second as judged this comment as vulgar, and the third could not reach a decision. Three people have given three types of judgments on this comment by looking at the word **Mangal** from a different religious point of view. Since this was a more difficult problem, surpassing the notion of vulgarity, we decided to not include it in the study this time but will consider it for separate research in the future. Figure 6 displays some of the most typical vulgar words in the dataset.

3.2.3. Inter-Rater Agreement Evaluation

Cohen's Kappa:

After annotating the data, we conducted an examination of the inter-rater agreement. The analysis, employing Cohen's Kappa [20], revealed an impressive average value of 0.91 Kappa. This indicates very strong agreement among the annotators, as demonstrated in Table 2.

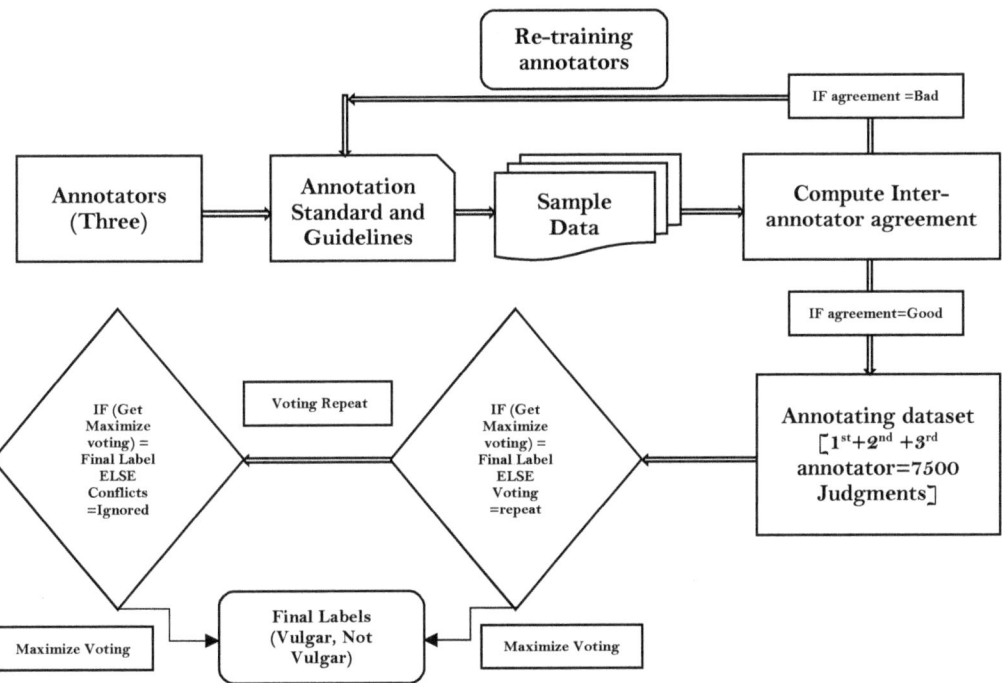

Figure 5. Data annotation process.

Chittagonian Dialect	English Translation	TF-IDF Score	Frequency
হানকি	Wh*re	0.003594	100
মাগি	Sl*t's	0.001797	40
হেডা	Female Genital Organ	0.001366	37
চুদি	F*ck	0.001078	35
মাদারচোদ	Mother F*cker	0.000791	34
চনু	P*nis	0.000647	20
চুইদ্দুম	F*cking	0.000575	18
বেশ্যা	H*oker	0.000575	18
দুধ	T*t	0.000431	14
কুত্তা	B*tch	0.000288	12

Figure 6. Top 10 most frequent vulgar words in the dataset.

Table 2. Cohen's Kappa score.

Annotator Pairs	Cohen's Kappa
1 and 2	0.92
1 and 3	0.90
2 and 3	0.91

Krippendorff's alpha: Using Cohen's Kappa score we can see that the annotators were in almost perfect agreement. To double-check the annotation agreements, we calculated the inter-rater agreement scores using Krippendorff's alpha [21]. As a result, we achieved an average agreement value of 0.914, demonstrating a significantly high agreement among annotators (refer to Table 3).

Table 3. Krippendorff's alpha.

Annotator Pairs	Krippendorff's Alpha
1 and 2	0.927
1 and 3	0.898
2 and 3	0.917

With the above two inter-rater agreements scores, we verified that there was a very high agreement between the annotators during the data annotation process. With the assurance of good quality guidelines and annotators with professional backgrounds, we can state that the created dataset can be considered high quality.

3.3. Baselines

It has been widely accepted that vulgar remark detection can be performed with sufficient accuracy using a basic keyword-matching method to find specific words or phrases of interest within a text [49]. By comparing the input text to a predetermined list of keywords, such a method searches for exact matches with a predetermined lexicon or a list of vulgar keywords [50]. Often, the input text is also preprocessed to remove any unnecessary words or information before keyword matching. Tokenization and removal of stop words or punctuation are a few potential strategies.

Numerous modifications of the keyword-matching method have been applied, including regular expressions [51], string matching algorithms [52], or pattern matching [53,54], all of which can be used to carry out the matching process. The algorithm analyzes each keyword in the text to determine whether it is used alone or as a component of a longer phrase.

Keyword matching has the advantage of being efficient, straightforward, and explainable [55]. Because the input text and keyword list are directly compared, the computational overhead is typically very low. The algorithm can handle a large number of keywords, making it suitable for tasks that call for the simultaneous identification of multiple specific terms.

However, one serious problem is that keyword lists are static, while the language on the Internet is constantly evolving. Therefore, the algorithm may have issues with newly emerging terms or terms that depend on context and are not on the keyword list. The keyword list, therefore, must be periodically updated if the method is to continue working effectively over time.

Despite these shortcomings, keyword matching is still a useful technique in many applications. It provides an efficient and flexible way to pick specific words or phrases of interest. When combined with other methods like machine learning, keyword matching can be a useful component of systems that perform more thorough content filtering [56].

To perform keyword matching, one first needs to prepare the list or lexicon of such keywords: in this case, a lexicon of vulgar words and phrases.

Although a variety of approaches can be designed for keyword extraction, such as rule-based linguistic approaches, statistical approaches, or even machine learning-based approaches, we can specify three general methods for extracting keywords or key phrases from text [55]. Namely, the required keywords can be extracted manually by a trained human designer or annotator, which assures high accuracy but is time-consuming and requires significant human effort. Additionally, the task of extracting vulgar words also poses a burden to the mental health of such human annotators. Secondly, the keyword list can be extracted fully automatically, which takes away all of the burden from the annotators. However, it is usually difficult to assure high accuracy of automatic extraction, since such methods rely on statistical properties of text (term occurrences, term frequencies, frequencies in the whole document, etc.). A third way is to improve the automatic extraction method to the point where it is as close to human judgment as possible and leave the remaining correcting work to the human annotator.

Consequently, in this paper firstly, we proposed a keyword-matching baseline method for vulgar remark detection, which we based on a vulgar keyword extraction. In the keyword-matching baseline, the lexicon of vulgar words provided to the method is considered as a list of features, while the matching procedure is treated as classification in the sense that if at least one vulgar word from the lexicon is matched in the input sentence, the sentence is considered vulgar.

Since the baseline method for classification is only based on simple keyword matching, we compared the three above-mentioned methods for keyword extraction to evaluate which of the keyword extraction methods would be the most effective and efficient and if we could find a method with sufficiently high efficiency and efficacy. For efficiency, we consider the amount of human effort put into preparing the lexicon, while for efficacy, we consider the method's accuracy in reclassifying the sentences into either vulgar or non-vulgar. The whole process of keyword extraction and comparison of all three methods is shown in Figure 7.

Firstly, in the purely manual extraction method, the human annotators read all sentences and manually extracted relevant vulgar expressions. This resulted in a total of 1010 vulgar words. The total accuracy was 0.648. Next, we aimed to propose a method capable of fully automatic (no human effort required) or semi-automatic (only a limited human effort required comparing to fully manual method) extraction of vulgar keywords.

Firstly, to initially extract vulgar keyword candidates, we applied *TF-IDF* and probability of occurrence. Here, *TF*, or term frequency, is calculated by dividing the occurrences of a specific word (or "term") in a document by the number of all terms in that document [57], as in Equation (1).

$$TF = \frac{\text{Frequency of a certain word in the document}}{\text{Word count in the entire document}} \quad (1)$$

Next, *IDF*, or inverse document frequency, determines the importance of keywords in a text and is calculated as a logarithmically scaled inverted division of the number of documents containing the term and total number of documents, as shown in Equation (2).

$$IDF = \log_2 \left(\frac{\text{Total documents}}{\text{Documents with a particular term}} \right) \quad (2)$$

Finally, in order to generate the *TF-IDF* measure, the *TF* and *IDF* are multiplied, as in Equation (3).

$$TF\text{-}IDF = TF \times IDF \quad (3)$$

By calculating *TF-IDF* for all words in the two groups of sentences (vulgar and non-vulgar) we obtain the list of words, where the higher *TF-IDF* score for the vulgar group represents a higher probability of the word being the most representative of the vulgar group, which by assumption should be equivalent to the word being potentially vulgar. In the evaluation, we test the efficacy of this purely TF-IDF-based method without any additional filtering. However, in reality it is not always true, and many non-vulgar words also become included in the list. Specifically, one can observe how many actual vulgar words are included in the first ten, twenty, etc., words on the list, as represented in the result and discussion section.

To solve this problem, we added an additional method to delete the words that have the highest probability of being non-vulgar from the list of top TF-IDF vulgar word candidate terms. The method is explained as follows. The idea for this step was borrowed from Ptaszynski and Yagahara's (2021) method for the automatic extraction of technical terms from larger corpora [58].

To perform this, firstly, we calculate the probability of occurrence (PoO) for each word in either vulgar or non-vulgar class according to Ptaszynski et al.'s (2019) pattern extortion method [54], represented in Equation (4), which is a simplified sigmoid function normalizing the weighted score between 1 (completely vulgar) and −1 (completely non-vulgar).

$$PoO = \left(\frac{\text{Occurrence of Vulgar word}}{(\text{Occurrence of Vulgar word} + \text{Occurrence of Non-vulgar word})} - 0.5 \right) * 2 \quad (4)$$

From that list, we then take all words for which the weight was −1 (appeared only in the non-vulgar group). We use this list to additionally filter out potential non-vulgar terms which might appear on the list of TF-IDF extracted terms. In this manner, we can to some extent automatically eliminate potential non-vulgar words included in the TF-IDF lists by mistake. To test the coverage and usability of this method, we evaluate to what extent were the non-vulgar words eliminated from the list by looking at the ratio of the number of vulgar words in all automatically extracted words. We also verify this for various extraction spans, namely, top ten, top twenty, etc.

In this manner, we end up with three baseline methods for vulgar remark detection based on simple keyword matching, each based on a different keyword extraction procedure, as follows.

1. Automatic keyword extraction method with no additional filtering of non-vulgar words,
2. Automatic keyword extraction method with manual filtering of non-vulgar words,
3. Automatic keyword extraction method with additional automatic filtering of non-vulgar words.

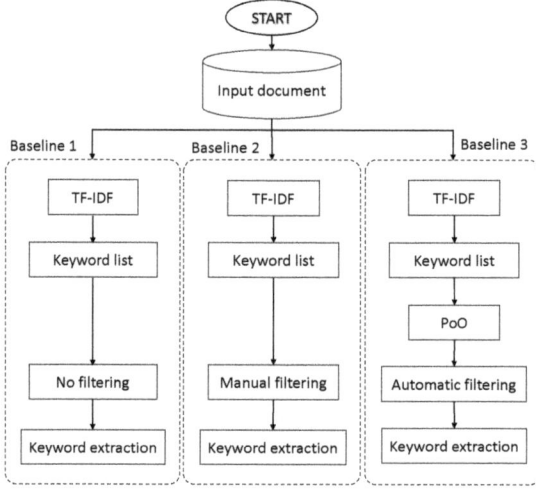

Figure 7. Mechanism of vulgar keyword extraction method.

3.4. Data Preprocessing

Apart from the baselines, we also applied classic machine learning (ML) algorithms to classify vulgar remarks. However, the ML algorithms required a number of specific data preprocessing and feature engineering techniques to be applied before the classification.

Regarding data preprocessing, it is important to consider that the dataset comprises SNS comments, which may contain a significant amount of irrelevant information for the analysis. To minimize the impact of such unwanted and redundant features, we have carried out the following data preprocessing steps. Figure 8 shows each preprocessing step.

Removing punctuation: When punctuation is removed from a text, all quotation marks and other special characters are also removed. Examples of punctuation used to denote pauses, emphasis, and other grammatical functions in written language include periods, commas, question marks, exclamation points, hyphens, parentheses, quotation marks, and other non-alphanumeric characters. The text data used in this study include a variety of punctuation and special characters, some of which may not have a discernible effect on the meaning of a sentence. In earlier research, Mahmud et al. [59] showed that removing punctuation can improve text classification, including automatic cyberbullying detection, especially when using traditional ML algorithms. As a result, we also adopted this strategy and removed all punctuation from the text under analysis. Therefore, we also followed this approach and eliminated all punctuation from the analyzed text. The list of punctuation we considered for removal includes the following characters: ' "!#()*+,-./:;<=>?@[]'| , etc.

Removing emoji and emoticons: Eliminating these graphical representations of emotions, expressions, or symbols from a text involves using emoji and emoticon removal. Emoticons and emoji are frequently used in written communication to convey emotions and reactions or to provide additional context in online communications. There are two ways to deal with this type of information: either completely removing them from the text or replacing them with corresponding text representations. While emojis have been shown to aid text classification in some cases [60], this research primarily focuses on testing the baseline performance of simple ML classifiers on the dataset. Hence, we opted not to consider emojis and emoticons during the classification process.

Removing English characters: Despite the fact that in Bangladesh Bengali is the primary official language, it is common for people to incorporate English words into their speech. However, since we were only focused on the Chittagonian dialect for this task, we eliminated any English characters. We did this in order to make sure that our analysis of the Chittagonian language was precise and focused.

Removing English digits: Upon thorough examination of the dataset samples, we observed the presence of digits and numbers that did not carry specific semantic meaning. In standard practice, named entity recognition (NER) [61] tools are employed to identify and categorize such numerical entities, such as phone numbers, percentages, and currencies. However, for the Chittagonian dialect, no NER tool is currently available. To address this limitation and facilitate the initial experiment, we opted to remove the digits and numbers from the dataset. Nevertheless, we acknowledge the importance of handling numerical entities accurately in future experiments. Therefore, our future plans include developing a dedicated NER tool specifically designed for Chittagonian. This tool will significantly enhance the processing capabilities and enable more effective handling of such cases in subsequent research and applications.

Removing stopwords: Stopwords are frequently eliminated when text data are processed for tasks involving natural language processing. Stop words are frequent words in a language that do not influence the meaning of a sentence. In the Chittagonian dialect, such words include, e.g., 'কি(What)','তুই(You)','ইতি(Him)', 'তুই(You)', 'তোর(Yours)' 'তোয়ারে(You)' আরার/আরীর(Our),কেইনে(How)', etc. They are often removed in various natural language processing tasks, like text classification, to reduce the dataset's dimensionality [62]. Therefore, in order to decrease the dimensionality of the text data and increase the effectiveness and efficiency of the subsequent analysis, we opted to eliminate stopwords from the dataset.

Tokenization: The process of tokenizing involves separating the text into tokens, or individual words. To facilitate further analysis for vulgar remark detection in the Chittagong dialect, we tokenized the text into useful linguistic units, such as words or subwords similarly to other previous research studying the Chittagonian dialect [59].

3.5. Feature Engineering

The transformation of lexical features (words) into numerical representations is necessary to enable the application of machine learning algorithms on textual data. In order to perform this, in this research we used four different feature extraction techniques, including CountVectorizer, *TF-IDF* Vectorizer, Word2vec, and fastText. We also followed the methodologies used in previous studies [31,35,59,63]. We successfully converted the textual data, which initially consisted of strings of characters (words), into numerical features using these feature extraction techniques.

3.5.1. Count Vectorizer

In tasks involving natural language processing, the widely used text preprocessing method CountVectorizer is employed [64]. A group of text documents are transformed into a matrix that shows the frequency of each word's (term's) occurrence in each document. The matrix's columns represent distinctive terms in the corpus, while the rows represent documents. In order to convert text data into a numerical format suitable for machine learning algorithms and enable further analysis and modeling based on the term frequencies in the documents, we used the CountVectorizer function of the Scikitlearn library [65].

3.5.2. TF-IDF Vectorizer

Information retrieval and natural language processing tasks often use the TF-IDF Vectorizer [66], a popular text processing method. It converts a collection of text documents into numerical feature vectors, where each feature denotes the weight of a term in a given document in relation to the corpus as a whole. It determines the term frequency (TF) and inverse document frequency (IDF) for every term in the documents, multiplies them, and then produces the *TF-IDF* score, which is a numerical representation appropriate for machine learning algorithms and other statistical analyses. In addition to assisting with a variety of text-based tasks, such as document classification, this vectorization process [63] also helps us to capture the significance of terms within documents. In this research, we specifically used the implementation of TF-IDF from the Scikit learn library [67].

3.5.3. Word2vec

By representing words as dense vectors in a high-dimensional space, Word2vec [68] is a popular classic word embedding technique. It is a shallow neural network model that learns to translate words into continuous vector spaces based on their patterns of co-occurrence in a significant body of text. Words with similar meanings can be placed closer together in the vector space thanks to the word embeddings that are created. This enables tasks related to word similarity, analogies, and text classification, among other natural language processing operations [69].

3.5.4. fastText

The widely used fastText library [70] was created by Facebook AI Research [71] for text representation and classification [72]. It is especially helpful for dealing with out-of-vocabulary words because it uses word embeddings and character n-grams to effectively encode words and capture subword information. As a result of the model's quick execution and scalability, massive text datasets can be trained and inferred with efficiency. It is widely used in the research and business communities thanks to its success in a variety of natural language processing tasks, such as text classification and language modeling.

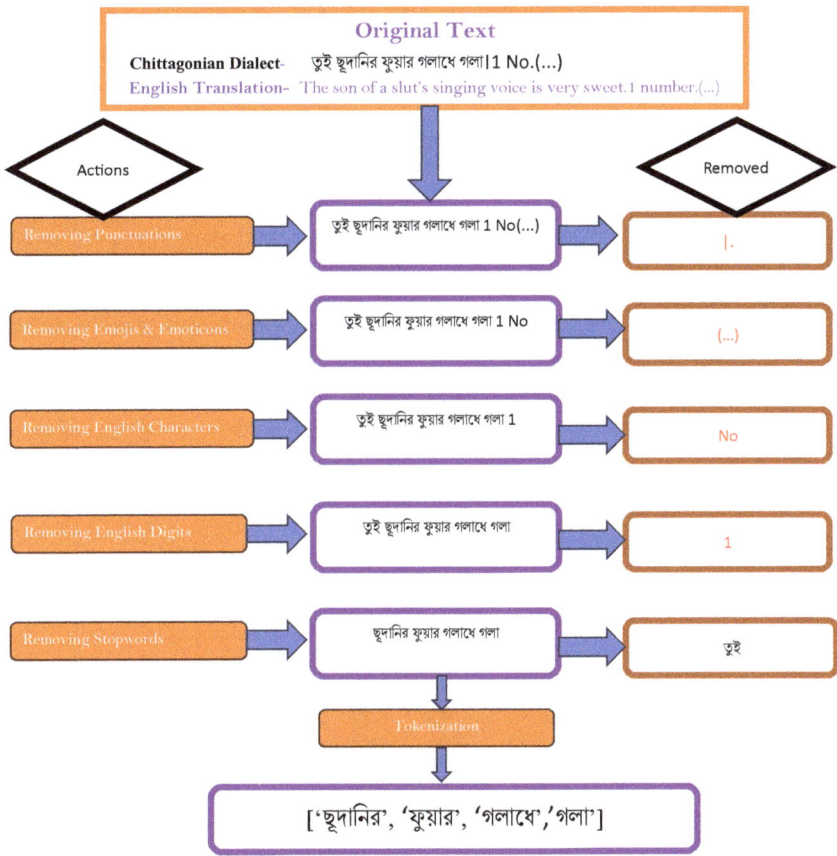

Figure 8. Step-by-step data preprocessing.

3.6. Classification

We investigated two deep learning (DL) algorithms and five machine learning (ML) algorithms in total during our experimentation. We used logistic regression (LR), support vector machine (SVM), decision tree (DT), random forest (RF), and multinomial naive Bayes (MNB) for the traditional ML approaches. We also used the long short-term memory network (LSTM) and the simple recurrent neural network (simpleRNN) in the context of deep learning.

3.6.1. Logistic Regression

A supervised machine learning algorithm called logistic regression (LR) is employed for binary classification tasks [63]. By fitting a logistic function, it models the association between a dependent binary variable and one or more independent variables. The logistic function converts the input features into probabilities, which are then used to categorize instances into one of the two classes. LR is widely used for a variety of applications, including cyberbullying detection [59], text classification [73,74], and sentiment analysis [75] because of its simplicity, interpretability, and efficiency.

3.6.2. Support Vector Machines

Generally used for classification tasks [59], a support vector machine (SVM) is a supervised machine learning algorithm. It looks for the best hyperplane to divide data points into distinct classes in a high-dimensional space. Due to SVM's robustness and effectiveness in handling complex decision boundaries, the margin (distance) between the closest data points of various classes is maximized. When the data cannot be linearly separated in the original feature space, it can also use kernel functions to transform the data into higher dimensions, allowing for effective classification [76].

3.6.3. Decision Tree

A popular supervised machine learning algorithm called a decision tree (DT) is often used for classification [59] and regression tasks. In order to produce a tree-like structure of choices, it operates by recursively partitioning the data into subsets according to the feature values. In classification or regression, each leaf node stands in for a class label while each internal node represents a feature test. Decision trees are used in many different applications because they can handle both categorical and numerical data [50].

3.6.4. Random Forest

An algorithm for collective learning called random forest (RF) is used for both classification [59] and regression tasks. In order to create more precise and reliable predictions, it builds multiple decision trees during training. In order to lessen overfitting and improve generalization, each tree is constructed using a random subset of features and data samples. Random forest creates a strong and adaptable machine learning model by combining the predictions of individual trees, which is widely used for numerous real-world applications [77].

3.6.5. Multinomial Naive Bayes

The probabilistic machine learning algorithm known as multinomial naive Bayes (MNB) is frequently employed for text classification tasks [59]. Based on the Bayes theorem [78,79], it makes the assumption that features are conditionally independent given the class label. MNB performs well when using features that represent word counts or term frequencies in the context of text classification. Despite its simplicity, MNB frequently performs surprisingly well in tasks like sentiment analysis and document categorization [80].

3.6.6. Simple Recurrent Neural Network

To process sequential data, such as time series or text, a simple recurrent neural network (RNN) is a type of neural network architecture [81]. It has a feedback loop that enables information to endure over time and can deal with inputs of varying length. A basic drawback of a simple RNN is the vanishing gradient problem [82], which makes it difficult for it to recognize long-term dependencies in sequences. Due to this problem, more sophisticated RNN variants, such as long short-term memory (LSTM) [83] and gated recurrent unit (GRU) [84], were created in order to solve the vanishing gradient issue and enhance performance on sequential tasks.

3.6.7. Long Short-Term Memory Network

Recurrent neural network (RNN) architectures with long short-term memory (LSTM) are made to handle sequential data [85]. It fixes the vanishing gradient issue [82] that prevents conventional RNNs from detecting long-range dependencies in sequences. For tasks like natural language processing, LSTMs use specialized memory cells with input, output, and forget gates that allow them to retain and forget information over time.

3.7. Performance Evaluation Metrics

Model evaluation involves assessing the performance of a model on test data. In this study, the following evaluation metrics were used: *precision* (PRE), *recall* (REC), *F1-score* (F1), *and accuracy* (ACC). These metrics are computed based on the counts of *true positives*

(*TPv*), *false positives* (*FPv*), *true negatives* (*TNv*), and *false negatives* (*FNv*). The calculations for these metrics are given by Equations (5)–(8).

$$Accuracy(ACC) = \frac{TPv + TNv}{TPv + TNv + FPv + FNv} \quad (5)$$

$$Precision(PRE) = \frac{TPv}{TPv + FPv} \quad (6)$$

$$Recall(REC) = \frac{TPv}{TPv + FNv} \quad (7)$$

$$F1 - score(F1) = 2 \times \left(\frac{Precision(PRE) \times Recall(REC)}{Precision(PRE) + Recall(REC)} \right) \quad (8)$$

In these equations:

True positives (*TPv*) are positive instances that were accurately predicted as positive. *False positives* (*FPv*) are instances that were predicted as positive when the actual label is negative. Observations that were accurately classified as negative are known as *true negative* (*TNv*) observations. Observations that were incorrectly classified as negative when they actually belonged to the positive class are known as *false negatives* (*FNv*).

4. Results and Discussion

4.1. Discussion on Performance of Keyword-Based Vulgarity Extraction and Classification Baselines

In this section, we present the results and analysis of three different methods for vulgar word extraction using keyword matching described in Section 3.3 and Figure 7. These methods involve various approaches to filtering non-vulgar words and determining the relevance of extracted words based on *TF-IDF* scores and probability of occurrences. The goal of this proposed method was to evaluate the effectiveness of these methods in identifying and extracting vulgar words from a given text.

4.1.1. Discussion on Ratio of Vulgar Words Extracted with TF-IDF Weighting

To estimate the potential of the method for automatic extraction of vulgar words, we first calculated the *TF-IDF* scores for each word in both the vulgar and non-vulgar parts of the dataset. Then, we manually determined the ratio of words that were actually vulgar within the top 10, 20, 30, etc., word spans from that list. As TF-IDF has the well-recognized potential to place words that are the most relevant for each compared group (here vulgar vs. non-vulgar) at the top of the list, this would show the accuracy of using only TF-IDF to extract vulgar words.

The results show that the method achieved the extraction ratios of 0.8, 0.75, 0.667, 0.6, 0.58, 0.583, 0.557, 0.525, 0.489, and 0.45 within the top 10, 20, 30, 40, 50, 60, 70, 80, 90, and 100 extracted words, respectively. The high accuracy at the beginning of the list indicates that the majority of extracted words with high *TF-IDF* scores were indeed vulgar. However, the lower accuracy in the later word spans suggests roughly half of the extracted words, although being statistically relevant to the vulgar group, were indeed not vulgar. Table 4 and Figure 9 represent the extraction accuracy for a wider span up to the first thousand words extracted with TF-IDF. As one can see, at the end of the list, only about sixteen percent of the extracted words (i.e., 161 in 1000 words precisely) were indeed vulgar. This suggests that using only TF-IDF without any additional filtering of non-vulgar words will not yield high scores and will not be practical in the long run. Therefore, we needed to improve the extraction method with an additional algorithm for filtering out the words that were most certainly non-vulgar from the group extracted with TF-IDF.

4.1.2. Baseline 1: Keyword-Matching Method Based on TF-IDF Term Extraction with No Additional Filtering

To verify the practical usability of the automatic vulgar term extraction method, we applied the automatically extracted terms as a lexicon (word list) in a simple keyword matching-based method for vulgar sentence detection.

In this method, we performed keyword matching without any additional filtering of the terms extracted with the TF-IDF scores. This means that in the list of the extracted terms there might be some non-vulgar words. As vulgar remarks can be expressed without specifically vulgar terms, this could either improve or impair vulgar sentence detection.

The results indicated an accuracy of the detection of vulgar sentences at levels of 0.197, 0.236, 0.255, 0.269, 0.280, 0.293, 032, 0.343, 0.352, and 0.36 for detection when only the words from the top 10, 20, 30, 40, 50, 60, 70, 80, 90, and 100 of the list of words extracted with TF-IDF (See Table 4 and Figure 9 for reference) were used, respectively. The accuracy suggests that the method successfully identified a significant number of vulgar sentences from the dataset. Interestingly, even using only the top 10 words allowed for a close to 20% accuracy. This is even more impressive if we acknowledge that some of those words were not specifically vulgar.

However, the overall low accuracy, reaching only 62%, indicates that (1) some non-vulgar sentences were incorrectly matched to the list of automatically extracted vulgar term keyword candidates (false positives), and (2) many vulgar sentences were not matched due to the limitations of purely keyword-matching-based method. This shows that there is a wide range of vulgar sentences where the vulgar meaning is not expressed with any specifically vulgar terms. Moreover, although the keyword-matching-based methods are advantageous in terms of processing speed, they have limited applicability, which confirms similar findings from previous research [86–88].

4.1.3. Baseline 2: Keyword-Matching Method Based on TF-IDF Term Extraction with Only Manual Filtering

In this method, after extracting the vulgar keyword candidates automatically using the TF-IDF, we filtered out the non-vulgar words manually. This was not a difficult task for the first several spans (top 10, 20, up to around 100 words), but as the extraction list became longer, it became apparent that continuing this task in the future would be time-consuming and unpractical, especially in the future when novel vulgar words are added to the everyday Internet vocabulary.

Despite the impracticality, the keyword-matching method based on this manually filtered list achieved somewhat satisfying results. For the lower spans of the top 10, 20, 30, 40, 50, 60, 70, 80, 90, and 100, the accuracy values were 0.211, 0.284, 0.327, 0.362, 0.381, 0.41, 0.432, 0.467, 0.489, and 0.523, respectively (see Table 4 and Figure 9). Even so, much higher accuracy for the longer word list was achieved, reaching even 73% for the first 1000 automatically extracted words, with 161 actually vulgar words. This can already be considered applicable in practice, especially since the number of vulgar words is very low. Achieving over 70% accuracy suggests that a large majority of vulgar sentences can in fact be detected with simple keyword-matching-based methods. The only problem remaining for this method thus would be the automation of the additional filtering of non-vulgar words to decrease the necessity of human effort in updating this method in the future, which refers to the final baseline method proposed in the following section.

Table 4. Ratios of extracted vulgar terms, automatic filtering, and performance of the keyword-matching baselines (no. of non-vulgar words deleted manually separately for each span (A); no. of non-vulgar words deleted by humans cumulatively (B); no. of vulgar words (C); ratio of vulgar words in extracted words (D); keyword-matching accuracy with no additional filtering of non-vulgar words (E); keyword-matching accuracy with manual filtering of non-vulgar words (F); keyword-matching accuracy with automatic filtering of non-vulgar words (G); no. of non-vulgar words deleted by automatic filtering separately for each span (H); no. of non-vulgar words deleted by automatic filtering cumulatively (I); accuracy of the automatic filtering method in the automatic filtering of non-vulgar words (J).)

Top # Extracted Words	A	B	C	D	E	F	G	H	I	J
10	2	2	8	0.800	0.197	0.211	0.200	1	1	0.50
20	3	5	15	0.750	0.236	0.284	0.245	3	4	0.80
30	5	10	20	0.667	0.255	0.327	0.300	0	4	0.40
40	6	16	24	0.600	0.269	0.362	0.324	2	6	0.38
50	5	21	29	0.580	0.280	0.381	0.363	2	8	0.38
60	4	25	35	0.583	0.293	0.410	0.385	0	8	0.32
70	6	31	39	0.557	0.320	0.432	0.427	2	10	0.32
80	7	38	42	0.525	0.343	0.467	0.449	1	11	0.29
90	8	46	44	0.489	0.352	0.489	0.467	1	12	0.26
100	9	55	45	0.450	0.360	0.523	0.475	4	16	0.29
200	71	126	74	0.370	0.44	0.66	0.553	30	46	0.37
300	77	203	97	0.323	0.48	0.665	0.571	33	79	0.39
400	79	282	118	0.295	0.510	0.681	0.59	28	107	0.38
500	85	367	133	0.266	0.541	0.685	0.626	42	149	0.41
600	91	458	142	0.237	0.576	0.689	0.637	46	195	0.43
700	96	554	146	0.209	0.597	0.691	0.649	52	247	0.45
800	97	651	149	0.186	0.606	0.698	0.681	50	297	0.46
900	90	741	159	0.177	0.618	0.71	0.689	41	338	0.46
1000	98	839	161	0.161	0.620	0.73	0.695	68	406	0.48

4.1.4. Baseline 3: Keyword-Matching Method Based on TF-IDF Term Extraction with Automatic Filtering of Non-Vulgar Words

Finally, to check to what extent the TF-IDF term extraction can be improved automatically as well as to decrease the human effort, we applied an additional automatic filtering of non-vulgar words.

In this method, for keyword matching we used a predefined list of vulgar words with automatically filtered-out words that had no probability of occurrence in vulgar context, as explained in Section 3.3. The method achieved accuracies of 0.2, 0.245, 0.3, 0.324, 0.363, 0.385, 0.427, 0.449, 0.467, and 0.475 within the top 10, 20, 30, 40, 50, 60, 70, 80, 90, and 100 extracted words, respectively (see Table 4 and Figure 9). Moreover, for the longer word lists, the method, despite filtering out on average only half of the actually non-vulgar words, achieved accuracies close to purely human-based filtering. This suggests the following: (1) the automatic filtering method can reduce human effort by half, and at the same time, (2) when no additional human effort is applied, the method still achieves near-human-level accuracy all of the time. For example, for the longest checked word list, namely, 1000 automatically extracted vulgar word candidates with 406 non-vulgar words additionally filtered out automatically, the method achieved 95% of the human-level accuracy (0.695/0.73 accuracy). Even after normalizing this by considering Baseline

1 as a starting point, the automatic filtering method still covered 68% of human level. The normalized human level is calculated as follows: (Baseline 3 accuracy − Baseline 1 accuracy)/(Baseline 2 accuracy − Baseline 1 accuracy). However, for a slightly shorter word list, namely the top 800 words, with 149 vulgar words and 297 automatically filtered out non-vulgar words overall, the method achieved a performance at 98% of the human level and 82% of the normalized human level.

The accuracy score, in general, suggests sufficiently accurate identification of vulgar sentences after automatic filtering based on the probability of occurrence.

These results indicate that a combination of keyword matching with additional filtering techniques, such as threshold-based or probability-based filtering, can improve the performance of vulgar word extraction. Although the choice of extraction and filtering methods could depend on the specific requirements and trade-offs between the accuracy and efficiency of the given application, the usefulness of such a simple yet effective keyword-based method can be of value both for vulgar sentence detection and especially for the extraction of new vulgar words in the future.

Further research could explore more advanced techniques, such as machine learning- or deep learning-based approaches for keyword extraction [35] to optimize the filtering process and improve the accuracy of vulgar word extraction. Additionally, the evaluation of these methods on larger and more diverse datasets would provide a better understanding of their generalizability and robustness in real-world scenarios.

Consequently, in the following Sections 4.2 and 4.3 we introduce a detailed description of such machine learning and deep learning frameworks for vulgar word detection.

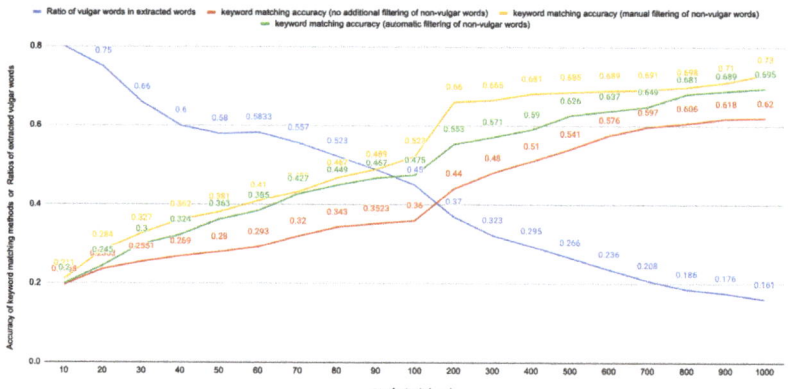

Figure 9. Comparison between different types of keyword-matching methods with the respective ratio of vulgar words in the lexicon.

4.2. Machine Learning Models for Vulgarity Detection

Detecting vulgar remarks in text data typically involves the application of machine learning (ML) and deep learning (DL) techniques. Figure 10 shows the main steps for detecting vulgar remarks.

In this study, we focused on detecting vulgarity in the Chittagonian language using classic ML algorithms. We used two different feature extraction techniques: CountVectorizer and *TF-IDF* Vectorizer. The dataset was divided into an 80–20 ratio for training and testing the models. In this data partitioning, a total of 1988 data points were designated for the training set, encompassing 80% of the entire dataset. The testing set, on the other hand, comprises 20% of the data, consisting of 497 data points.

Table 5 and Figure 11 presents the overall performance of the machine learning models using CountVectorizer. The logistic regression (LR) model performed the best, achieving the highest accuracy of 0.91. It also outperformed other models in this study and current research [35] in terms of recall for both classes. The random forest (RF) model was the second-best performer with an accuracy of 0.87, while multinomial naive Bayes (MNB) demonstrated good precision and recall compared to LR.

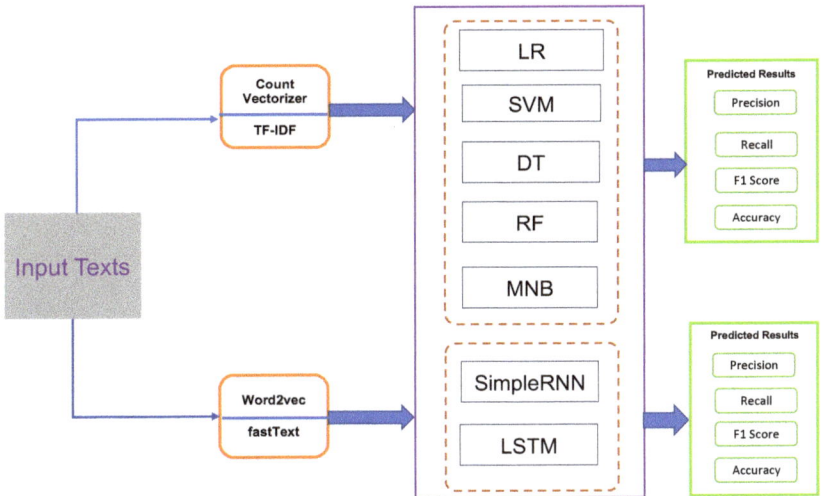

Figure 10. Layout of the experimental procedure for selecting optimal ML/DL model for vulgarity detection.

On the other hand, TF-IDF Vectorizer assesses the relevance of words within the dataset. LR also achieved the highest accuracy of 0.91 with good recall, and performed better than state-of-the-art methods [35]. MNB had an accuracy of 0.83, lower than RF, but it showed a balanced performance across other metrics, as depicted in Table 6 and the corresponding Figure 12.

Overall, the study suggests that using CountVectorizer with the LR model yields the best results for vulgarity detection in Chittagonian, while TF-IDF Vectorizer also performed well with LR and MNB models showing competitive performance.

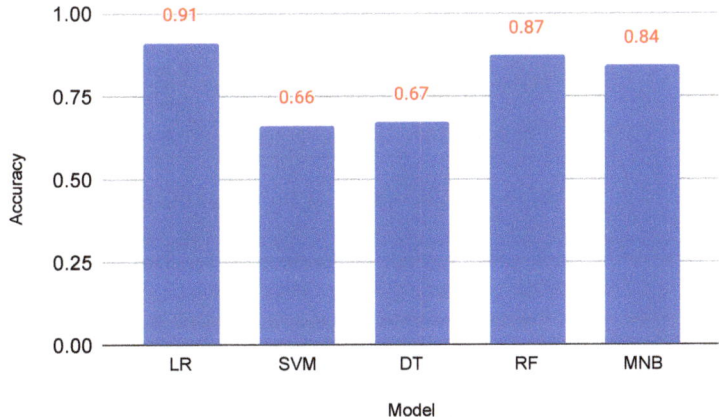

Figure 11. Model accuracy using CountVectorizer.

Table 5. Model performance utilizing CountVectorizer.

Model	Vulgar			Non Vulgar			ACC
	PRE	REC	F1	PRE	REC	F1	
Logistic Regression (LR)	0.800	0.921	0.860	0.910	0.761	0.833	0.910
Support Vector Machines (SVM)	0.654	0.721	0.682	0.680	0.600	0.631	0.660
Decision Tree (DT)	0.623	0.861	0.722	0.771	0.470	0.583	0.671
Random Forest (RF)	0.670	0.942	0.791	0.900	0.534	0.673	0.871
Multinomial Naive Bayes (MNB)	0.811	0.910	0.863	0.902	0.791	0.842	0.842

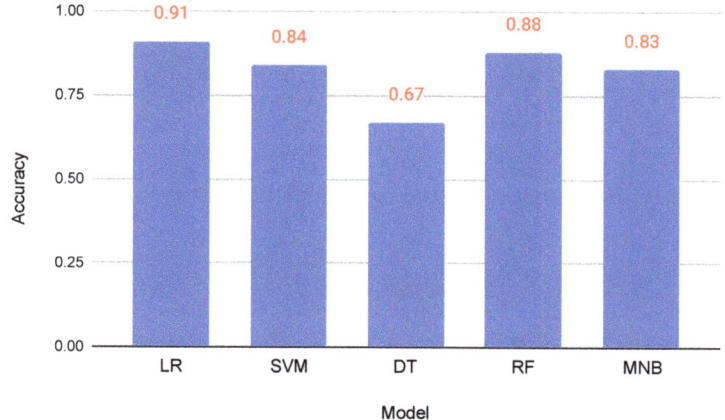

Figure 12. Model accuracy using TF-IDF Vectorizer.

Table 6. Model performance utilizing *TF-IDF* Vectorizer.

Model	Vulgar			Non Vulgar			ACC
	PRE	REC	F1	PRE	REC	F1	
Logistic Regression (LR)	0.820	0.921	0.870	0.901	0.802	0.853	0.911
Support Vector Machines (SVM)	0.810	0.890	0.853	0.881	0.792	0.832	0.843
Decision Tree (DT)	0.561	0.963	0.712	0.852	0.211	0.341	0.671
Random Forest (RF)	0.643	0.971	0.770	0.942	0.453	0.612	0.881
Multinomial Naive Bayes (MNB)	0.801	0.913	0.854	0.891	0.770	0.832	0.832

4.3. Deep Learning Models for Vulgarity Detection

The hyperparameter tuning section for the paper outlines the key choices made in configuring the neural network model, particularly the SimpleRNN and LSTM architectures. In this study, the hyperparameter tuning process was designed to optimize the performance of the text classification models. The first critical hyperparameter is the input dimension of the embedding layer, which directly corresponds to the size of the vocabulary in the corpus. The choice of this parameter is pivotal, as it determines the richness of the word representations. In our experiments, the input dimension was set based on the vocabulary size extracted from the dataset, ensuring that the model could effectively capture the lexical diversity present in the text data. Additionally, the output dimensions of the embedding layer were set to 64, determining the length of the word vectors. This value was selected through experimentation to strike a balance between model expressiveness and computational efficiency. The maximum length of a sequence was set to 100, aligning with the nature of the text data, ensuring that sequences of text were standardized to this length during preprocessing. Another crucial set of hyperparameters pertained to the optimization process. We employed the Adam optimizer, a popular choice for gradient-based optimization, known for its adaptive learning rate capabilities. The learning rate, set at 0.001, is a hyperparameter that can significantly impact training dynamics. The choice of batch size was set to 32, balancing the trade-off between computational efficiency and model convergence speed. The training duration spanned ten epochs, allowing the model to iteratively update its parameters while monitoring convergence. Lastly, it is noteworthy that the dataset comprised 2500 data samples with balanced labels, ensuring that the model's performance was evaluated on a representative and unbiased dataset.

In this study, we explored the effectiveness of deep learning-based models, specifically RNN and LSTM, using Word2vec and fastText word embeddings for various NLP tasks. To ensure robust evaluation, we divided the dataset into three parts: the training set, the validation set (to check for overfitting), and the test set (to evaluate the model's performance). In this data split, a total of 1741 data points have been allocated to the training set, which constitutes 70% of the entire dataset. The validation set comprises 15% of the data, containing 372 data points, while the testing set also consists of 15% of the data, encompassing another 372 data points.

When employing Word2vec word embeddings, we observed that the SimpleRNN model outperformed the LSTM model and performed better than state-of-the-art methods [35], achieving an accuracy of 0.84 compared to LSTM's 0.63. Furthermore, SimpleRNN exhibited superior precision and recall for both classes, as illustrated in Table 7.

Moreover, we explored the application of fastText word embedding technique for both deep learning models. In this scenario, SimpleRNN achieved an impressive accuracy of 0.90, demonstrating its proficiency at the task. Notably, both models displayed excellent performance in detecting both classes, as shown in Table 7 and Figure 13.

4.4. Comparison with Keyword-Matching Baseline

Each of the four previously mentioned keyword extraction methods had an accuracy rating of roughly up to 70% with varying degrees of success. The results were influenced by the keyword list, the filtering methods, and the presence of vulgar word variations or misspellings. We could locate specific keywords within a text quickly by using these straightforward, effective keyword-matching techniques. However, the results have shown these methods also had problems with keyword variations and the need for frequent keyword list updating remained. Another problem not solvable by simple keyword-matching methods was the use of language that is context-sensitive. This shows that the ability of keyword-matching methods to handle complex linguistic structures or comprehend the context and meaning of the text is limited.

On the other hand, machine learning and deep learning methods leverage algorithms and models trained on labeled datasets to classify text segments as vulgar or non-vulgar. The results are typically measured using metrics such as accuracy, precision, recall, and F1-score. Machine learning and deep learning approaches achieved a high accuracy of approximately 84–91%, or roughly twenty percentage points above the keyword-matching baselines, which shows such algorithms can learn patterns and features from the data more effectively. They are also capable of capturing complex linguistic nuances, identifying context-sensitive vulgar words, and handling variations. A disadvantage of these methods is that they require significant amounts of labeled training data and computational resources for training and inference, which means that unless the training data are constantly updated, their performance will also degrade with time. However, this could be to some extent mitigated by (1) applying the automatic vulgar term extraction method proposed in this paper to update and expand the vulgar term lexicon of keyword-matching baselines and (2) using those baselines to collect potential vulgar and non-vulgar sentence candidates. This would allow for initial information triage [89] and ensure more efficiency in the vulgar remark data collection and annotation process.

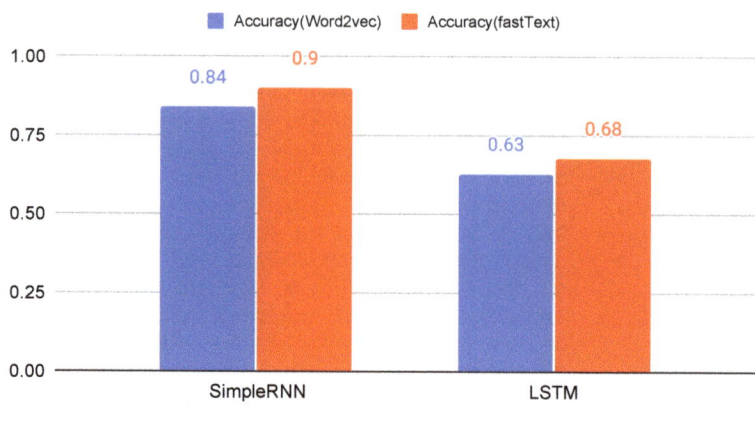

Figure 13. Model accuracy using Word2vec and fastText.

Table 7. Performance of models using Word2Vec and FastText.

Word2vec	Vulgar			Non Vulgar			ACC
	PRE	REC	F1	PRE	REC	F1	
SimpleRNN	0.784	0.983	0.863	0.972	0.704	0.812	0.842
Long Short-Term Memory(LSTM)	0.612	0.811	0.703	0.682	0.451	0.544	0.631
FastText	Vulgar			Non Vulgar			ACC
	PRE	REC	F1	PRE	REC	F1	
SimpleRNN	0.943	0.872	0.901	0.872	0.941	0.903	0.902
Long Short-Term Memory(LSTM)	0.632	0.893	0.744	0.792	0.452	0.573	0.681

4.5. Comparison with Previous Studies

To our knowledge, this is the first study to identify vulgar words in the Chittagonian dialect of Bangla. For that, we could not find any previous research to compare with our study. Since there was a study [35] to find vulgar words in the Bengali language, we compared it to our approaches. Below are the main differences and findings compared to the previous study to facilitate our research evaluation.

1. Our research language domain is the Chittagong dialect of Bangla, while previous work focused on Bengali/Bangla. Working with dialects has many challenges such as data collection, data annotation, data processing, dataset validation, model creation, etc. By overcoming all these challenges, we successfully completed the research.
2. We carried out the research in two steps. Firstly, we reported the performance of the three keyword-matching baselines. No previous research has tried this type of method.
3. Then, we built machine learning and deep learning models and compared them with baseline methods. We observed that our models gave comparatively better results then previous studies [35].

4.6. Limitations of Study

There were several limitations of this study: Firstly, the dataset size of 2500 comments, while comprehensive, is relatively small for training deep learning models, potentially limiting their generalizability. Secondly, the focus on the Chittagonian dialect of Bangla narrows the applicability of findings to Bangla and related languages. This means that the results also need to be confirmed in other widely spoken languages and different language families. Additionally, the initial reliance on a lexicon-based approach for vulgar language detection may not effectively capture nuanced or context-specific variations. Resource constraints in deep learning models, like SimpleRNN, can also impact their competitive performance compared to traditional methods.

4.7. Ethical Considerations

Because this study concerns the human population (social media users), especially the unethical use of language on the Internet, an important part was using an ethical approach to research, beginning with data collection [90]. In order to collect data for this study, only public user posts and comments were collected. Facebook's open access policy permits data collection of public posts. Since we collected only public comments/posts, they were no longer regarded as private, and therefore, no special agreement was necessary for the collection of data and research [91]. Moreover, we adhered to at the following ethical concerns while collecting the data:

1. As the source for our dataset, we primarily used social media groups. Therefore, while gathering the data, we verified and complied with those groups' terms and conditions.
2. We performed anonymization of posts containing such sensitive information as names of private persons, organizations, religious groups, institutions, and states.
3. We deleted personal information such as phone numbers, home addresses, etc.

5. Conclusions and Future Work

5.1. Conclusions

This study primarily centered on identifying vulgar language in social media posts. As the common approach to vulgar remark detection is using simple lists of vulgar keywords, we firstly proposed a method to automatically extract such vulgar keywords from raw data and used those keywords in simple keyword-matching baseline classification methods. The automatic keyword extraction method was able to successfully extract vulgar keywords and additionally successfully filter out half of the non-vulgar words, which allowed the method to reach a satisfying approximately 70% accuracy in detecting vulgar sentences. Next, we also employed machine learning (ML) and deep learning (DL) classifiers, including logistic regression (LR), support vector machine (SVM), decision tree (DT), random forest (RF), multinomial naive Bayes (MNB), simple recurrent neural network (simpleRNN), and long short-term memory (LSTM). These classifiers were coupled with various feature extraction techniques like CountVectorizer, TF-IDF Vectorizer, Word2Vec, and fastText. Our dataset consisted of 2485 comments, balanced between vulgar and non-vulgar remarks. To evaluate the performance of the proposed methods, we ran experiments on this dataset. The results indicated that LR with CountVectorizer or TF-IDF Vectorizer, as well as simpleRNN with Word2Vec and fastText, were particularly effective in detecting vulgar comments.

5.2. Contributions of this Study

In summary, this research makes substantial contributions to the field of vulgar remark detection in the Chittagonian dialect, as follows:

1. Gathered a dataset of 2500 comments and posts from publicly accessible Facebook accounts.
2. Ensured dataset reliability through rigorous manual annotation and validated the annotations using Cohen's Kappa statistics and Krippendorff's alpha.
3. Introduced a keyword-matching-based baseline method using a hand-crafted vulgar word lexicon.
4. Developed an automated method for augmenting the vulgar word lexicon, ensuring adaptability to evolving language.
5. Introduced various sentence-level vulgar remark detection methods, from lexicon matching to advanced techniques.

6. Conducted comprehensive comparisons between keyword-matching and machine learning (ML) and deep learning (DL) models to achieve high detection accuracy.
7. Achieved over 90% accuracy in detecting vulgar remarks in Chittagonian social media posts, demonstrating a performance acceptable for real-world applications.

5.3. Future Work

Building upon the outcomes of this study, our future research directions will focus on devising resource-constrained strategies for vulgar remark recognition with a specific emphasis on the Chittagonian dialect. While the results of our current study are promising for vulgar remark detection in the Chittagonian dialect, there exist several avenues for further investigation and development. One crucial aspect involves expanding the size of our training dataset. A larger, more diversified dataset that encompasses a broader spectrum of vulgar words and contextual nuances will be meticulously gathered and annotated. This expansion aims to enhance the robustness and generalization capabilities of our models, enabling them to effectively identify vulgar language in a wider array of real-world scenarios.

Furthermore, our future research agenda will explore the realm of multi-modal approaches for vulgar remark detection. This entails analyzing not only textual content but also visual elements if available. By combining text and image analysis, we aim to gain a more profound understanding of vulgar remarks, especially in multimedia contexts where visual cues play a significant role. Additionally, we plan to advance our classification methods by incorporating more sophisticated machine learning techniques. Models such as bidirectional LSTM and transformers, renowned for their ability to capture intricate language patterns, will be leveraged to further elevate the accuracy and effectiveness of vulgar remark detection. This holistic approach to future research endeavors aims to refine and expand the capabilities of vulgar remark detection systems, ultimately contributing to a cleaner and safer online environment and more effective content moderation practices.

Author Contributions: Conceptualization, T.M. and M.P.; methodology, T.M.; validation, T.M., M.P. and F.M.; formal analysis, T.M. and M.P.; investigation, T.M. and M.P.; resources, T.M. and M.P.; data curation, T.M. and M.P.; writing—original draft preparation, T.M.; writing—review and editing, T.M. and M.P.; visualization, T.M.; supervision, M.P. and F.M. All authors have read and agreed to the published version of the manuscript.

Funding: This research received no external funding.

Institutional Review Board Statement: Not applicable.

Informed Consent Statement: All necessary informed consent was obtained from all subjects involved in the study.

Data Availability Statement: The data used to support the findings of this study are available upon reasonable request to the corresponding author.

Conflicts of Interest: The authors declare no conflicts of interest.

Abbreviations

The following abbreviations are used in this manuscript:

NLP	Natural Language Processing
ML	Machine Learning
DL	Deep Learning
RNN	Recurrent Neural Networks
BTRC	Bangladesh Telecommunication Regulatory Commission
NGO	Non-governmental Organization
TF-IDF	Term Frequency-Inverse Document Frequency
LR	Logistic Regression
SVM	Support Vector Machine
DT	Decision Tree

RF	Random Forest
MNB	Multinomial Naive Bayes
LSTM	Long Short-Term Memory network
BiLSTM	Bidirectional Long Short-Term Memory network
BERT	Bidirectional Encoder Representations from Transformers
ELECTRA	Pre-training Text Encoders as Discriminators Rather Than Generators

References

1. Bangladesh Telecommunication Regulatory Commission. Available online: http://www.btrc.gov.bd/site/page/347df7fe-409f-451e-a415-65b109a207f5/- (accessed on 15 January 2023).
2. United Nations Development Programme. Available online: https://www.undp.org/bangladesh/blog/digital-bangladesh-innovative-bangladesh-road-2041 (accessed on 20 January 2023).
3. Chittagong City in Bangladesh. Available online: https://en.wikipedia.org/wiki/Chittagong (accessed on 1 April 2023).
4. StatCounter Global Stats. Available online: https://gs.statcounter.com/social-media-stats/all/bangladesh/#monthly-202203-202303 (accessed on 24 April 2023).
5. Facebook. Available online: https://www.facebook.com/ (accessed on 28 January 2023).
6. imo. Available online: https://imo.im (accessed on 28 January 2023).
7. WhatsApp. Available online: https://www.whatsapp.com (accessed on 28 January 2023)
8. Addiction Center. Available online: https://www.addictioncenter.com/drugs/social-media-addiction/ (accessed on 28 January 2023).
9. Prothom Alo. Available online: https://en.prothomalo.com/bangladesh/Youth-spend-80-mins-a-day-in-Internet-adda (accessed on 28 January 2023).
10. United Nations. Available online: https://www.un.org/en/chronicle/article/cyberbullying-and-its-implications-human-rights (accessed on 28 January 2023).
11. ACCORD—African Centre for the Constructive Resolution of Disputes. Available online: https://www.accord.org.za/conflict-trends/social-media/ (accessed on 28 January 2023).
12. Cachola, I.; Holgate, E.; Preoţiuc-Pietro, D.; Li, J.J. Expressively vulgar: The socio-dynamics of vulgarity and its effects on sentiment analysis in social media. In Proceedings of the 27th International Conference on Computational Linguistics, Santa Fe, NM, USA, 20–26 August 2018; pp. 2927–2938.
13. Wang, N. An analysis of the pragmatic functions of "swearing" in interpersonal talk. *Griffith Work. Pap. Pragmat. Intercult. Commun.* **2013**, *6*, 71–79.
14. Mehl, M.R.; Vazire, S.; Ramírez-Esparza, N.; Slatcher, R.B.; Pennebaker, J.W. Are women really more talkative than men? *Science* **2007**, *317*, 82. [CrossRef] [PubMed]
15. Wang, W.; Chen, L.; Thirunarayan, K.; Sheth, A.P. Cursing in English on twitter. In Proceedings of the 17th ACM Conference on Computer Supported Cooperative Work & Social Computing, Baltimore, MD, USA, 15–19 February 2014; pp. 415–425.
16. Holgate, E.; Cachola, I.; Preoţiuc-Pietro, D.; Li, J.J. Why swear? Analyzing and inferring the intentions of vulgar expressions. In Proceedings of the 2018 Conference on Empirical Methods in Natural Language Processing, Brussels, Belgium, 31 October–4 November 2018; pp. 4405–4414.
17. Chittagonian Language. Available online: https://en.wikipedia.org/wiki/Chittagonian_language (accessed on 11 February 2023).
18. Lewis, M.P. *Ethnologue: Languages of the World*, 16th ed.; SIL International: Dallax, TX, USA, 2009.
19. Masica, C.P. *The Indo-Aryan Languages*; Cambridge University Press: Cambridge, UK, 1993.
20. Cohen, J. A coefficient of agreement for nominal scales. *Educ. Psychol. Meas.* **1960**, *20*, 37–46. [CrossRef]
21. Krippendorff, K. Measuring the reliability of qualitative text analysis data. *Qual. Quant.* **2004**, *38*, 787–800. [CrossRef]
22. Sazzed, S. A lexicon for profane and obscene text identification in Bengali. In Proceedings of the International Conference on Recent Advances in Natural Language Processing (RANLP 2021), Online, 1–3 September 2021; pp. 1289–1296.
23. Das, S.; Mahmud, T.; Islam, D.; Begum, M.; Barua, A.; Tarek Aziz, M.; Nur Showan, E.; Dey, L.; Chakma, E. Deep Transfer Learning-Based Foot No-Ball Detection in Live Cricket Match. *Comput. Intell. Neurosci.* **2023**, *2023*, 2398121. [CrossRef]
24. Mahmud, T.; Barua, K.; Barua, A.; Das, S.; Basnin, N.; Hossain, M.S.; Andersson, K.; Kaiser, M.S.; Sharmen, N. Exploring Deep Transfer Learning Ensemble for Improved Diagnosis and Classification of Alzheimer's Disease. In Proceedings of the 2023 International Conference on Brain Informatics, Hoboken, NJ, USA, 1–3 August 2023; Springer: Cham, Switzerland, 2023; pp. 1–12.
25. Wu, Z.; Luo, G.; Yang, Z.; Guo, Y.; Li, K.; Xue, Y. A comprehensive review on deep learning approaches in wind forecasting applications. *CAAI Trans. Intell. Technol.* **2022**, *7*, 129–143. [CrossRef]
26. Gasparin, A.; Lukovic, S.; Alippi, C. Deep learning for time series forecasting: The electric load case. *CAAI Trans. Intell. Technol.* **2022**, *7*, 1–25. [CrossRef]
27. Pinker, S. *The Stuff of Thought: Language as a Window into Human Nature*; Penguin: London, UK, 2007.
28. Andersson, L.G.; Trudgill, P. *Bad Language*; Blackwell/Penguin Books: London, UK, 1990.
29. Eshan, S.C.; Hasan, M.S. An application of machine learning to detect abusive bengali text. In Proceedings of the 2017 20th International Conference of Computer and Information Technology (ICCIT), Dhaka, Bangladesh, 22–24 December 2017; pp. 1–6.

30. Akhter, S.; Abdhullah-Al-Mamun. Social media bullying detection using machine learning on Bangla text. In Proceedings of the 2018 10th International Conference on Electrical and Computer Engineering (ICECE), Dhaka, Bangladesh, 20–22 December 2018; pp. 385–388.
31. Emon, E.A.; Rahman, S.; Banarjee, J.; Das, A.K.; Mittra, T. A deep learning approach to detect abusive bengali text. In Proceedings of the 2019 7th International Conference on Smart Computing & Communications (ICSCC), Sarawak, Malaysia, 28–30 June 2019; pp. 1–5.
32. Awal, M.A.; Rahman, M.S.; Rabbi, J. Detecting abusive comments in discussion threads using naïve bayes. In Proceedings of the 2018 International Conference on Innovations in Science, Engineering and Technology (ICISET), Chittagong, Bangladesh, 27–28 October 2018; pp. 163–167.
33. Hussain, M.G.; Al Mahmud, T. A technique for perceiving abusive bangla comments. *Green Univ. Bangladesh J. Sci. Eng.* **2019**, *4*, 11–18.
34. Das, M.; Banerjee, S.; Saha, P.; Mukherjee, A. Hate Speech and Offensive Language Detection in Bengali. *arXiv* **2022**, arXiv:2210.03479.
35. Sazzed, S. Identifying vulgarity in Bengali social media textual content. *PeerJ Comput. Sci.* **2021**, *7*, e665. [CrossRef]
36. Jahan, M.; Ahamed, I.; Bishwas, M.R.; Shatabda, S. Abusive comments detection in Bangla-English code-mixed and transliterated text. In Proceedings of the 2019 2nd International Conference on Innovation in Engineering and Technology (ICIET), Dhaka, Bangladesh, 23–24 December 2019; pp. 1–6.
37. Ishmam, A.M.; Sharmin, S. Hateful speech detection in public facebook pages for the bengali language. In Proceedings of the 2019 18th IEEE International Conference on Machine Learning and Applications (ICMLA), Boca Raton, FL, USA, 16–19 December 2019; pp. 555–560.
38. Karim, M.R.; Dey, S.K.; Islam, T.; Sarker, S.; Menon, M.H.; Hossain, K.; Hossain, M.A.; Decker, S. Deephateexplainer: Explainable hate speech detection in under-resourced bengali language. In Proceedings of the 2021 IEEE 8th International Conference on Data Science and Advanced Analytics (DSAA), Porto, Portugal, 6–9 October 2021; pp. 1–10.
39. Sazzed, S. Abusive content detection in transliterated Bengali-English social media corpus. In Proceedings of the Fifth Workshop on Computational Approaches to Linguistic Code-Switching, Online, 11 June 2021; pp. 125–130.
40. Faisal Ahmed, M.; Mahmud, Z.; Biash, Z.T.; Ryen, A.A.N.; Hossain, A.; Ashraf, F.B. Bangla Text Dataset and Exploratory Analysis for Online Harassment Detection. *arXiv* **2021**, arXiv:2102.02478.
41. Romim, N.; Ahmed, M.; Talukder, H.; Islam, S. Hate speech detection in the bengali language: A dataset and its baseline evaluation. In Proceedings of the International Joint Conference on Advances in Computational Intelligence, Dhaka, Bangladesh, 20–21 November 2020; Springer: Singapore, 2021; pp. 457–468.
42. Islam, T.; Ahmed, N.; Latif, S. An evolutionary approach to comparative analysis of detecting Bangla abusive text. *Bull. Electr. Eng. Inform.* **2021**, *10*, 2163–2169. [CrossRef]
43. Aurpa, T.T.; Sadik, R.; Ahmed, M.S. Abusive Bangla comments detection on Facebook using transformer-based deep learning models. *Soc. Netw. Anal. Min.* **2022**, *12*, 24. [CrossRef]
44. Devlin, J.; Chang, M.W.; Lee, K.; Toutanova, K. Bert: Pre-training of deep bidirectional transformers for language understanding. In Proceedings of the 2019 Conference of the North American Chapter of the Association for Computational Linguistics: Human Language Technologies, Volume 1 (Long and Short Papers), Minneapolis, MI, USA, 2–7 June 2019; Association for Computational Linguistics: Minneapolis, MI, USA, 2019; pp. 4171–4186.
45. Clark, K.; Luong, M.T.; Le, Q.V.; Manning, C.D. Electra: Pre-training text encoders as discriminators rather than generators. In Proceedings of the International Conference on Learning Representations, ICLR 2020, Virtual, 26 April–1 May 2020.
46. List of Non-Governmental Organisations in Bangladesh. Available online: https://en.wikipedia.org/wiki/List_of_non-governmental_organisations_in_Bangladesh (accessed on 15 February 2023).
47. Pradhan, R.; Chaturvedi, A.; Tripathi, S.; Sharma, D.K. A review on offensive language detection. In *Advances in Data and Information Sciences: Proceedings of ICDIS 2019, Agra, India, 29–30 March 2019*; Springer: Singapore, 2020; pp. 433–439.
48. Khan, M.M.; Shahzad, K.; Malik, M.K. Hate speech detection in roman urdu. *ACM Trans. Asian Low-Resour. Lang. Inf. Process. (TALLIP)* **2021**, *20*, 1–19. [CrossRef]
49. Novitasari, S.; Lestari, D.P.; Sakti, S.; Purwarianti, A. Rude-Words Detection for Indonesian Speech Using Support Vector Machine. In Proceedings of the 2018 International Conference on Asian Language Processing (IALP), Bandung, Indonesia, 15–17 November 2018; pp. 19–24. [CrossRef]
50. Kim, S.N.; Medelyan, O.; Kan, M.Y.; Baldwin, T. Automatic keyphrase extraction from scientific articles. *Lang. Resour. Eval.* **2013**, *47*, 723–742. [CrossRef]
51. Li, J.; Jiang, G.; Xu, A.; Wang, Y. The Automatic Extraction of Web Information Based on Regular Expression. *J. Softw.* **2017**, *12*, 180–188.
52. Alqahtani, A.; Alhakami, H.; Alsubait, T.; Baz, A. A survey of text matching techniques. *Eng. Technol. Appl. Sci. Res.* **2021**, *11*, 6656–6661. [CrossRef]
53. Califf, M.E.; Mooney, R.J. Bottom-up relational learning of pattern matching rules for information extraction. *J. Mach. Learn. Res.* **2003**, *4*, 177–210.
54. Ptaszynski, M.; Lempa, P.; Masui, F.; Kimura, Y.; Rzepka, R.; Araki, K.; Wroczynski, M.; Leliwa, G. Brute-force sentence pattern extortion from harmful messages for cyberbullying detection. *J. Assoc. Inf. Syst.* **2019**, *20*, 1075–1127. [CrossRef]

55. Beliga, S. *Keyword Extraction: A Review of Methods and Approaches*; University of Rijeka, Department of Informatics: Rijeka, Croatia, 2014; Volume 1.
56. Su, G.y.; Li, J.h.; Ma, Y.h.; Li, S.h. Improving the precision of the keyword-matching pornographic text filtering method using a hybrid model. *J. Zhejiang Univ.-Sci. A* **2004**, *5*, 1106–1113. [CrossRef]
57. Liu, F.; Pennell, D.; Liu, F.; Liu, Y. Unsupervised approaches for automatic keyword extraction using meeting transcripts. In Proceedings of the Human Language Technologies: The 2009 Annual Conference of the North American Chapter of The association for Computational Linguistics, Boulder, CO, USA, 31 May–5 June 2009; pp. 620–628.
58. Ptaszynski, M.; Yagahara, A. Senmon Yogo Chushutsu Sochi, Senmon yogo Chushutsu hoho Oyobi Puroguramu (Technical Term Extraction Device, Technical Term Extraction Method and Program). 16 December 2021. Available online: https://jglobal.jst.go.jp/en/detail?JGLOBAL_ID=202103002313491840 (accessed on 29 January 2023). (In Japanese)
59. Mahmud, T.; Ptaszynski, M.; Eronen, J.; Masui, F. Cyberbullying detection for low-resource languages and dialects: Review of the state of the art. *Inf. Process. Manag.* **2023**, *60*, 103454. [CrossRef]
60. Li, D.; Rzepka, R.; Ptaszynski, M.; Araki, K. HEMOS: A novel deep learning-based fine-grained humor detecting method for sentiment analysis of social media. *Inf. Process. Manag.* **2020**, *57*, 102290. [CrossRef]
61. Haque, M.Z.; Zaman, S.; Saurav, J.R.; Haque, S.; Islam, M.S.; Amin, M.R. B-NER: A Novel Bangla Named Entity Recognition Dataset with Largest Entities and Its Baseline Evaluation. *IEEE Access* **2023**, *11*, 45194–45205. [CrossRef]
62. Eronen, J.; Ptaszynski, M.; Masui, F.; Smywiński-Pohl, A.; Leliwa, G.; Wroczynski, M. Improving classifier training efficiency for automatic cyberbullying detection with feature density. *Inf. Process. Manag.* **2021**, *58*, 102616. [CrossRef]
63. Mahmud, T.; Das, S.; Ptaszynski, M.; Hossain, M.S.; Andersson, K.; Barua, K. Reason Based Machine Learning Approach to Detect Bangla Abusive Social Media Comments. In *Intelligent Computing & Optimization, Proceedings of the 5th International Conference on Intelligent Computing and Optimization 2022 (ICO2022), Virtual, 27–28 October 2022*; Springer: Cham, Switzerland, 2022; pp. 489–498.
64. Ahmed, T.; Mukta, S.F.; Al Mahmud, T.; Al Hasan, S.; Hussain, M.G. Bangla Text Emotion Classification using LR, MNB and MLP with TF-IDF & CountVectorizer. In Proceedings of the 2022 26th International Computer Science and Engineering Conference (ICSEC), Sakon Nakhon, Thailand, 21–23 December 2022; pp. 275–280.
65. sklearn.feature_extraction.text.CountVectorizer. Available online: https://scikit-learn.org/stable/modules/generated/sklearn.feature_extraction.text.CountVectorizer.html (accessed on 23 February 2023).
66. Chakraborty, M.; Huda, M.N. Bangla document categorisation using multilayer dense neural network with tf-idf. In Proceedings of the 2019 1st International Conference on Advances in Science, Engineering and Robotics Technology (ICASERT), Dhaka, Bangladesh, 3–5 May 2019; pp. 1–4.
67. sklearn.feature_extraction.text.TfidfVectorizer. Available online: https://scikit-learn.org/stable/modules/generated/sklearn.feature_extraction.text.TfidfVectorizer.html (accessed on 23 February 2023).
68. Rahman, R. Robust and consistent estimation of word embedding for bangla language by fine-tuning word2vec model. In Proceedings of the 2020 23rd International Conference on Computer and Information Technology (ICCIT), Dhaka, Bangladesh, 19–21 December 2020; pp. 1–6.
69. Ma, L.; Zhang, Y. Using Word2Vec to process big text data. In Proceedings of the 2015 IEEE International Conference on Big Data (Big Data), Santa Clara, CA, USA, 29 October–1 November 2015; pp. 2895–2897.
70. facebookresearch/fastText: Library for Fast Text. Available online: https://github.com/facebookresearch/fastText (accessed on 25 February 2023).
71. Research—Meta AI. Available online: https://ai.meta.com/research/ (accessed on 25 February 2023).
72. Mojumder, P.; Hasan, M.; Hossain, M.F.; Hasan, K.A. A study of fasttext word embedding effects in document classification in bangla language. In Proceedings of the Cyber Security and Computer Science: Second EAI International Conference, ICONCS 2020, Dhaka, Bangladesh, 15–16 February 2020; Proceedings 2; Springer: Cham, Switzerland, 2020; pp. 441–453.
73. Shah, K.; Patel, H.; Sanghvi, D.; Shah, M. A comparative analysis of logistic regression, random forest and KNN models for the text classification. *Augment. Hum. Res.* **2020**, *5*, 12. [CrossRef]
74. Mahmud, T.; Ptaszynski, M.; Masui, F. Vulgar Remarks Detection in Chittagonian Dialect of Bangla. *arXiv* **2023**, arXiv:2308.15448.
75. Hasanli, H.; Rustamov, S. Sentiment analysis of Azerbaijani twits using logistic regression, Naive Bayes and SVM. In Proceedings of the 2019 IEEE 13th International Conference on Application of Information and Communication Technologies (AICT), Baku, Azerbaijan, 23–25 October 2019; pp. 1–7.
76. Hussain, M.G.; Hasan, M.R.; Rahman, M.; Protim, J.; Al Hasan, S. Detection of bangla fake news using mnb and svm classifier. In Proceedings of the 2020 International Conference on Computing, Electronics & Communications Engineering (iCCECE), Southend, UK, 17–18 August 2020; pp. 81–85.
77. Alam, M.R.; Akter, A.; Shafin, M.A.; Hasan, M.M.; Mahmud, A. Social Media Content Categorization Using Supervised Based Machine Learning Methods and Natural Language Processing in Bangla Language. In Proceedings of the 2020 11th International Conference on Electrical and Computer Engineering (ICECE), Dhaka, Bangladesh, 17–19 December 2020; pp. 270–273.
78. Joyce, J. Bayes' theorem. In *Stanford Encyclopedia of Philosophy*; Stanford University: Stanford, CA, USA, 2003.
79. Berrar, D. Bayes' theorem and naive Bayes classifier. *Encycl. Bioinform. Comput. Biol. ABC Bioinform.* **2018**, *403*, 412.
80. Islam, T.; Prince, A.I.; Khan, M.M.Z.; Jabiullah, M.I.; Habib, M.T. An in-depth exploration of Bangla blog post classification. *Bull. Electr. Eng. Inform.* **2021**, *10*, 742–749. [CrossRef]

81. Haydar, M.S.; Al Helal, M.; Hossain, S.A. Sentiment extraction from bangla text: A character level supervised recurrent neural network approach. In Proceedings of the 2018 International Conference on Computer, Communication, Chemical, Material and Electronic Engineering (IC4ME2), Rajshahi, Bangladesh, 8–9 February 2018; pp. 1–4.
82. Hu, Z.; Zhang, J.; Ge, Y. Handling vanishing gradient problem using artificial derivative. *IEEE Access* **2021**, *9*, 22371–22377. [CrossRef]
83. Mumu, T.F.; Munni, I.J.; Das, A.K. Depressed people detection from bangla social media status using lstm and cnn approach. *J. Eng. Adv.* **2021**, *2*, 41–47. [CrossRef]
84. Dam, S.K.; Turzo, T.A. Social Movement Prediction from Bangla Social Media Data Using Gated Recurrent Unit Neural Network. In Proceedings of the 2021 5th International Conference on Electrical Information and Communication Technology (EICT), Khulna, Bangladesh, 17–19 December 2021; pp. 1–6.
85. Uddin, A.H.; Bapery, D.; Arif, A.S.M. Depression analysis from social media data in Bangla language using long short term memory (LSTM) recurrent neural network technique. In Proceedings of the 2019 International Conference on Computer, Communication, Chemical, Materials and Electronic Engineering (IC4ME2), Rajshahi, Bangladesh, 11–12 July 2019; pp. 1–4.
86. Ptaszynski, M.; Dybala, P.; Shi, W.; Rzepka, R.; Araki, K. A system for affect analysis of utterances in Japanese supported with web mining. *J. Jpn. Soc. Fuzzy Theory Intell. Inform.* **2009**, *21*, 194–213. [CrossRef]
87. Ptaszynski, M.; Masui, F.; Dybala, P.; Rzepka, R.; Araki, K. Open source affect analysis system with extensions. In Proceedings of the 1st International Conference on Human–Agent Interaction, iHAI, Sapporo, Japan, 7–9 August 2013.
88. Ptaszynski, M.; Dybala, P.; Rzepka, R.; Araki, K.; Masui, F. ML-Ask: Open source affect analysis software for textual input in Japanese. *J. Open Res. Softw.* **2017**, *5*, 16. [CrossRef]
89. Ptaszynski, M.; Masui, F.; Fukushima, Y.; Oikawa, Y.; Hayakawa, H.; Miyamori, Y.; Takahashi, K.; Kawajiri, S. Deep Learning for Information Triage on Twitter. *Appl. Sci.* **2021**, *11*, 6340. [CrossRef]
90. Gray, D.E. *Doing Research in the Real World*; Sage: Newcastle upon Tyne, UK, 2021; pp. 1–100.
91. Mahoney, J.; Le Louvier, K.; Lawson, S.; Bertel, D.; Ambrosetti, E. Ethical considerations in social media analytics in the context of migration: Lessons learned from a Horizon 2020 project. *Res. Ethics* **2022**, *18*, 226–240. [CrossRef]

Disclaimer/Publisher's Note: The statements, opinions and data contained in all publications are solely those of the individual author(s) and contributor(s) and not of MDPI and/or the editor(s). MDPI and/or the editor(s) disclaim responsibility for any injury to people or property resulting from any ideas, methods, instructions or products referred to in the content.

Article

Social Media Opinion Analysis Model Based on Fusion of Text and Structural Features

Jie Long [1], Zihan Li [2,3], Qi Xuan [2,3], Chenbo Fu [2,3], Songtao Peng [2,3] and Yong Min [4,5,*]

1 College of Computer Science, Zhejiang University of Technology, Hangzhou 310023, China
2 Institute of Cyberspace Security, Zhejiang University of Technology, Hangzhou 310023, China
3 College of Information Engineering, Zhejiang University of Technology, Hangzhou 310023, China
4 Center for Computational Communication Research, Beijing Normal University, Zhuhai 519087, China
5 School of Journalism and Communication, Beijing Normal University, Beijing 100875, China
* Correspondence: myong@bnu.edu.cn

Abstract: The opinion recognition for comments in Internet media is a new task in text analysis. It takes comment statements as the research object, by learning the opinion tendency in the original text with annotation, and then performing opinion tendency recognition on the unannotated statements. However, due to the uncertainty of NLP (natural language processing) in short scenes and the complexity of Chinese text, existing methods have some limitations in accuracy and application scenarios. In this paper, we propose an opinion tendency recognition model HGAT (heterogeneous graph attention network) that integrates text vector and context structure methods to address the above problems. This method first trains a text vectorization model based on annotation text content, then constructs an isomorphic graph with annotation, news, and theme as its apex, and then optimizes the feature vectors of all nodes using an isomorphic graph neural network model with attention mechanism. In addition, this article collected 1,684,318 news items and 57,845,091 comments based on Toutiao, sifted through 511 of those stories and their corresponding 103,787 comments, and tested the impact of HGAT on this dataset. Experiments show that this method has stable improvement effect on different NLP methods, increasing accuracy by 2–10%, and provides a new perspective for opinion tendency recognition.

Keywords: social network; natural language process; opinion tendency recognition; graph embedding; graph neural network

1. Introduction

With the rapid development of the Internet, the public can participate in and discuss a wide range of topics on the Internet according to their background, positions and viewpoints. The development of online media has gone through three generations, including a one-way e-newspaper model represented by a portal, a social network and media model represented by Weibo and WeChat, and an intelligent model based on personalized recommendation information represented by Toutiao and Douyin. In the current internet ecology, different types of internet media complement each other, leading to the diversification of motivations, paths and modes of interaction in internet information dissemination and topic discussion. People can express their views on major social media platforms at any time, leading to a flood of content with a personal bias. This phenomenon of open comment, on the one hand, increases the motivation and initiative of public to participate in the management of society, on the other hand, it raises problems of the proliferation of fake news, the polarization of perceptions and the intensification of public opinion conflicts [1,2].

The study of sentiment analysis has received extensive attention and research from interdisciplinary researchers, especially in the fields of fake news detection brought about by online media, social media bot/dong-army recognition, filter bubble identification and countermeasures [3,4]. The most direct way to reflect public's attitude to these real news events is to analyze the content of their published comments. The comment text on social media represents the views of each individual on the current issue, and accurate identification of the sentiment orientation in these comments can help observe and judge the value trend of the public towards hot events. However, comments in the online media were mainly short stories, which made the study difficult. Compared with ordinary texts, short Chinese web texts tend to have the following characteristics: Comments tend to be short in length, and most people do not form complex sentences when commenting. There is a lot of noisy data, and short text posted on the internet often contains misspellings due to the lack of strict writing requirements. There are many "new words" with special meanings, and social media platforms often have original words with special meanings. Traditional text recognition methods struggle to understand the sentiment orientation these words represent, and Chinese participles are ambiguous and polysemous. Because of the above characteristics, traditional text recognition models cannot accurately recognize the sentiment orientation of network short text.

Research on recognizing sentiment orientation has accumulated to some extent, and most current methods rely on text content for identifying and judging sentiment orientation. These methods are based on natural language processing models to quantify Chinese words or characters [5,6], and then map the entire sentence or paragraph to a vector space followed by the recognition of sentiment orientation and judgement via classification methods. While NLP methods are able to identify sentiment orientation information in text to some degree, existing models are not effective in these problems due to the semantic diversity of Chinese texts and the emergence of new words on the Web. In recent years, the rise of graph convolutional neural networks and the significant advantages of graph embedding algorithms have been accompanied by: Ability to handle large-scale graph data. Graph embedding algorithms typically have efficient computational complexity and can handle large-scale graph data with millions or billions of nodes. Ability to capture similarity and association between nodes. Graph embedding algorithms can map adjacent nodes to similar low-dimensional vector spaces, thereby preserving the similarity and association between nodes. Ability to support various application scenarios. Graph embedding algorithms can be used in social network analysis, recommendation systems, bioinformatics, and other fields. Graph embedding algorithms have gradually been applied in the field of sentiment orientation classification [6,7]. This paper thus design a model based on word vectors and methods for embedding heterogeneous network graphs for comment response structures and textual representations in news stories. Using NLP methods, the model converts short text comments to word vectors and combines the feature vectors of neighbouring comments in the current comment domain (the neighboring comments include the first-order neighborhood and higher-order neighborhood range of neighboring nodes). Attention mechanisms can weight different parts of input, allowing the model to pay closer attention to important information and better understand the context of the input. Therefore, our model incorporates an attention mechanism for extracting important features and ultimately uses a classifier to classify the sentiment orientation of comments.

Given the paucity of datasets containing feedback-reply relationships, 511 public news items about Huawei and corresponding comments were obtained from Today's Headline media platform as the dataset, covering a range of hot topics such as finance and technology. All comments in this article have been manually annotated for sentiment orientation and performed some exploratory research on sentiment analysis of user comments on this dataset, with the following key contributions:

1. Given the special nature of Chinese short texts, a method of extracting features from the comment structure to modify text vectors was adopted in order to achieve more accurate vector mappings. To improve the classification accuracy of sentiment orientation for Chinese short texts, we combined the dynamic comment representation vectors generated by the text vectorization model with features of the comment network structure.

2. In response to the heterogeneity of the comment structure network (the sentiment orientation expressed by comments with reply relationships is often opposite), in this paper, we proposed the HGAT model for efficient feature fusion, and we also incorporated an attention mechanism to aid in feature embedding. The proposed model can further improve text features, and even with a relatively simple classifier, it can obtain good prediction results for the sentiment orientation of the comments.
3. For the proposed HGAT model, a Toutiao dataset is proposed in this paper, and it is verified on the Toutiao dataset that the model in this paper performs better on the dataset of Toutiao compared to the text-only classification method and the graph embedding method based on isomorphism network.

The remainder of this paper is organized as follows. Section 2 is devoted to related work, including opinion tendency recognition, opinion conflict detection, and graph embedding algorithms. Section 3 details the proposed opinion tendency recognition method based on heterogeneous network graph embedding fused attention mechanism in this paper, where the related designs of word embedding layer, input layer, coding layer and output layer are explicitly given. Section 4 conducts experiments on the Toutiao dataset based on the method proposed in this paper, and shows the comparison experimental results between this method and NLP-based, structure-based methods and combined NLP and structure-based algorithms, through which it is verified that the model in this paper has better classification effect and better stability. Section 5 summarizes the various advantages of our proposed model and points out some directions for future work.

2. Related Work

In this section, the paper reviews the work related to this study, including opinion tendency recognition, opinion conflict detection, and graph embedding algorithms.

2.1. Opinion Tendency Recognition

Opinion tendency recognition is a subtask of sentiment classification, which is primarily concerned with the problem of text classification. The purpose of this task is to judge whether a statement about a target expresses a positive, negative, or neutral attitude toward the target. Early traditional methods relied primarily on dictionary models to perform feature counting on textual phrases, such as computing feature values via text decomposition, keyword extraction, and so on, and then determining the opinion-orientation of the text based on the feature values. Early research not only used dictionaries to count feature information in text, but also applied machine learning methods like Support Vector Machines (SVM) [8], decision trees [9,10] and other methods, which are primarily statistical methods for the analysis of opinion guidance. Most of these methods first compute sentence features based on text features and then combine all of the features into a vector. Lastly, they train classifiers such as SVMs and decision trees to achieve opinion-driven text classification.

The aforementioned methods, however, require manual design of feature computation methods and statistical analysis, which is time-consuming and labor-intensive. In contrast, neural network methods compensate for the shortcomings of the above methods. The success of neural networks in image and speech recognition has led to the gradual application of some related models to position detection. For example, in recent years, a variety of methods have been proposed that incorporate textual information into CNN and RNN models for sentiment analysis [11–14]. Apart from CNN and RNN models, LSTM has also been shown to have significant effects in opinion tendency recognition. Siddiqua et al. [15] proposed a nested BiLSTM and LSTM model structure in order to learn information from a larger set of contextual texts. Furthermore, they used an attention mechanism to magnify the impact of salient information content, which further improved the accuracy of the results. Mohtarami et al. [16] augmented the model structure of memristive neural networks using LSTM and CNNs, and introduced a similarity matrix for comparing content similarity across context, improving the prediction accuracy of the model.

With the introduction of the BERT model in recent years [17], a large number of Transformer based models have demonstrated significant performance on NLP tasks. Some recent work has also investigated the effectiveness of the BERT model for opinion tendency recognition. For example, Ghosh et al. [18] compared the performance of the BERT model with other methods on the SemEval2016 dataset and demonstrated that the BERT model has the best performance. Li et al. [19] also studied the effectiveness of the BERT model for data augmentation, and found that the BERT model performs exceptionally well in opinion tendency recognition.

For sentiment classification in the Chinese domain, scholars have optimized different aspects based on the mainstream methods. Chinese text opinion mining research started relatively late, and the sentiment lexicon and deactivation lexicon are not as rich as English. To solve this problem, Xu et al. [20] constructed an extended sentiment dictionary. The extended sentiment dictionary contains basic sentiment words, domain sentiment words and multisense sentiment words, which improves the accuracy of sentiment analysis. A plain Bayesian classifier is used to determine the text domain in which the multisense sentiment words are located. To address the problem that neural network models cannot accurately capture sentiment information in sentiment analysis tasks, Li et al. [21] proposed a sentiment information-based network model (SINM) that uses converter encoders and LSTM as model components to automatically find sentiment knowledge in Chinese texts with the help of Chinese sentiment dictionaries. Sheng et al. [22] to better solve the problem of sentiment analysis of long Chinese texts, proposed a bert-based fusion model and further used the attention mechanism to obtain the effective core sentiment of long Chinese texts.

Opinion tendency recognition by NLP is relatively intuitive and straightforward, but for social networks, NLP methods tend to ignore information related to the content of contextual comment texts, and there may be some correlation between the opinion orientation of contextual comments and the opinion orientation of current comment texts. There is a lack of effective methods in the field of Chinese opinion tendency detection to fuse different granularity information in models. Therefore, we consider integrating text information with other structural information to explore whether this method will improve the effectiveness of opinion tendency detection.

2.2. Opinion Conflict Detection

One aspect of opinion tendency recognition is the prediction of conflicts and controversies, and relevant research on detecting controversies on web pages and social media platforms has been going on for a long time. Recent work by Garimella et al. [23] explained a series of graph structure characteristics of feedback responses under a variety of topics. In addition, they proposed a feature-based graph structure algorithm for measuring the level of controversy on a topic. Research on micro-level conflict at the post or comment level, however, is still not well developed because there are many typos or special terms in the text of posts or comments. The focus of research in this area is on the use of linguistic features of comments (such as the number of appearances of statistically opinionated and topic-related phrases, and some Twitter-specific feature statistics) for controversy detection. Coletto et al. [24], for example, devised a method based on pattern feature extraction to extract features from a Twitter data set and to determine which of the responses in the data set are controversial. Coletto et al. primarily constructed graphs based on responses between comments and relationships between the users' friends in the data set, and then extracted a set of structural features using the motifs algorithm. Lastly, we applied a classifier to classify the features obtained. Zhong et al. [25] also designed TPC-GCN and DTPC-GCN based on the GCN model in order to distinguish whether the post content is controversial or not. Both the The TPC-GCN and DTPC-GCN methods efficiently incorporate structural information from the heterogeneous network and introduce attention mechanisms, achieving higher recognition accuracies than the NLP methods.

2.3. Graph Embedding

Graph embedding methods are now a common approach to network analysis and research. Graph embedding methods map nodes, edges or the entire network of graphs into a low dimensional vector space for representation, and then use machine learning, deep learning, and other methods for downstream tasks such as classification of nodes, prediction of links, and classification of graphs.

DeepWalk [26] is considered to be the first work in this area in graph embedding problems. DeepWalk collects a series of sequences of nodes from the graph via random walks and then vectorizes each representation of the nodes using the Skip-Gram [27] model. The node2vec [28] method expands the search space of DeepWalk's sampling strategy by combining breadth-first search and depth-first search in order to obtain more global and local structural features in a better way, which is referred to as biased random walks, improving the representational learning capability of the network. The development of this typical research field has been accelerated by the emergence of deep learning methods. Variants of Graph Neural Networks (GNN) include GCN (Graph Convolution Network) [29], which provides a simplified method for computing graph embeddings. Based on the structure of the CNN model, the GCN model captures the graph structure and corresponding feature information of each node through graph convolution calculation, and converts the structural information between nodes into vector representations.

Finally, the emergence of attention mechanisms provides a novel method for making accurate predictions based on the weighted combination of all encoded input feature vectors. Li et al. [30] proposed an alternative method to utilize attention mechanisms in dynamic heterogeneous networks. The system employs three types of attention: structural, semantic and temporal, and obtains a better performance.

All of the aforementioned graph embedding methods based on message passing and attention mechanisms are based on the assumption of network isomorphism, i.e., nodes in the network are primarily connected with nodes belonging to the same attributes or classes. Real world networks, however, do not always satisfy the isomorphism hypothesis. To address highly heterogeneous networks, Zhu et al. [31] proposed the H_2GCN model, which aggregates neighborhood node characteristics and uses node degree values for normalization to update node characteristics, thus improving the efficiency of graph embedding algorithms in heterogeneous networks. Fu et al. [32] proposed a method to improve the performance of cross domain classification tasks using network embedding similarity metrics.

With respect to the application of graph embedding algorithms in sentiment analysis of opinions, Zhang et al. [7] used a dependency tree to build simple syntactic dependency relations and used a graph convolutional neural network to fuse the syntactic information, learn text-to-word vector representations, and derive the final vector representation weighted by the importance of the context content. The algorithm introduces graph embedding methods to sentiment classification, but there are few methods that combine feedback response relationships with feedback text for exploration and search.

To summarize, feature-based algorithms of text or network structure have achieved some effects in sentiment classification problems. For text vectorization methods, on the other hand, they ignore the structural information of the feedback response context in social media, although pure network methods fall short for information mining of textual information. Motivated by the above situation, this paper proposes a model that integrates textual information with structural response information from comments for the opinion tendency recognition task.

3. Proposed HGAT Model

This article mainly performs the task of identifying the opinion tendency of comments on social media. The sentence that needs to be predicted can be visually described as a triplet $T = <S, P, C>$ composed of text content, network attributes, and emotional polarity, $S = w_1, w_2, w_3, \cdots, w_n$ represents a sentence consisting of n words w_i, where $w_i, 0 < i < n$,

$P = T_1, T_2, T_3, \cdots, T_n$ represents n triplet entities that have a direct reply relationship with the current sentence. $C = -1, 0, 1$ represents the opinion polarity of the sentence, where -1 indicates that the comment statement has a negative opinion on the topic of the current news, 0 indicates a neutral opinion, and 1 indicates a positive opinion. The main task of this article is to predict the opinion tendency of a specified comment based on the text information content of the comment and the reply relationships between comments. The overall architecture of the HGAT model consists of four layers, as shown in Figure 1.

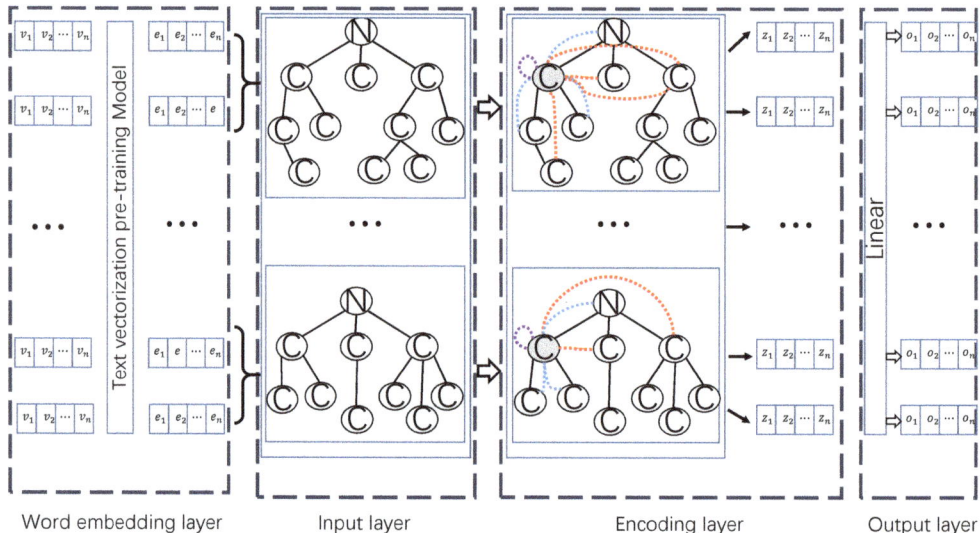

Figure 1. Logical structure of HGAT model. N is the news node, and C is the comment node. The red and blue lines represent the process of attention coefficient calculation between the target node and each node in its different neighborhood, and feature fusion is performed based on the attention coefficient. The red line represents the second-order domain, while the blue line represents the first-order domain. The purple line indicates that the encoding layer concatenates the input features of nodes into the output vector.

1. Word Embedding Layer: The text of the comment is tokenized and fed into a pre-trained text vectorization model using the corresponding phrase table. In this case, The input text is the text from the training set news, and the initial vector representations of all comment phrases and news content are obtained based on the output of the model after secondary training;
2. Inputting Layer: The feature vectors of the text obtained from the previous layer are combined with the social media commentary network, and text feature vectors are used as input to node features;
3. Encoding Layer: Training is carried out on the structure of each news story. The attention mechanism is used to merge the features of each neighbor node of the comment node at various levels with the node's own features, and the feature vector of the comment node is changed accordingly;
4. Output Layer: Given the output vector of the previous layer, the polarity category of the predicted opinion is obtained via the softmax function.

3.1. Word Embedding Layer

The HGAT model first divides and filters the raw data at the word embedding layer, and the division tool used in this paper is the tokenizer package under transformer, and the deactivation table uses the data from the deactivation table published by the Harbin Institute of Technology [33]. First of all, based on the stopwords in the stopword list, words that are not highly relevant to opinion orientation information or expression of position information in the corpus are filtered out for subsequent opinion orientation analysis. In terms of representation, the tokenize method splits the text into tokens by word. Assuming a statement $S = w_1, w_2, w_3, \cdots, w_n$ represents a statement composed of n words w_i, each word is converted into its corresponding numerical value through a tokenizer-predefined dictionary to form the corresponding vector of the sentence: $V = v_1, v_2, v_3, \cdots, v_n$, where $v_i, i \in [1, n]$ is the numerical value after each word is converted. Then, after the training set is processed and numbered, all the numbered comment texts of the training set are input into the pre-trained text vectorization model, and the dynamic word vector text E is obtained through calculation: $E = [e_1, e_2, e_3, \cdots, e_n]^T$. Where e represents the vectorized output of each comment utterance after being trained by the text vectorization model. This paper added a fully connected layer to the original text vector model after the output of the model to reduce the dimension of the output vector to 128 dimensions, i.e., $n = 128$. Furthermore, based on the trained model described above, this paper also outputs a corresponding 128 dimensional vector for the textual content of all news stories, representing news content.

3.2. Input Layer

After obtaining the comment text vectors, the HGAT model fuses the text vector features with the network structure features in the input layer and passes them to the subsequent coding layer. In the case of the news commentary network constituted by the dataset in this paper: it consists primarily of two types of nodes: news nodes and comment nodes. News nodes represent the specific content of news releases, and comment nodes are the content of comments posted by users. The composition presented in this paper is based on the response relationship between comments, and the results are presented in Figure 2. Since the news nodes and the comment nodes in this paper's method are both represented by textual feature vectors, and the dimensions of both of these vector types are 128, the network containing the two types of nodes is treated as a homogeneous network when the network is input to the input layer. In this paper, the network is constructed based on the comment response structure: where N represents a news node, C_1, C_2, C_3 represent three independent first level comments under N news stories, this means that all three of these comments are directly commented on under the new. Similarly, $C_{(1-1)}$, $C_{(1-2)}$ represent two comments replying to C_1, and $C_{(3-1)}$, $C_{(3-2)}$ represent two comments replying to C_3. This paper uses the above method of graph construction to convert the set of response relations in 511 news content into network structures for the purpose of representation: $G = g_1, g_2, \cdots, g_n$, where $g_i, i \in [1, n]$ represents the network consisting of all comments below a news item and the news node itself.

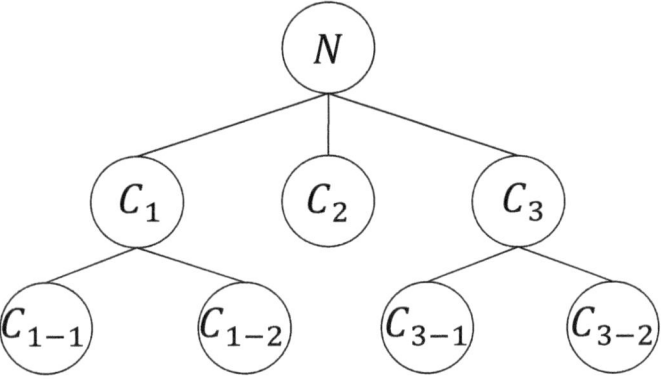

Figure 2. News-Comment Network Architecture.

3.3. Encoding Layer

At this layer, a graph embedding method based on heterogeneous networks is used to extract and fuse contextual feature information from the feedback. In the dataset of this paper, the polarity of sentiment expressed in contextual feedback is often opposite, i.e., responses to feedback under the "Huawei" topic news often have conflicting relationships with one another. Unfortunately, most graph embedding algorithms, such as GCN and GAT, often merge neighboring node features based on the assumption of node homogeneity, which is contrary to the purpose of the present paper. Social media opinion prediction requires an algorithm that can perform graph embeddings on strongly heterogeneous graphs and can aggregate features from neighboring nodes with strong heterogeneity in order to update the node's own feature vector.

The H_2GCN model is a graph embedding vector representation method for heterogeneous networks, which re-designs the feature fusion strategy for highly heterogeneous networks. The H_2GCN model first aggregates the features of the neighboring nodes of the target node and combines the calculation results with the target node's own feature vector to update the target node's feature. In this paper, the HGAT model is modified based on the H_2GCN model, and the node representation vector obtained by the input layer is multiplied by an initial trainable weight matrix $W_{emb} \in \mathbb{R}^{(128 \times \rho)}$ as the initial vector feature r_v^0 of the node.

$$r_v^0 = relu(F_{fc}(X) * W_{emb}) \tag{1}$$

relu represents the Rectified Linear Unit activation function, F_{fc} represents the fully connected function, ρ represents the specified output vector dimension of the hidden layer in HGAT, and X represents the matrix of input vector representations in the encoding layer. In addition, in this paper, an attention mechanism is introduced to aggregate the characteristics of the neighboring nodes for the representation vector $r_v^{(k)}$ of the node v. The attention mechanism calculates the attention coefficients of the neighboring nodes of the target node and updates the feature vector of the target node based on the feature vectors of the neighboring nodes. A key point of the attention mechanism is to determine the influence of the surrounding nodes on the attention coefficients of the current node. To calculate the attention coefficients, the correlation coefficient between the target node and its neighboring nodes needs to be calculated. The input feature matrix $h = \{\vec{h_0}, \vec{h_1}, \vec{h_2}, \cdots, \vec{h_n}\}, \vec{h_i} \in \mathbb{R}^F$, then the calculation formula of the correlation coefficient $e_{i,j}$ between node i and node j is as follows:

$$e_{i,j} = LeakyReLu(\vec{a}(W\vec{h_i} \| W\vec{h_j})) \tag{2}$$

in the formula, $W \in \mathbb{R}^{(F' \times F)}$ represents the weight matrix, where F' represents the dimension of the specified output features. W is applied to each node to ensure that each node can perform a self-attention operation to obtain an attention coefficient. The $\|$ symbol represents the defined attention operation function. The vector $\vec{a}$ in the formula represents a feedforward neural network with a dimension of $\vec{a} \in \mathbb{R}^{(2 \times F')}$, and the result calculated by the neural network is non-linearized by the *LeakyReLU* function to obtain the correlation coefficient $e_{i,j}$ between node i and node j.

Based on Equation (3), the attention coefficient between node i and node j can be obtained by normalization, denoted as $\alpha_{(i,j)}$, which is given by:

$$\alpha_{i,j} = \frac{exp(e_{i,j})}{\sum_{k \in N_i} exp(e_{i,k})} \tag{3}$$

where N_i represents all neighboring nodes of the target node i. After obtaining the node attention coefficient, the updated feature vector of node i can be obtained by summing up the weighted features of all its neighboring nodes.

$$h_i = \sigma(\sum_{j \in N_i} \alpha_{i,j} W \vec{h}_j) \tag{4}$$

where, σ represents a non-linear activation function. Since this paper needs to aggregate the features of the second-order neighboring nodes of node i, during the process of updating the original network node features using the attention mechanism, feature updates need to be performed through different adjacency matrices. Then, the feature vectors updated by the second-order neighborhood are concatenated to obtain the nth-order vector feature.

$$r_v^n = [r1, r2] \tag{5}$$

$$r1 = att(r_v^{n-1}, A - E), r2 = att(r_v^{n-1}, A*A - A - E) \tag{6}$$

where att denotes the process of obtaining the node update vector based on the attention mechanism. After stacking n layers through the above process, the vector r_v^n of the output vertex v will be used as the node vector input of the next output layer. In this paper, n is set to 2 in the experiments.

The output vector after the feature vector update can be represented as:

$$r_v^{(final)} = [r_v^{(1)}, r_v^{(2)}, \cdots, r_v^{(K)}, F_{fc}(X)] \tag{7}$$

$r_v^{(final)}$ is the output vector of the embedding layer, which is the node embedding vector output by the graph embedding model. The benefit of doing this is to separate the target node's own features from the features of its neighboring nodes, and calculate them separately, emphasizing the heterogeneity between the node and its surrounding nodes. Therefore, when dealing with heterogeneous networks, H$_2$GCN often has better performance than graph embedding algorithms based on the homogeneity assumption, such as GCN.

3.4. Output Layer

This layer mainly obtains the node embedding vector representation of the previous layer's encoding layer, smoothly calculates the result through the *softmax* function, and predicts the sentiment polarity of the review text, as shown in the equation:

$$Y = softmax(r_v^{(final)} W_c) \tag{8}$$

where $W_c \in \mathbb{R}^{((2^{K+1}-1)*\rho \times \lceil Y \rceil)}$ is a weight matrix, $\lceil Y \rceil$ is the number of sentiment polarities that need to be classified, and $\lceil Y \rceil$ is set to 3 in this paper. The model in this paper is trained by minimizing the cross-entropy loss value between the predicted value and the true value, as shown below:

$$l_n = -\omega_{Y_n} X_{n,Y_n} \tag{9}$$

$$l(X, Y) = \begin{cases} \sum_{n=1}^{N} \frac{1}{\sum_{n=1}^{N} \omega_{Y_n}} l_n, & \text{if } reduction = \text{'mean'} \\ \sum_{n=1}^{N} l_n, & \text{if } reduction = \text{'sum'} \end{cases} \tag{10}$$

where N represents the batch data $D(X, Y)$ containing N samples, where X is the output of the neural network and has been normalized and logarithmically processed, and Y is the category label corresponding to the sample. l_n is the loss corresponding to the nth sample, which can be obtained from Formula (9). The constant ω is used to deal with the problem of sample imbalance among multiple categories. Formula (10) represents the loss result calculated for the batch data containing N samples, where $reduction = $ 'mean' and *reduction* = '*sum*' represent two ways of calculating the loss by taking the mean and directly accumulating, respectively.

4. Experimental Section

4.1. Datasets and Evaluation Metric

Given the paucity of Chinese datasets with comment-reply structures and sentiment labels at the comment level, data for this article were obtained from the Toutiao media platform. The site includes news articles, reviews, and information on the corresponding users of reviews. The news articles focus on multiple areas such as technology, finance and entertainment, with a particular focus on Huawei-related news published between March and December 2019. Multiple comments are included in each news article, and these comments reflect users' attitudes toward the content of the news article. For this reason, this paper categorizes the sentiment of comments into positive, neutral, and negative. An example of a conflict occurring in a news article is shown in Figure 3. In this example, the news article N belongs to the topic T, and multiple comments in the news article express different views of the users. Figure 3 shows that comments are labelled as positive, neutral or negative depending on the user's attitude towards the information. There is a case where the comment C_{3-1} expresses a negative sentiment, but in fact supports the content of the news article N. This is because in the structure of the comment tree, C_{3-1} refutes the viewpoint of C_3 towards news article N, while C_3 expresses an opposing opinion towards news article N. Therefore, we consider comments like C_{3-1} to have a positive sentiment towards news article N. In data preprocessing, we manually confirmed each comment's label based on its content and contextual feedback. If one of the two comments expresses a positive attitude toward the news while the other one expresses a negative attitude, we consider this to be a conflict of interest. Edges between comments with no conflict indicate that both comments have the same sentiment. Furthermore, this paper does not consider edges between neutral comments and other types of comments in the news. As shown in Table 1, we collected a total of 511 news articles and 103,787 reviews for the entire dataset. In some special cases, there are comments without content or publication time. We conjecture that these comments have been removed by users. In the case of such comments, their labels can only be determined on the basis of the labels of their child comments.

Topic: Huawei News

Current Topic T
A Groups of news about Huawei including technology, Product launch and so on.

Current News N Attached to T
Huawei's flagship tablet MatePad conference preview–use a pen to define an office tablet.

Comments under the news
(Positive) C_1: Huawei didn't make a tablet before, now it's starting to work hard.
(Neutral) C_2: Does this pen come with it or need to be purchased separately?
(Negative) C_3: It's imitating the iPad again! Why don't I go far to iPad for the same price.
  ↳ (Positive) C_{3-1}: You can only buy low-end Apple Air at this price.

Figure 3. News and related Comments.

Depending on the needs of the task, the focus of this paper is on the conflicting content in hot news among all news stories. For this purpose, we extracted three data subsets for experimentation. In particular, we first found the two most active users who made the most comments under different news stories, denoted by u_1 and u_2, respectively. To simplify the notation, we represent the actuality and corresponding comments commented by u_1 as

Toutiao#1, and the other subset consisting of all the news items commented by u_2 such as Toutiao#2. Toutiao#1 is connected to each news item in Toutiao#2 in this paper to form a larger news commentary dataset. The text-feature vector of the Topic node is the vectorised representation of the word "Huawei". Table 1 shows the statistics for the two subsets and the entire dataset.

Table 1. Toutiao and three subsets Dataset Statics.

	Toutiao	Toutiao#1	Toutiao#2
Number of news	511	11	11
Number of users	71,579	3496	5940
Number of comments	103,787	5570	10,580
Positive comments	54,994	2224	3622
Neutral comments	23,236	1647	4095
Negative comments	25,557	1699	2863

The evaluation criteria used in this article are *accuracy*, *macro-precision*, *macro-recall*, and *macro-F1*. The calculation formulas are shown below:

$$accurancy = \frac{TP + FN}{TP + TN + FP + FN} \quad (11)$$

$$recall_i = \frac{TP_i}{TP_i + FN_i} \quad (12)$$

$$precision_i = \frac{TP_i}{TP_i + FP_i} \quad (13)$$

$$F1_i = 2\frac{recall_i precision_i}{recall_i + precision_i} \quad (14)$$

$$macro\text{-}recall = \frac{\sum_{i=1}^{N} recall_i}{N} \quad (15)$$

$$macro\text{-}precision = \frac{\sum_{i=1}^{N} precision_i}{N} \quad (16)$$

$$macro\text{-}F1 = \frac{\sum_{i=1}^{N} F1_i}{N} \quad (17)$$

where, TP_i represents the number of samples that were predicted as positive and actually are positive for the i-th opinion polarity category, FP_i represents the number of samples that were predicted as positive but actually are negative for the i-th opinion polarity category, FN_i represents the number of samples that were predicted as negative but actually are positive for the i-th opinion polarity category, and TN_i represents the number of samples that were predicted as negative and actually are negative for the i-th opinion polarity category. N represents the total number of opinion polarity categories.

4.2. Baseline Method

A comparison of our proposed model with other text vectorization methods including RoBERTa, Ernie, BERT, CPT, GPT2, and so on. The description of these models is as follows.

RoBERTa [34]: RoBERTa is an enhanced version of BERT, achieving better performance by improving training tasks as well as data generation methods, by training for a longer period of time, using larger batch sizes, and by using more data.

Ernie [35]: ERNIE is a large scale knowledge enhanced model published by Baidu, covering NLP and cross-modal models. ERNIE performed further optimisations based on the BERT model and performed better on Chinese NLP tasks. The main improvement of this technique is the masking mechanism.

BERT [17]: BERT is a pre-trained language representation model that emphasizes not using traditional unidirectional language models or shallowly concatenating two unidirectional language models for pre-training, but by using a novel masked language model (MLM) to generate a bidirectional deep language representation.

CPT [36]: The CPT model primarily makes changes to the structure of the encoding and decoding portions of the Transformer and adds three portions: the shared encoder (S-Enc), the decoder comprehension (U-Dec), and the decoder generation (G-Dec). It improves language comprehension and generation skills by using multi-task articulatory training.

GPT2 [37]: The GPT2 model is a new pre-trained model published by the OpenAI organisation on the basis of the GPT model in 2018. Unlike the BERT model, GPT2 is built using the transformer decoder module, while BERT is built using the transformer encoder module. The auto-regression mechanism allows GPT2 to better capture information about the context content.

GCN [29]: GCN was first proposed by Bruna et al. Each node in the network structure updates its own node state by exchanging information with one another, and uses convolutional computations to extract spatial features to learn the node representation, thus performing classification tasks.

4.3. Model Parameter Setting

The experiments in this article were conducted using the PyTorch framework and the A-100 GPU for training.

The number of epoch iterations was set to 50, and the best model was saved based on the accuracy score.

4.4. Comparison of Experimental Results

In addition to the aforementioned text vectorization models, several comparative models were added for ablation experiments in this article: the HGAT + ones method represents the use of an identity matrix instead of node feature vectors as input to the HGAT model in this article. We use this method to test the effect of pure network methods on the true opinion classification after ablation of the feature vectors of the text vectors. The GCN + BERT method represents the use of the BERT model to extract comment node text vector representations and updating comment node feature vectors through the GCN model. We use this method to test the effectiveness of the heterogeneity graph embedding model in HGAT for feature extraction of network structure. In this paper, we build on the above models by comparing several text classification methods, network structure-based methods, and two methods that combine text features with structural features.

The same set of training and test samples was used during implementation so that the experimental results from different algorithms could be compared. Likewise, the training and test sets used for training with the text methods and updating the node feature vectors using the GCN and HGAT models were the same set of nodes. To balance the training and test samples, this paper divided the three comment types equally, and the split results are presented in Table 2. In this paper, the ratio of training set to test set is 90/10, we also did the experiment under 80/20 ratio, which has little effect on the result, here we choose the result under 90/10 ratio.

Table 2. The division result of datasets.

Data Set	Toutiao#1			Toutiao#2		
	Positive	Neutral	Negative	Positive	Neutral	Negative
Training set	904	899	899	1689	1689	1692
Test set	97	102	102	188	191	185

Table 3 shows the experimental results based on the training and test sets above, where Acc. is the precision, Pre. represents the macro-precision, Rec. represents the macro-recall, and F1 represents the macro-F1. The experimental results show that:

1. On both the Toutiao#1 and Toutiao#2 datasets, the text and structural feature combination method proposed in this paper performs better than the original text vectorisation algorithm, which states that the algorithm proposed in this paper, which optimizes and updates the text vectors based on the structural characteristics of the network, has some effect.
2. The HGAT + BERT model proposed in this paper outperforms all text vectorization methods as well as all structure-based algorithms on both datasets.
3. On the Toutiao#1 dataset, the HGAT proposed in this paper has a slightly weaker effect than the GCN, which could be because there are more neutral comments in the structure of the network. This paper's model uses graph embeddings of heterogeneous networks to update node vectors, as shown in Figures 4 and 5, but responses to neutral feedback are often still neutral feedback, making prediction of neutral feedback vectors slightly worse for HGAT than for GCN.

In order to test whether the proposed HGAT model is universally applicable to a variety of text vectorization algorithms, in this paper, five text vectorization algorithms have been combined and compared with two graph embedding algorithms, GCN and HGAT, and the experimental results can be seen in Table 4. As can be seen in Table 4, the algorithm that updates the node based on the word embedding vectors Ernie, CPT, and GPT2 performs better than GCN, although it is weaker than GCN in terms of updating RoBERTa word embedding vectors but still has better performance than the original RoBERTa algorithm. Furthermore, while the RoBERTa text vectorization method does not perform as well as BERT in the pure text vectorization representation, the improvement is largest after updating the nodes by the network structure method. In Toutiao#2 dataset, the value of ACC. was increased from the initial value of 54.96% to 61.82% by HGAT algorithm, and to 68.66% by the GCN algorithm. Figures 4 and 5 show that while the RoBERTa algorithm has the lowest accuracy for predicting neutral feedback, it performs best overall in predicting both negative and positive feedback. As a result, the overall result obtained after updating the node feature vector by the HGAT and GCN graph neural network models is better than that obtained by the BERT, Ernie, CPT and GPT2 algorithms. Furthermore, when compared to all text vectorization algorithms, the algorithm combined with HGAT and text vectorization algorithms has superior performance, who show that the heterogeneous network structure-based method for graph embeddings can efficiently optimize the original text vector and obtain better results with better overall performance. The main reason for the improvement, in our view, is that the HGAT model can effectively capture the correlation between the structural features of sentences with different views and their context, which makes the identification of the opinion tendency more accurate.

In order to investigate the accuracy of the text vectorization methods, as well as the combination of structure and text vectorization methods for the identification of three types of comment nodes, in this paper, we have compared five text vectorization models and models that combine textual and structural features. Figures 4 and 5 show the results. It can be seen from the results that.

1. For the most part, the model that combines HGAT and text vectorization has better accuracy in identifying both positive and negative reviews than the model that combines GCN and text vectorization, and also has a larger improvement over the text vectorization methods.
2. The combination model of GCN and text vectorisation has an advantage in terms of identifying neutral comments. We believe this is due to the fact that the context of neutral comments often involves neutral comments in response comments, thus, the node homogeneity-based GCN method has an advantage for identifying neutral comments.
3. The GCN and text vectorization combination models are relatively unstable, and their accuracy is greatly impacted by the text vectorization. Of these, in the Toutiao#1

dataset, based on the CPT vectorisation method, the GCN + CPT model even had an accuracy of around 20% lower in the identification of negative comments when compared to the original method. The proposed HGAT model in this paper has better stability than the GCN model. The reason for the improvement, in our view, is that the HGAT model introduces attention mechanisms, which may better quantify the influence of neighbouring nodes on the current one. Furthermore, the network heterogeneity-based model is best suited to environments with extreme opinion polarization in social networks.

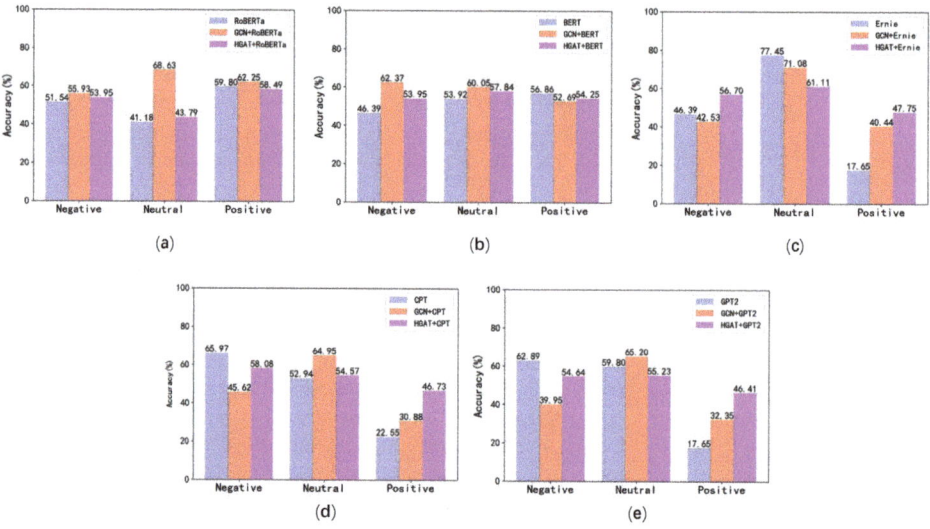

Figure 4. Comparison of the recognition results of comments with different opinions by different methods on the Toutiao#1 dataset. (**a**) Based on the RoBERTa model; (**b**) Based on the BERT model; (**c**) Based on the Ernie model; (**d**) Based on the CPT model; (**e**) Based on the GPT2 model.

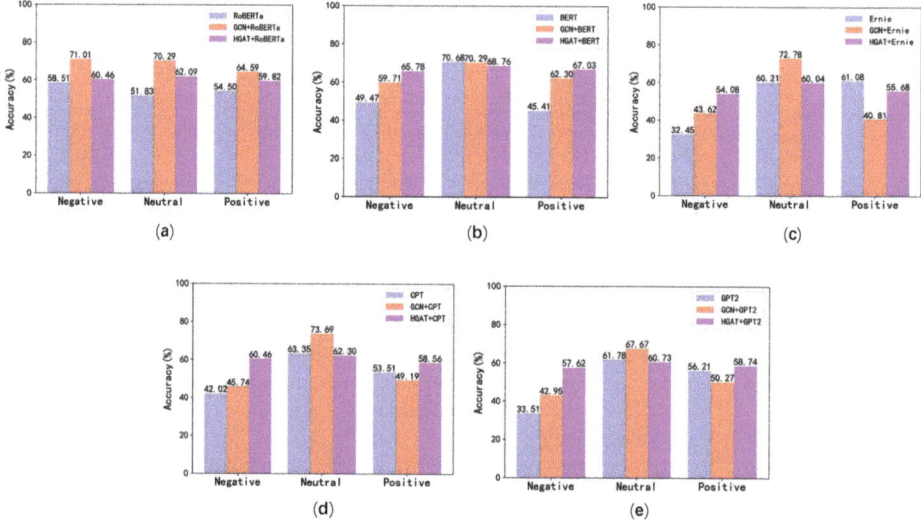

Figure 5. Comparison of the recognition results of comments with different opinions by different methods on the Toutiao#2 dataset. (**a**) Based on the RoBERTa model; (**b**) Based on the BERT model; (**c**) Based on the Ernie model; (**d**) Based on the CPT model; (**e**) Based on the GPT2 model.

Table 3. Comparison of evaluation for different models.

Model		Toutiao#1				Toutiao#2			
		Acc. (%)	Pre. (%)	Rec. (%)	F1 (%)	Acc. (%)	Pre. (%)	Rec. (%)	F1 (%)
NLP-based	Roberta	50.83 ± 0.06	51.15 ± 0.60	50.84 ± 0.07	50.55 ± 0.05	54.96 ± 0.51	54.99 ± 0.44	54.98 ± 0.50	54.94 ± 0.69
	Bert	52.49 ± 0.20	52.77 ± 2.07	52.39 ± 0.22	52.38 ± 0.99	55.32 ± 0.70	55.27 ± 1.13	55.18 ± 0.71	54.72 ± 0.81
	Ernie	47.17 ± 0.66	48.70 ± 0.33	47.16 ± 0.60	43.83 ± 0.82	51.24 ± 0.81	52.15 ± 1.07	51.25 ± 0.82	50.34 ± 1.28
	CPT	46.51 ± 0.66	46.73 ± 1.15	46.78 ± 0.62	43.80 ± 0.45	50.53 ± 0.20	50.38 ± 1.63	50.50 ± 0.35	49.70 ± 1.60
	GPT2	46.84 ± 0.60	47.77 ± 0.49	47.16 ± 0.65	44.95 ± 0.79	53.01 ± 1.78	53.08 ± 1.86	52.96 ± 1.14	52.63 ± 0.94
Structure-based	HGAT + ones	44.43 ± 3.58	48.03 ± 7.56	44.68 ± 3.65	42.07 ± 5.16	48.23 ± 1.93	48.36 ± 2.00	48.28 ± 1.93	48.16 ± 1.94
	GCN + ones	53.15 ± 1.16	55.18 ± 1.48	53.32 ± 1.21	52.61 ± 1.28	66.49 ± 1.00	67.34 ± 0.32	66.46 ± 0.91	66.32 ± 0.91
Combined	GCN + Bert	58.31 ± 1.17 [1]	59.39 ± 1.48 [1]	58.37 ± 1.39 [1]	58.38 ± 1.28 [1]	64.14 ± 0.84	64.82 ± 0.32	64.09 ± 0.90	64.00 ± 0.91
	HGAT + Bert	55.82 ± 1.48	56.04 ± 1.45	55.79 ± 1.46	55.84 ± 1.46	67.20 ± 0.82 [1]	67.38 ± 0.88 [1]	67.19 ± 0.82 [1]	67.22 ± 0.83 [1]

[1] The bold numbers represent the best experimental results.

Table 4. Experimental results comparison of different NLP methods and structural combination.

Model	Toutiao#1				Toutiao#2			
	Acc. (%)	Pre. (%)	Rec. (%)	F1 (%)	Acc. (%)	Pre. (%)	Rec. (%)	F1 (%)
Bert	52.49 ± 0.20	52.77 ± 2.07	52.39 ± 0.22	52.38 ± 0.99	55.32 ± 0.70	55.27 ± 1.13	55.18 ± 0.71	54.72 ± 0.81
GCN + BERT	58.31 ± 1.32 [1]	59.39 ± 1.12 [1]	58.37 ± 1.36 [1]	58.38 ± 1.34 [1]	64.14 ± 0.64	64.82 ± 0.36	64.09 ± 0.68	64.00 ± 0.73
HGAT + BERT	55.82 ± 1.48	56.04 ± 1.45	55.79 ± 1.46	55.84 ± 1.46	67.20 ± 0.82 [1]	67.38 ± 0.88 [1]	67.19 ± 0.82 [1]	67.22 ± 0.83 [1]
Roberta	50.83 ± 0.06	51.15 ± 0.60	50.84 ± 0.07	50.55 ± 0.05	54.96 ± 0.51	54.99 ± 0.44	54.98 ± 0.50	54.94 ± 0.69
GCN + Roberta	62.37 ± 3.10 [1]	64.67 ± 2.04 [1]	62.27 ± 3.07 [1]	62.44 ± 3.23 [1]	68.66 ± 0.30 [1]	69.23 ± 0.48 [1]	68.63 ± 0.36 [1]	68.58 ± 0.48 [1]
HGAT + Roberta	52.41 ± 1.17	52.65 ± 1.24	52.44 ± 1.19	52.24 ± 1.13	61.82 ± 0.48	61.89 ± 0.56	61.79 ± 0.49	61.79 ± 0.52
Ernie	47.17 ± 0.66	48.70 ± 0.33	47.16 ± 0.60	43.83 ± 0.82	51.24 ± 0.81	52.15 ± 1.07	51.25 ± 0.82	50.34 ± 1.28
GCN + Ernie	51.49 ± 0.70	52.44 ± 0.93	51.35 ± 0.74	50.54 ± 0.97	52.57 ± 0.54	52.84 ± 0.52	52.41 ± 0.51	51.49 ± 0.51
HGAT + Ernie	54.41 ± 3.33 [1]	54.34 ± 3.36 [1]	54.43 ± 3.35 [1]	54.23 ± 3.39 [1]	56.62 ± 0.47 [1]	56.60 ± 0.46 [1]	56.60 ± 0.46 [1]	56.59 ± 0.46 [1]
CPT	46.51 ± 0.66	46.73 ± 1.15	46.78 ± 0.62	43.80 ± 0.45	50.53 ± 0.20	50.38 ± 1.63	50.50 ± 0.35	49.70 ± 1.60
GCN + CPT	47.17 ± 2.68	52.84 ± 2.92	47.15 ± 2.81	43.56 ± 4.09	56.34 ± 0.57	56.91 ± 0.55	56.21 ± 0.61	55.39 ± 0.45
HGAT + CPT	52.99 ± 1.20 [1]	53.01 ± 1.05 [1]	53.06 ± 1.18 [1]	52.95 ± 1.22 [1]	60.46 ± 0.95 [1]	60.64 ± 0.92 [1]	60.44 ± 0.94 [1]	60.47 ± 0.92 [1]
GPT2	46.84 ± 0.60	47.77 ± 0.49	47.16 ± 0.65	44.95 ± 0.79	53.01 ± 1.78	53.08 ± 1.86	52.96 ± 1.14	52.63 ± 0.94
GCN + GPT2	45.93 ± 1.23	46.39 ± 1.05	45.83 ± 1.25	44.53 ± 1.47	53.72 ± 0.43	53.66 ± 0.46	53.63 ± 0.42	53.16 ± 0.28
HGAT + GPT2	51.24 ± 2.21 [1]	51.28 ± 2.20 [1]	51.30 ± 2.19 [1]	51.16 ± 2.25 [1]	59.04 ± 0.78 [1]	59.09 ± 0.80 [1]	59.03 ± 0.79 [1]	59.05 ± 0.79 [1]

[1] The bold numbers represent the best experimental results.

5. Conclusions

The HGAT model proposed in this paper, which combines semantic and structural features, first extracts the semantic feature vector from the text of the comment by means of NLP, obtaining the text feature vector of the comment. Based on the graph embedding method of heterogeneous networks, the text vector is then further modified according to the context response relation of the comment, effectively ameliorating the problem caused by the short length of the Chinese comments and expressing the comment vector accurately. The attention mechanism is simultaneously combined with the graph embedding method of heterogeneous networks to assign attention probabilities to the salient features during learning of the input features, improving the final classification accuracy. Overall, the experimental results demonstrate that the proposed model has better performance on the Toutiao News dataset when compared to both pure text classification and graph embedding methods based on homogenous networks. Future analysis will focus on a more in-depth exploration of the textual representation and properties of the comment-news network, by combining deeper semantic recognition models with heterogeneous network graph embedding models to improve the effectiveness of opinion trend classification. In summary, the proposed method is easy to transfer, and the trained model can be used extensively in practical social media platforms.

Author Contributions: Conceptualization, J.L., Z.L. and Y.M.; methodology, J.L. and Z.L.; data curation and formal analysis J.L. and Z.L.; writing—original draft preparation, J.L. and Z.L.; writing—review and editing, Q.X., C.F. and S.P.; funding acquisition, Q.X. and Y.M. All authors have read and agreed to the published version of the manuscript.

Funding: This work was partially supported by the National Natural Science Foundation of China under Grant 61973273 and by the Zhejiang Provincial Natural Science Foundation of China under Grant LGF21G010003.

Institutional Review Board Statement: Not applicable.

Informed Consent Statement: Informed consent was obtained from all subjects involved in the study.

Data Availability Statement: Data will be made available upon reasonable request to the corresponding author.

Conflicts of Interest: The authors declare no conflict of interest.

References

1. Conover, M.; Ratkiewicz, J.; Francisco, M.; Goncalves, B.; Menczer, F. Political Polarization on Twitter. In Proceedings of the International AAAI Conference on Web and Social Media, Barcelona, Spain, 17–21 July 2011.
2. Linden, S.; Roozenbeek, J.; Compton, J. Inoculating against fake news about COVID-19. *Front. Psychol.* **2020**, *11*, 566790. [CrossRef] [PubMed]
3. Sharma, K.; Feng, Q.; He, J.; Ruchansky, N.; Liu, Y. Combating Fake News: A Survey on Identification and Mitigation Techniques. *ACM Trans. Intell. Syst. Technol.* **2019**, *10*, 1–42. [CrossRef]
4. Vicario, M.; Vivaldo, G.; Bessi, A.; Zollo, F.; Quattrociocchi, W. Echo Chambers: Emotional Contagion and Group Polarization on Facebook. *Sci. Rep.* **2016**, *6*, 37825 . [CrossRef] [PubMed]
5. Zhang, D.; Xu, H.; Su, Z.; Xu, Y. Chinese comments sentiment classification based on word2vec and SVMperf. *Expert Syst. Appl.* **2015**, *42*, 1857–1863. [CrossRef]
6. Qiao, Q.; Bo, T.; Huang, M.; Yang, L.; Zhu, X. Learning Tag Embeddings and Tag-specific Composition Functions in Recursive Neural Network. In Proceedings of the 53rd Annual Meeting of the Association for Computational Linguistics and the 7th International Joint Conference on Natural Language Processing, Beijing, China, 26–31 July 2015; Volume 1: Long Papers.
7. Zhang, Y.; Qi, P.; Manning, C.D. Graph Convolution over Pruned Dependency Trees Improves Relation Extraction. *arXiv* **2018**, arXiv:1809.10185.
8. Bouchlaghem, R.; Elkhelifi, A.; Faiz, R. SVM based approach for opinion classification in Arabic written tweets. In Proceedings of the 2015 IEEE/ACS 12th International Conference of Computer Systems and Applications (AICCSA), Marrakech, Morocco, 17–20 November 2015.
9. Es-Sabery, F.; Es-Sabery, K.; Qadir, J.; Sainz-De-Abajo, B.; Torre-Diez, I. A MapReduce Opinion Mining for COVID-19-Related Tweets Classification Using Enhanced ID3 Decision Tree Classifier. *IEEE Access* **2021**, *9*, 58706–58739. [CrossRef]

10. Sanjay, K.S.; Danti, A. Detection of fake opinions on online products using Decision Tree and Information Gain. In Proceedings of the 2019 3rd International Conference on Computing Methodologies and Communication (ICCMC), Erode, India, 27–29 March 2019.
11. Wan, W.; Xiao, Z.; Liu, X.; Wei, C.; Wang, T. pkudblab at SemEval-2016 Task 6: A Specific Convolutional Neural Network System for Effective Stance Detection. In Proceedings of the 10th International Workshop on Semantic Evaluation (SemEval-2016), San Diego, CA, USA, 16–17 June 2016.
12. Du, J.; Xu, R.; He, Y.; Lin, G. Stance Classification with Target-specific Neural Attention. In Proceedings of the 26th International Joint Conference on Artificial Intelligence, IJCAI 2017, Melbourne, Australia, 19–25 August 2017.
13. Zhou, Y.; Cristea, A.I.; Shi, L. Connecting Targets to Tweets: Semantic Attention-Based Model for Target-Specific Stance Detection. In *Lecture Notes in Computer Science, Proceedings of the Web Information Systems Engineering—WISE 2017—18th International Conference, Puschino, Russia, October 7–11. 2017, Proceedings, Part I*; Bouguettaya, A., Gao, Y., Klimenko, A., Chen, L., Zhang, X., Dzerzhinskiy, F., Jia, W., Klimenko, S.V., Li, Q., Eds.; Springer: Berlin/Heidelberg, Germany, 2017; Volume 10569, pp. 18–32.
14. Augenstein, I.; Rocktäschel, T.; Vlachos, A.; Bontcheva, K. Stance Detection with Bidirectional Conditional Encoding. *arXiv* **2016**, arXiv:1606.05464.
15. Siddiqua, U.A.; Chy, A.N.; Aono, M. Tweet Stance Detection Using an Attention based Neural Ensemble Model. In Proceedings of the 2019 Conference of the North American Chapter of the Association for Computational Linguistics: Human Language Technologies (NAACL-HLT 2019), Minneapolis, MO, USA, 2–7 June 2019; Long and Short Papers; Association for Computational Linguistics: Minneapolis, MO, USA, 2019; Volume 1.
16. Mohtarami, M.; Baly, R.; Glass, J.R.; Nakov, P.; Moschitti, A. Automatic Stance Detection Using End-to-End Memory Networks. *arXiv* **2018**, arXiv:1804.07581.
17. Devlin, J.; Chang, M.W.; Lee, K.; Toutanova, K. BERT: Pre-training of Deep Bidirectional Transformers for Language Understanding. *arXiv* **2018**, arXiv:1810.04805.
18. Ghosh, S.; Singhania, P.; Singh, S.; Rudra, K.; Ghosh, S. Stance Detection in Web and Social Media: A Comparative Study. In *Experimental IR Meets Multilinguality, Multimodality, and Interaction: Proceedings of the 10th International Conference of the CLEF Association, CLEF 2019, Lugano, Switzerland, 9–12 September 2019, Proceedings 10*; Springer International Publishing: Berlin/Heidelberg, Germany, 2020.
19. Li, Y.; Caragea, C. Target-Aware Data Augmentation for Stance Detection. In Proceedings of the Association for Computational Linguistics 2021, Virtual, 1–6 August 2021; pp. 1850–1860.
20. Xu, G.; Yu, Z.; Yao, H.; Li, F.; Meng, Y.; Wu, X. Chinese Text Sentiment Analysis Based on Extended Sentiment Dictionary. *IEEE Access* **2019**, *7*, 43749–43762. [CrossRef]
21. Li, G.; Zheng, Q.; Zhang, L.; Guo, S.; Niu, L. Sentiment Infomation based Model For Chinese text Sentiment Analysis. In Proceedings of the 2020 IEEE 3rd International Conference on Automation, Electronics and Electrical Engineering (AUTEEE), Shenyang, China, 20–22 November 2020; pp. 366–371. [CrossRef]
22. Sheng, D.; Yuan, J. An Efficient Long Chinese Text Sentiment Analysis Method Using BERT-Based Models with BiGRU. In Proceedings of the 2021 IEEE 24th International Conference on Computer Supported Cooperative Work in Design (CSCWD), Dalian, China, 5–7 May 2021; pp. 192–197. [CrossRef]
23. Garimella, K.; Morales, G.D.; Gionis, A.; Mathioudakis, M. Quantifying Controversy on Social Media. *ACM Trans. Soc. Comput.* **2018**, *1*, 1–27. [CrossRef]
24. Coletto, M.; Garimella, K.; Gionis, A.; Lucchese, C. Automatic controversy detection in social media: A content-independent motif-based approach. *Online Soc. Netw. Media* **2017**, *3–4*, 22–31. [CrossRef]
25. Zhong, L.; Cao, J.; Sheng, Q.; Guo, J.; Wang, Z. Integrating Semantic and Structural Information with Graph Convolutional Network for Controversy Detection. *arXiv* **2020**, arXiv:2005.07886.
26. Perozzi, B.; Al-Rfou, R.; Skiena, S. DeepWalk: Online Learning of Social Representations. In Proceedings of the 20th ACM SIGKDD International Conference on Knowledge Discovery and Data Mining, New York, NY, USA, 24–27 August 2014.
27. Mikolov, T.; Sutskever, I.; Kai, C.; Corrado, G.; Dean, J. Distributed Representations of Words and Phrases and their Compositionality. In *Advances in Neural Information Processing Systems*; Curran Associates, Inc.: Red Hook, NY, USA, 2013.
28. Grover, A.; Leskovec, J. node2vec: Scalable Feature Learning for Networks. In Proceedings of the 22nd ACM SIGKDD International Conference on Knowledge Discovery and Data Mining, New York, NY, USA, 13–17 August 2016.
29. Kipf, T.N.; Welling, M. Semi-Supervised Classification with Graph Convolutional Networks. *arXiv* **2016**, arXiv:1609.02907.
30. Li, Q.; Shang, Y.; Qiao, X.; Dai, W. Heterogeneous Dynamic Graph Attention Network. In Proceedings of the 2020 IEEE International Conference on Knowledge Graph (ICKG), Nanjing, China, 9–11 August 2020.
31. Zhu, J.; Yan, Y.; Zhao, L.; Koutra, D. Beyond Homophily in Graph Neural Networks: Current Limitations and Effective Designs. *Adv. Neural Inf. Process. Syst.* **2020**, *33*, 7793–7804.
32. Fu, C.; Zheng, Y.; Liu, Y.; Xuan, Q.; Chen, G. NES-TL: Network Embedding Similarity-Based Transfer Learning. *IEEE Trans. Netw. Sci. Eng.* **2020**, *7*, 1607–1618. [CrossRef]
33. Datatang丨StopWordsSet. Available online: http://www.datatang.com/data/19300 (accessed on 18 March 2021).
34. Liu, Y.; Ott, M.; Goyal, N.; Du, J.; Joshi, M.; Chen, D.; Levy, O.; Lewis, M.; Zettlemoyer, L.; Stoyanov, V. RoBERTa: A Robustly Optimized BERT Pretraining Approach. *arXiv* **2019**, arXiv:1907.11692.
35. Sun, Y.; Wang, S.; Li, Y.; Feng, S.; Wu, H. ERNIE: Enhanced Representation through Knowledge Integration. *arXiv* **2019**, arXiv:1904.09223.

36. Shao, Y.; Geng, Z.; Liu, Y.; Dai, J.; Yang, F.; Zhe, L.; Bao, H.; Qiu, X. CPT: A Pre-Trained Unbalanced Transformer for Both Chinese Language Understanding and Generation. *arXiv* **2021**, arXiv:2109.05729.
37. Radford, A.; Wu, J.; Child, R.; Luan, D.; Amodei, D.; Sutskever, I. Language models are unsupervised multitask learners. *OpenAI Blog* **2019**, *1*, 9.

Disclaimer/Publisher's Note: The statements, opinions and data contained in all publications are solely those of the individual author(s) and contributor(s) and not of MDPI and/or the editor(s). MDPI and/or the editor(s) disclaim responsibility for any injury to people or property resulting from any ideas, methods, instructions or products referred to in the content.

Article

Emotional State Detection Using Electroencephalogram Signals: A Genetic Algorithm Approach

Rosa A. García-Hernández [1], José M. Celaya-Padilla [1,*], Huizilopoztli Luna-García [1], Alejandra García-Hernández [1], Carlos E. Galván-Tejada [1], Jorge I. Galván-Tejada [1], Hamurabi Gamboa-Rosales [1], David Rondon [2] and Klinge O. Villalba-Condori [3]

[1] Unidad Académica de Ingeniería Eléctrica, Universidad Autónoma de Zacatecas, Jardín Juárez 147, Centro, Zacatecas 98000, Mexico; adriana.gar.her@uaz.edu.mx (R.A.G.-H.)
[2] Departamento Estudios Generales, Universidad Continental, Arequipa 04001, Peru
[3] Vicerrectorado de Investigación, Universidad Católica de Santa María, Arequipa 04001, Peru
* Correspondence: jose.celaya@uaz.edu.mx

Abstract: Emotion recognition based on electroencephalogram signals (EEG) has been analyzed extensively in different applications, most of them using medical-grade equipment in laboratories. The trend in human-centered artificial intelligence applications is toward using portable sensors with reduced size and improved portability that can be taken to real life scenarios, which requires systems that efficiently analyze information in real time. Currently, there is no specific set of features or specific number of electrodes defined to classify specific emotions using EEG signals, and performance may be improved with the combination of all available features but could result in high dimensionality and even worse performance; to solve the problem of high dimensionality, this paper proposes the use of genetic algorithms (GA) to automatically search the optimal subset of EEG data for emotion classification. Publicly available EEG data with 2548 features describing the waves related to different emotional states are analyzed, and then reduced to 49 features with genetic algorithms. The results show that only 49 features out of the 2548 can be sufficient to create machine learning (ML) classification models with, using algorithms such as k-nearest neighbor (KNN), random forests (RF) and artificial neural networks (ANN), obtaining results with 90.06%, 93.62% and 95.87% accuracy, respectively, which are higher than the 87.16% and 89.38% accuracy of previous works.

Keywords: emotion recognition; electroencephalogram; affective computing; genetic algorithms; RF; KNN; ANN

1. Introduction

In recent years, research in emotion detection has become increasingly important. The development of user-centric artificial intelligence-based technologies has been one of the main reasons for the growth in different application areas such as healthcare, education, entertainment, robotics, marketing, security, and surveillance. Physical expressions such as facial gestures, speech or postures have been used to identify human emotions [1–6], but in some cases, this can be ineffective because people may purposely or unconsciously mask their true feelings, which is why physiological signals can provide a more precise and objective recognition of emotions [7]. For this reason, many of the approaches in affective computing research have turned their attention to analysis through physiological signals [8–15].

Emotion recognition and human-centered research efforts have provided improved achievements thanks to the use of machine learning and deep learning algorithms [16,17]; these have evolved rapidly in recent decades and an important aspect to consider is the selection of features used to create prediction or classification models.

Genetic algorithms are optimization algorithms based on natural selection. These algorithms are executed through an iterative process of selection, crossing and mutation of chromosomes from an initial population; during this process, only the best-adapted chromosomes survive [18]. GAs have proven to be very efficient in terms of parameter optimization and dimensionality reduction [19–21]. In a study of emotion recognition through pulse signals and the SVM classifier, the authors evaluated the effect of using genetic algorithms for feature selection and compared the recognition rate without a selector and with a GA as the feature selector. The recognition rate increased from 52.5% to 90% when using genetic algorithms [22].

In addition, an advantage of GAs over deep network training algorithms in backpropagation is that GAs do not suffer from an evanescent gradient, as they do not use gradients for optimization. Instead, they use search techniques based on natural selection and reproduction to find optimal solutions to complex problems [23]. Algorithms that have proven to be very effective in solving classification problems are random forest, KNN and ANN, which are described in more detail in Section 2. Electroencephalogram (EEG) signals are types of physiological signals that are very sensitive to changes in emotional states. The electroencephalogram registers the potential differences of neurons when they are active, recording the electrical information from the activity of the autonomic nervous system and central nervous system [15]. One of the disadvantages of emotion analysis based on electroencephalogram signals is the fact that it involves the use of many electrodes attached to the head of the analyzed subjects, as well as a large amount of data to be analyzed. For this reason, the trend in the development of sensors for emotion recognition has been leaning towards reducing size and improving portability in order to be useful for real-life scenarios, such as wearable wireless sensors [24]. An important aspect to consider for the development of this type of sensors and systems is a reduction in the dimensionality of the data to be processed.

Previous work on EEG feature extraction has shown that there are many useful temporal and statistical features, which have proven to be effective for the recognition of different emotions. In their study, Bird J. et al. [25] explored five different feature selection methods and seven classification models, comparing their predictive ability and the number of features used. For each test, 10-fold cross validation was used to train the model. For feature selection, the best result was obtained with the OneR algorithm with 44 features selected from a set of 2100, which were used with classification models such as Naïve Bayes, Bayesian networks, random tree, support vector machine, multilayer perceptron and random forest, the last one being the most accurate with a prediction accuracy of 87.16%.

In a following study, Bird J. et al. [26] compared single and ensemble methods for classifying emotions from an EEG brainwave database, and they also compare feature selection methods such as OneR, Bayes network, information gain, and symmetrical uncertainty. From 2548 characteristics, 63 were chosen with the information gain method and their higher results of classification were found with the random forest ensemble classifier with an overall accuracy of 97.89%. The best single classifier was a deep neural network with an accuracy of 94.89%. Ten-fold cross validation was used for training the models.

In their study, Jodie Ashford and collaborators [27] classified EEG signals with an approach based on the representation of the statistical features from the signals as images. In their work, they reduce a large set of features containing 2479 features to a set only containing 256 using the algorithm of information gain measure; then, they used the resulting features to reshape them and express them as grayscale images which were then used in a deep convolutional neural network to train the data and classify the mental state of the subject with an accuracy of 89.38%. For the validation of this study, the data were split in a 70/30 ratio for training and testing, respectively.

Moreover, Liu Z. et al. [28] proposed a special feature extraction methodology using the empirical mode decomposition (EMD) domain mixed with sequence backward selection (SBS) in order to remove the reductant features. Additionally they compared different temporal windows' lengths and the kinds of rhythms of EEG signals, the one with a length of 1 s being the one achieving the highest recognition accuracy of 86.46% in valence and

84.90% in arousal. In this work, the dataset samples were split into an 80/20 test and training ratio.

On the other hand, Xu H. and collaborators [29], using their approach, studied the effects of selecting different channels and frequency bands of EEG signals on the precision of emotion recognition. Initially, they used the discrete wavelet transform method to separate the signals into bands such as gamma, beta, alpha, and theta; then, the entropy and energy were extracted from each band as the class features. Afterwards, they evaluated channel combinations and compared them using three methods, the first one based on experience, the second based on the indirect minimal redundancy maximal relevance (mRMR-FS) algorithm, and the third one based on the direct channel selection method mRMR-CS algorithm, using each channel as a whole. In their results, of the three methods used, mRMR-CS had the best ability to reduce channels, reaching an accuracy of 79.46% and reducing the number of channels from 32 to 22.

This study proposes the use of intelligent algorithms to perform temporal and statistical feature selection from EEG brain waves related to different emotional states, with the use of genetic algorithms (GA) and KNN, RF and ANN machine learning models to classify three emotional states with improved accuracy. Section 2 discusses the materials and methods used in this research work. Section 3 provides the results. Section 4 shows the discussion, and Section 5 reports the conclusions.

The main contributions of this study are outlined below.

The proposed model enabled the dimensionality of the data to be reduced by 98.08%. Compared to similar studies, this dimensionality reduction is superior to others, and could improve the feasibility of developing specific-purpose devices for real-time applications in feature jobs.

The 49 optimal features to which the proposed model reduced the initial total of 2548 features were used to create machine learning (ML) classification models with algorithms such as k-nearest neighbor (KNN), random forests (RF) and artificial neural networks (ANN), obtaining results with 90.06%, 93.62% and 95.87% accuracy, respectively, which are higher than the 87.16% and 89.38% accuracy reported in previous works.

2. Materials and Methods

This section describes the methodology applied for data analysis using different selection and classification algorithms, as well as giving a general description of the data analyzed. Figure 1 illustrates the methodology applied, which consists of four main stages: (1) data description, (2) feature selection, (3) implementation of ML models and then (4) validation of the models. Details of the methodology are described below.

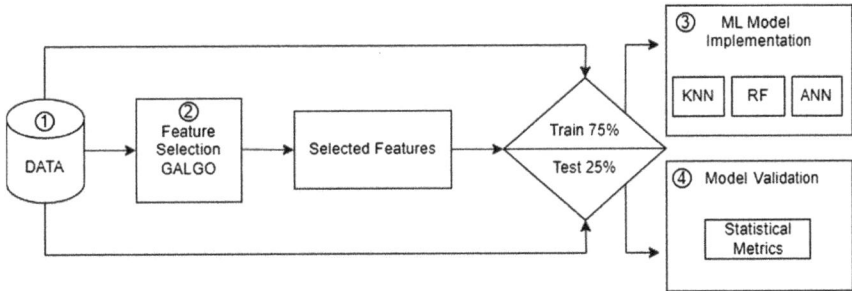

Figure 1. Methodology.

2.1. Data Description

A publicly available EEG database describing the waves related to different emotional states was used. Data from 1 male and 1 female were recorded during 3 min sessions for positive, neutral, and negative states using a Muse EEG headset. Locations TP9, AF7, AF8 and TP10 of the dry electrode were selected. Emotions were evoked using negative, positive, and neutral emotional movie clips. From the EEG brain waves, a static dataset was created using a sliding window approach. The dataset included 2548 features and 2132 observations related to an emotional state [10,25]. It was not necessary to process or remove any data.

2.2. Feature Selection

The database used to carry out this work included 2548 statistical and time-dependent features to describe three mental states. All these features may have the potential to include important information to describe part of the emotional state classification problem; however, this might be computationally costly. The genetic algorithm introduced by Goldberg and Holland [30] is based on the natural selection mechanism. The algorithm's objective is to discover the best solution for chromosome survival, with a statistical optimization approach. Genetic algorithms are based on natural behavior.

There are different types of GAs; some authors classify the variants into five main categories: real- and binary-coded, multi-objective, parallel, chaotic and hybrid [18]. The type of GA used in this study is a multi-objective statistical model-building approach with a specific search strategy for variable selection. The general operation of the GA is based on the random creation of a population called (Y) of n chromosomes; the fitness of each chromosome in the population is evaluated, according to this evaluation two chromosomes, C1 and C2 are selected, and a crossover operator with crossover probability (Cp) is applied to those chromosomes C1 and C2 to produce an offspring, O. Then, a mutation operator with mutation probability (Mp) is applied on the offspring O to generate O'. Finally, the offspring O' is placed in the new population. From here, the selection, crossover and mutation operations are repeated iteratively until the new population is complete. Figure 2 represents the GA flow chart with the following listed steps:

- Step 1: creation of a random initial population of chromosomes, which in this case are sets of 5 genes or features.
- Step 2: consisting in the evaluation of the capability of the chromosomes to predict the different emotional states, with this creating a statistical model, the GA assigns a score to each chromosome, and this score is proportional to the resulting accuracy of the model. In this study, the nearest centroid classifier is used in the model.
- Step 3: if the score in the previous step is higher than that of the defined fitness goal, the chromosome is selected; if it is not, the process continues.
- Step 4: the chromosomes best suited to the problem are replicated; the higher the score, the bigger the offspring.
- Step 5: the crossover consists of a recombination of pairs of good chromosomes from the genetic information of the replicated parents.
- Step 6: the mutations created in Step 5 are now included in the new population, allowing new genes to be included in the chromosomes.
- Step 7: the process is repeated from Step 2 onwards until a solution is found; each cycle from Step 4 to Step 6 is referred to as a generation.

In this study, an R package named GALGO is used [30,31].

Table 1 describes the parameters selected for the GALGO analysis. The classification method selected was nearest centroid, which is a very simple and fast classifier capable of classifying data without feature selection and is also computationally inexpensive.

Table 1. GALGO parameters.

Parameter	Value
Classification method	Nearcent
Chromosome size	5
Solutions	4000
Generations	200
Goal fitness	1

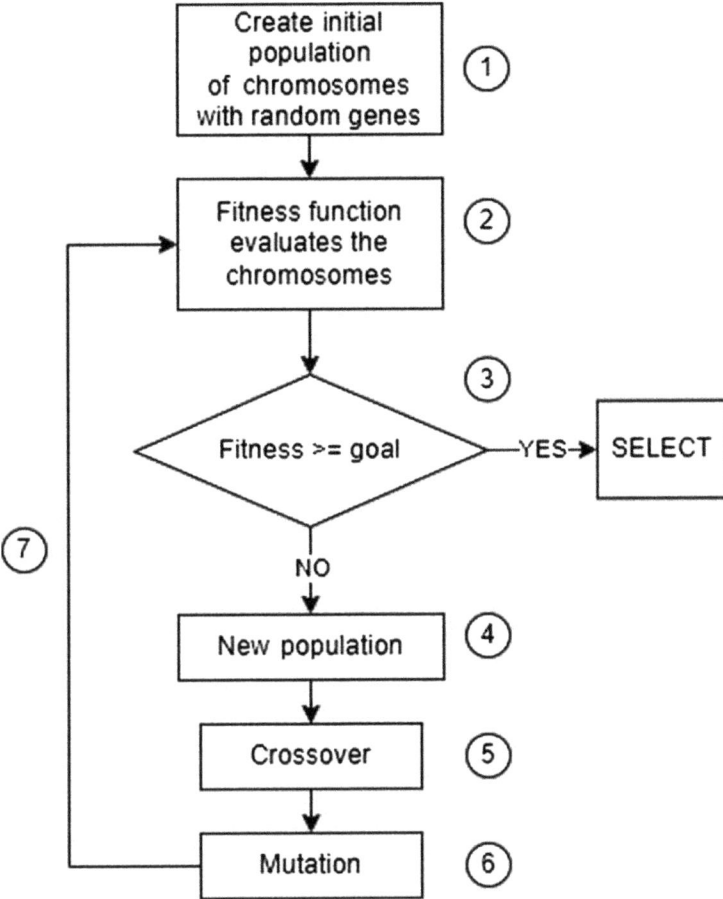

Figure 2. GA procedure flow chart.

2.3. ML Model Implementation

After the feature selection process, in order to avoid bias towards a specific algorithm, three different classification algorithms were implemented to evaluate the efficiency of the models; for this stage, random forest, KNN and ANN were used, and these models are described below. The experiments in this study were carried out on a Dell G15 Ryzer Edition machine with a Windows 11 operating system. The models and methodologies were implemented in R, and the open source software was validated by the scientific community. The libraries used for the ML models were "caret" and "neuralnet".

2.3.1. Random Forest

Random forest is a complex machine learning algorithm that combines the results of different decision trees in parallel, as shown in Figure 3, on different subsamples of datasets using the majority or average voting for the final result to perform a classification or regression problem [32].

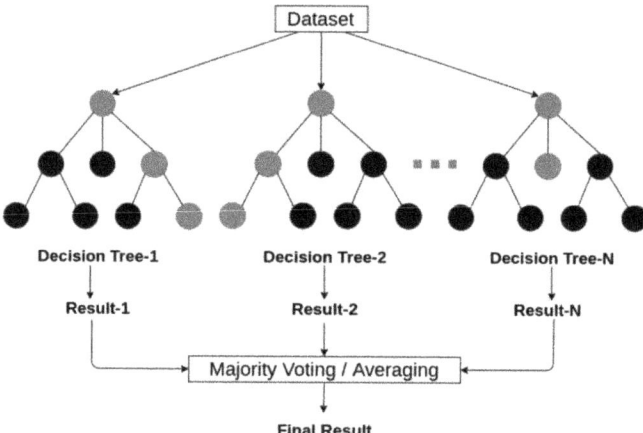

Figure 3. Structure of random forest with multiple decision trees [32].

2.3.2. k-Nearest Neighbor

Many classification and regression problems are described with the k-nearest neighbor algorithm, a supervised statistical algorithm which measures the similarity of the training set and the test set; it defines the class of an object based on the level of similarity between it and the training examples. The first step is to select a sample for training and the second step is to define "k", the number of neighbors; the higher the number of k, the higher the accuracy. The algorithm utilizes the Euclidean distance formula in order to find the k-nearest samples of the training data for each test data as shown in Equation (1). The training samples that each class has are counted and compared with "k", and then the class of the test event is selected as the most popular class from the k training samples [33].

$$D(A,B) = \sqrt{\sum\nolimits_{i=1}^{n} (a_i - b_i)^2} \tag{1}$$

where A is the feature vector of a new sample, B is the feature vector of a single training sample, n is the total number of features used for prediction, and a and b are the i-th components of A and B, respectively.

2.3.3. Artificial Neural Networks

ANNs are a machine learning method used for classification or prediction that emulates learning from prior information, such as that from neurons in the human brain. A neural network is defined by the connections between the neurons, while the training algorithm is used to determine the weights on the connections, and the activation function, e.g., the logistic sigmoid function shown in Equation (2). Figure 4 shows the structure of an ANN, with three essential parts: the input layers, hidden layers and output layer. The data are received in the input layer and then used in the hidden layers to perform mathematical calculations and recognize patterns. The result obtained from the calculation is the output layer [34]

$$f(x) = \frac{1}{1+e^{-x}}, \tag{2}$$

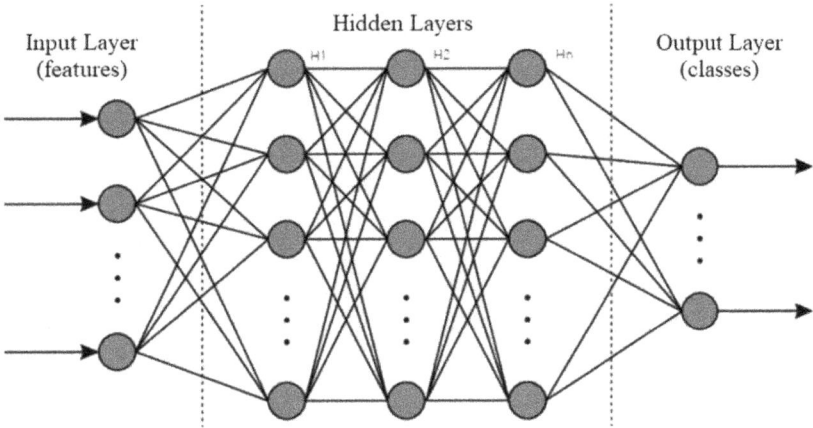

Figure 4. Structure of an artificial neural network.

2.4. Model Validation

To evaluate the results of the algorithms, the database was split in a 75:25 ratio. The models were trained with 75% of the data and the remaining 25% was used to test them. The performance of the models in emotion recognition was evaluated using the following statistical metrics: overall accuracy, confusion matrix, sensitivity, and specificity. From the information provided by the confusion matrix shown in Table 2, sensitivity, specificity and overall accuracy can be computed. Overall accuracy as shown in Equation (3) denotes the proportion of correctly predicted classifications over the total number of instances; in some cases, this could be a deceptive measure, and for this reason sensitivity and specificity are computed. Sensitivity describes the capability of an algorithm to predict a positive class when the actual one is positive, and specificity describes the capability of an algorithm to not predict a positive class when the actual one is not positive [35].

$$\text{Overall Accuracy} = TP + TN/(TP + TN + FP + FN) \tag{3}$$

$$\text{Sensitivity} = TP/(TP + FN) \tag{4}$$

$$\text{Specificity} = TN/(TN + FP) \tag{5}$$

Table 2. Confusion matrix.

	Actually Positive	**Actually Negative**
Predicted Positive	True positives (TP)	False positives (FP)
Predicted Negative	False negatives (FN)	True negatives (TN)

3. Results

With the results of the GA analysis in the R software (Version 3.6.3), Figure 5 shows the fitness scores over generations; the ordinate axis indicates the fitness score, and the abscissa axis represents the generations. This graph represents the chromosomes' capacity to accurately detect each emotional state, with the blue line indicating all models' mean fitness. As can be deduced from Figure 5, after 150 generations, there was no appreciable change in the fitness score, demonstrating that the previously established value of 200 was congruent.

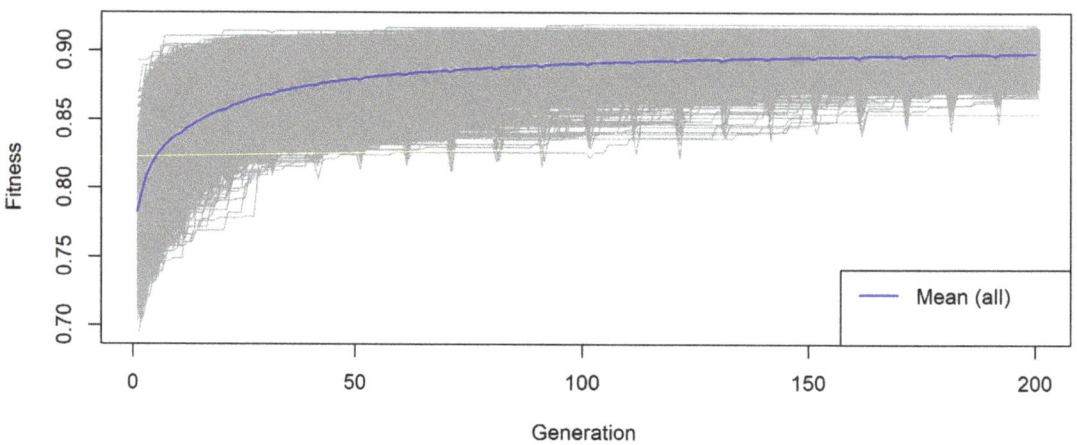

Figure 5. Fitness scores over generations.

Figure 6 showing the Gene Rank Stability graph displays the ranked stability of the genes showing the stability of the features in the models in order. The first features are the most stable and their color is solid, which means these features contribute more to the classification.

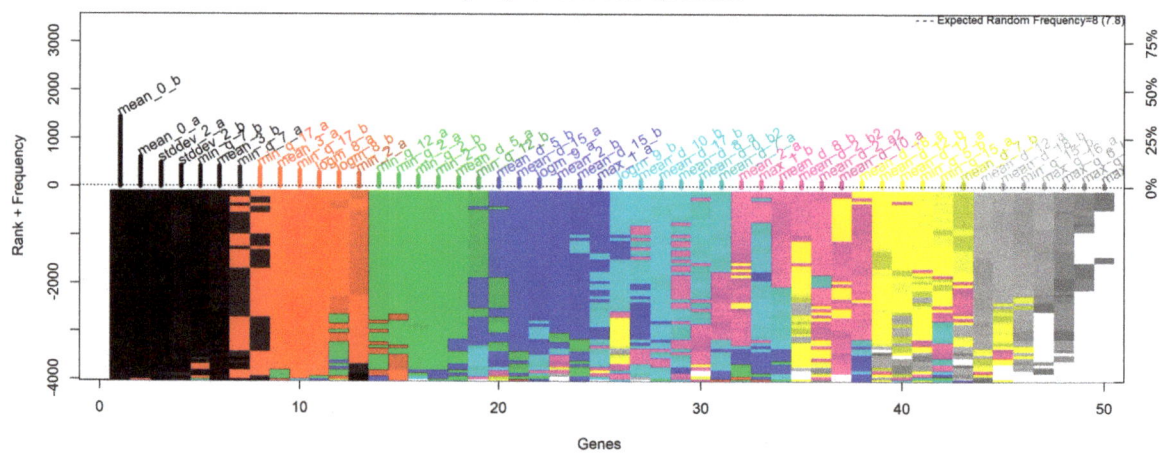

Figure 6. Gene rank stability. Graph shows the genes in order of importance from left to right according to the frequency of occurrence in the solutions, but also in order of stability, which is defined by the amount of rank changes they suffered through the evaluation of all solutions. This rank change is represented in the graph by colors; if the genes had many rank changes, the graph shows a mixture of colors. Specifically, the darker tones such as black and red tones represent the most stable genes, which stabilized over a few hundred solutions, while the genes indicated by light tones such as gray and yellow tones required thousands of solutions to stabilize.

After ranking the most significant features, the forward selection method outputs 49 features, as shown in Figure 7. The ordinate axis indicates the accuracy in terms of classification, and the abscissa axis presents the features in order of importance, while the black line is the best-performing model which in this case is model number 23 with an accuracy of 0.9019. With regard to this, it could be said that the 49 features listed in Table 3 include important information from the EEG signal with which to classify the emotional states, and this represent a 98% reduction in the total number of features. Figure 8 shows the data distribution related to each emotional state from the 49 selected features.

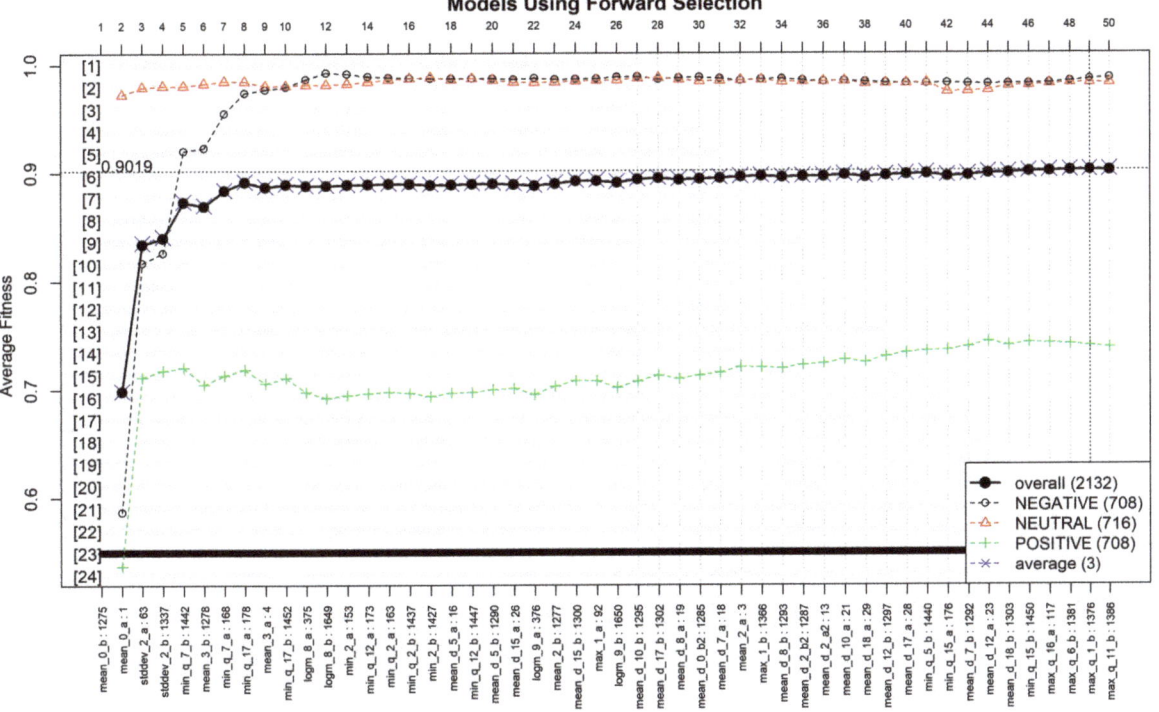

Figure 7. Models Average Fitness from the forward selection method.

Table 3. Resulting features.

Model 23's Features
"mean_0_b", "mean_0_a", "stddev_2_a", "stddev_2_b", "min_q_7_b", "mean_3_b", "min_q_7_a", "min_q_17_a", "mean_3_a", "min_q_17_b", "logm_8_a", "logm_8_b", "min_2_a", "min_q_12_a", "min_q_2_a", "min_q_2_b", "min_2_b", "mean_d_5_a", "min_q_12_b", "mean_d_5_b", "mean_d_15_a", "logm_9_a", "mean_2_b", "mean_d_15_b", "max_1_a", "logm_9_b", "mean_d_10_b", "mean_d_17_b", "mean_d_8_a", "mean_d_0_b2", "mean_d_7_a", "mean_2_a", "max_1_b", "mean_d_8_b", "mean_d_2_b2", "mean_d_2_a2", "mean_d_10_a", "mean_d_18_a", "mean_d_12_b", "mean_d_17_a", "min_q_5_b", "min_q_15_a", "mean_d_7_b", "mean_d_12_a", "mean_d_18_b", "min_q_15_b", "max_q_16_a", "max_q_6_b", "max_q_1_b".

GA-selected features listed in order of importance according to their rank from Figure 6.

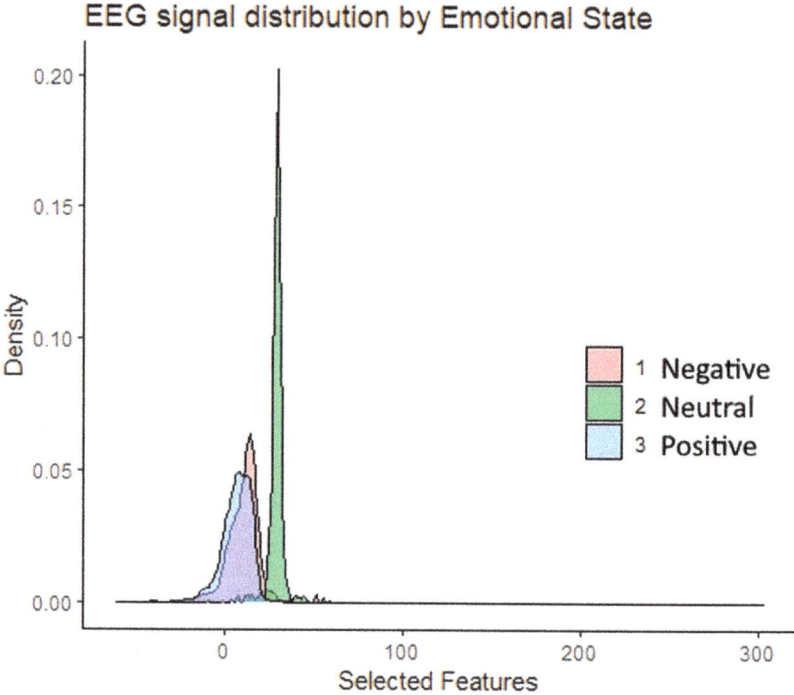

Figure 8. Data Distribution Graph.

With the 49 features selected within the best-accuracy model during the genetic algorithm's implementation, the performance of three machine learning methods was assessed for the prediction of the negative, neutral, and positive emotional states in the test proportion of the data. The algorithms used are; KNN, RF and ANN. Table 4 show the Overall accuracy of the three models. The results have shown a higher overall accuracy of ANN model over KNN and RF. For more detailed information the Confusion Matrixes are presented in Table 5. KNN, RF and ANN Model's confusion matrix for the test data.

Table 4. ML model's overall accuracy for the test data.

	KNN	RF	ANN
Overall Accuracy	90.43%	93.43%	95.87%

Table 5. KNN, RF and ANN model's confusion matrix for the test data.

	KNN	Reference				RF	Reference				ANN	Reference		
		Neg.	Neu.	Pos.			Neg.	Neu.	Pos.			Neg.	Neu.	Pos.
Prediction	Neg.	160	10	1	Prediction	Neg.	170	8	0	Prediction	Neg.	166	5	6
	Neu.	0	184	1		Neu.	0	170	0		Neu.	0	177	2
	Pos.	22	17	138		Pos.	1	26	150		Pos.	4	5	168

Table 5 shows the confusion matrixes for the three models. In the case of KNN, from this visualization, it can be inferred that the best performance of the model is for classifying the positive class, since according to the reference positive column, only two instances were wrongly predicted which correspond to just 1.4% of the total positive instances, and this can be confirmed with the sensitivity result of 0.9857 shown in Table 6. In contrast, in the negative and neutral classes, 22 and 27 predictions were wrong, which correspond to 12.08% and 12.7% of the total negative and neutral instances.

Table 6. KNN, RF and ANN model's sensitivity and specificity for the test data.

	Neg.	Neu.	Pos.
KNN Sensitivity	0.8791	0.8720	0.9857
RF Sensitivity	0.9942	0.8396	1.0000
ANN Sensitivity	0.9765	0.9465	0.9545
	Neg.	Neu.	Pos.
KNN Specificity	0.9687	0.9969	0.9008
RF Specificity	0.9779	1.0000	0.9295
ANN Specificity	0.9697	0.9942	0.9748

In the same way, the confusion matrix of the Random Forest model shows its best performance classifying the positive class with 0% of mistakes, and it shows a very good performance classifying the negative class, since there was only 1 instance wrongly predicted which corresponds to only 0.58% of the total Negative classes. In contrast, the neutral class has 34 predictions wrongly classified out of 204, which corresponds to the 16% of the total Neutral classes. These results can be correlated with their corresponding values of sensitivity and specificity shown in Table 6.

In the case of the ANN model, the overall accuracy of which was the highest, it can be said that its best performance was for recognizing the negative emotional state, since there were only four instances wrongly predicted. In the second place was its performance for the positive class with eight instances being wrongly predicted. Finally, in the third place was its performance for the neutral class with ten instances being wrongly predicted. In overall terms, ANN had the best performance, and this can be correlated with the corresponding values of sensitivity and specificity, where all results are over 0.9465 and close to 1, as shown in Table 6.

4. Discussion

In this analysis, we developed a GA model based on the nearest center classification method, in order to reduce the data dimensionality from a publicly available EEG signal dataset to classify three emotional states. After reducing the data from 2548 features to only 49 features and reaching an accuracy of 90.19%, three ML classification algorithms (KNN, RF and ANN) were evaluated to classify negative, neutral, and positive emotional states. The results show that the 49 features selected by the GA from the total of 2548 features are sufficient to create ML classification models with, obtaining results of 90.43% and 93.43% and 95.87% for KNN, RF and ANN, respectively, the last one being the most accurate. Table 7 shows a comparison of this study with previous studies of the same database using different methodologies to improve feature selection and emotional state recognition via EEG signals. Although our results present the second highest overall performance, as shown in Table 7, our work uses 14 fewer features than the report [26] which has 2.02 points higher overall performance. This means that in terms of dimensionality reduction our method is still better.

Table 7. Comparison of this study with previous similar studies.

Autor's	Overall Accuracy	Observations
This Study	95.87%	49 out of 2548 features selected with genetic algorithms and ANN ML model for classification.
Bird J. et al. [25]	87.16%	44 out of 2100 features were used with classification models such as Bayesian networks, support vector machine and random forest, the last one being the most accurate.
Bird J. et al. [26]	97.89%	63 out of 2548 features selected via information gain measurement. Classification method: random forest ensemble classifier.
Ashford et al. [27]	89.38%	256 out of 2479 features selected based on information gain measurement from gray-scale image representation of statistical features. Classification method: deep convolutional neural network.

Furthermore, considering Figure 7 (models using the forward selection), it can be observed that with only the first eight features, listed as mean_0_b, mean_0_a, stddev_2_a, stddev_2_b, min_q_7_b, mean_3_b, min_q_7_a and min_q_17_a in the graph, very similar performances close to 90% accuracy could be achieved with only 0.31% of the total 2548 features contained in the database, which is the percentage represented by these eight features. This is a much greater reduction in the dimensionality of data than that previously reported in similar studies.

Moreover, in order compare our methodology with standard method for feature selection in Python, called sequential feature selection (SFS) [36], an additional experiment was performed. The SFS algorithm starts by selecting an initial set of features and evaluating their predictive power. The algorithm then proceeds iteratively, evaluating the performance of the model with an increasing number of features, adding or removing one feature at a time, until the desired number of features is reached or no further improvement in performance is observed. To compare our proposed feature selection methodology, GALGO, to SFS, the same number of optimal features (49) were set as the parameter for the SFS methodology, and the resulting SFS features were tested with the same ML models as those used in our study. As shown in Table 8, all the overall accuracies of the ML models analyzed with the selected features from the SFS model were below our accuracies, so it can be said that the multi-objective genetic algorithms outperform SFS. For this test, the Python library used was SciKit-Learn, version 0.24 [37].

Table 8. GA and SFS feature selection comparison.

GA Feature Selection	KNN	RF	ANN
Overall Accuracy	90.43%	93.43%	95.87%
Sequential Feature Selection	KNN	RF	ANN
Overall Accuracy	74.11%	92.68%	84.98%

5. Conclusions

In this work, a multi-objective GA statistical model with a specific search strategy for variable selection was proposed and tested in a high-dimensional EEG signal database for emotional classification.

The proposed model allowed a reduction in the dimensionality of the data of 98.08%. This model, compared to that used in similar studies, outperforms the dimensionality reduction sought, which could enhance the feasibility of developing specific-purpose devices for real-time applications in future works.

The proposed model reduced the 2548 features describing the waves related to different emotional states, to only 49 optimal features, which were used to create machine learning (ML) classification models with algorithms such as k-nearest neighbor (KNN), random forests (RF) and artificial neural networks (ANN), obtaining results with 90.06%, 93.62% and 95.87% accuracy, respectively, which are higher than the 87.16% and 89.38% accuracy of previous works.

Furthermore, according to our forward selection analysis, it can be inferred that by sacrificing a little accuracy without going below 90%, emotional states could be described with only eight features or 0.31% of the total of 2548 features contained in the database used, which was a reduction unlike any other published in similar studies.

Additionally, a comparison was made between the selection method used and the widely validated method: sequential feature selection of the SciKit-Learn library (https://scikit-learn.org/stable/, accessed on 9 May 2023). From this comparison, it can be observed that GALGO has better results. As shown in Table 8, the three algorithms that used the features selected using GALGO obtained better results than the algorithms that used the features selected with SFS did, so we can validate the feature selection methodology proposed in this research and affirm that GALGO works very well and obtains better models than the traditional ones do.

The proposed methodology generated a model with a promising approach for detecting emotions in real time using only 49 features extracted from EEG signals. With the increasing interest in affective computing and the need for non-invasive methods of emotion recognition, this model could have significant applications in fields such as psychology, neuroscience, and human–computer interactions. The potential for real-time emotion detection represents a significant step forward in our understanding of the complex relationship between brain activity and emotions.

In future work, we plan to create our own databases in order to increase the number of test subjects and validate the architecture with more support. Additionally, the research can be further extended by focusing on identifying specific features and electrodes describing each class of emotion to improve the feasibility of developing purpose-built devices for real-time applications.

Author Contributions: Conceptualization, R.A.G.-H. and A.G.-H.; data curation, R.A.G.-H.; formal analysis, R.A.G.-H.; funding acquisition, J.M.C.-P., H.L.-G., J.I.G.-T. and H.G.-R.; methodology, R.A.G.-H., J.M.C.-P., A.G.-H. and C.E.G.-T.; project administration, R.A.G.-H.; resources, J.I.G.-T., H.G.-R., D.R. and K.O.V.-C.; supervision, J.M.C.-P., H.L.-G., A.G.-H., C.E.G.-T., J.I.G.-T. and H.G.-R.; validation, J.M.C.-P.; visualization, R.A.G.-H. and A.G.-H.; writing—original draft, R.A.G.-H.; writing—review and editing, R.A.G.-H. All authors have read and agreed to the published version of the manuscript.

Funding: This research received no external funding.

Institutional Review Board Statement: Not applicable.

Informed Consent Statement: Not applicable.

Data Availability Statement: Not applicable.

Acknowledgments: The authors thank CONACYT for the support granted by their national scholarship program, CVU number 307551 to Rosa Adriana García Hernández.

Conflicts of Interest: The authors declare no conflict of interest.

References

1. Kim, J.H.; Poulose, A.; Han, D.S. The Extensive Usage of the Facial Image Threshing Machine for Facial Emotion Recognition Performance. *Sensors* **2021**, *21*, 2026. [CrossRef] [PubMed]
2. Canal, F.Z.; Müller, T.R.; Matias, J.C.; Scotton, G.G.; de Sa Junior, A.R.; Pozzebon, E.; Sobieranski, A.C. A Survey on Facial Emotion Recognition Techniques: A State-of-the-Art Literature Review. *Inf. Sci.* **2022**, *582*, 593–617. [CrossRef]
3. Karnati, M.; Seal, A.; Bhattacharjee, D.; Yazidi, A.; Krejcar, O. Understanding Deep Learning Techniques for Recognition of Human Emotions Using Facial Expressions: A Comprehensive Survey. *IEEE Trans. Instrum. Meas.* **2023**, *72*, 1–31. [CrossRef]

4. Kakuba, S.; Poulose, A.; Han, D.S. Deep Learning-Based Speech Emotion Recognition Using Multi-Level Fusion of Concurrent Features. *IEEE Access* **2022**, *10*, 125538–125551. [CrossRef]
5. Yan, Y.; Shen, X. Research on Speech Emotion Recognition Based on AA-CBGRU Network. *Electronics* **2022**, *11*, 1409. [CrossRef]
6. Soman, G.; Vivek, M.V.; Judy, M.V.; Papageorgiou, E.; Gerogiannis, V.C. Precision-Based Weighted Blending Distributed Ensemble Model for Emotion Classification. *Algorithms* **2022**, *15*, 55. [CrossRef]
7. Lin, W.; Li, C. Review of Studies on Emotion Recognition and Judgment Based on Physiological Signals. *Appl. Sci.* **2023**, *13*, 2573. [CrossRef]
8. Awais, M.; Raza, M.; Singh, N.; Bashir, K.; Manzoor, U.; Islam, S.U.; Rodrigues, J.J.P.C. LSTM-Based Emotion Detection Using Physiological Signals: IoT Framework for Healthcare and Distance Learning in COVID-19. *IEEE Internet Things J.* **2021**, *8*, 16863–16871. [CrossRef]
9. AlZoubi, O.; D'Mello, S.K.; Calvo, R.A. Detecting Naturalistic Expressions of Nonbasic Affect Using Physiological Signals. *IEEE Trans. Affect. Comput.* **2012**, *3*, 298–310. [CrossRef]
10. Albraikan, A.; Tobon, D.P.; El Saddik, A. Toward User-Independent Emotion Recognition Using Physiological Signals. *IEEE Sens. J.* **2019**, *19*, 8402–8412. [CrossRef]
11. Chao, H.; Dong, L. Emotion Recognition Using Three-Dimensional Feature and Convolutional Neural Network from Multichannel EEG Signals. *IEEE Sens. J.* **2021**, *21*, 2024–2034. [CrossRef]
12. Egger, M.; Ley, M.; Hanke, S. Emotion Recognition from Physiological Signal Analysis: A Review. *Electron. Notes Theor. Comput. Sci.* **2019**, *343*, 35–55. [CrossRef]
13. Santamaria-Granados, L.; Munoz-Organero, M.; Ramirez-Gonzalez, G.; Abdulhay, E.; Arunkumar, N. Using Deep Convolutional Neural Network for Emotion Detection on a Physiological Signals Dataset (AMIGOS). *IEEE Access* **2019**, *7*, 57–67. [CrossRef]
14. Saganowski, S.; Perz, B.; Polak, A.; Kazienko, P. Emotion Recognition for Everyday Life Using Physiological Signals from Wearables: A Systematic Literature Review. *IEEE Trans. Affect. Comput.* **2022**, *12*, 1. [CrossRef]
15. Bota, P.J.; Wang, C.; Fred, A.L.N.; Placido Da Silva, H. A Review, Current Challenges, and Future Possibilities on Emotion Recognition Using Machine Learning and Physiological Signals. *IEEE Access* **2019**, *7*, 140990–141020. [CrossRef]
16. Sepúlveda, A.; Castillo, F.; Palma, C.; Rodriguez-Fernandez, M. Emotion Recognition from ECG Signals Using Wavelet Scattering and Machine Learning. *Appl. Sci.* **2021**, *11*, 4945. [CrossRef]
17. Sedik, A.; Marey, M.; Mostafa, H. WFT-Fati-Dec: Enhanced Fatigue Detection AI System Based on Wavelet Denoising and Fourier Transform. *Appl. Sci.* **2023**, *13*, 2785. [CrossRef]
18. Katoch, S.; Chauhan, S.S.; Kumar, V. A Review on Genetic Algorithm: Past, Present, and Future. *Multimed. Tools Appl.* **2021**, *80*, 8091–8126. [CrossRef]
19. Salih, O.; Duffy, K.J. Optimization Convolutional Neural Network for Automatic Skin Lesion Diagnosis Using a Genetic Algorithm. *Appl. Sci.* **2023**, *13*, 3248. [CrossRef]
20. Al-Tawil, M.; Mahafzah, B.A.; Al Tawil, A.; Aljarah, I. Bio-Inspired Machine Learning Approach to Type 2 Diabetes Detection. *Symmetry* **2023**, *15*, 764. [CrossRef]
21. Lin, Z.-H.; Woo, J.-C.; Luo, F.; Chen, Y.-T. Research on Sound Imagery of Electric Shavers Based on Kansei Engineering and Multiple Artificial Neural Networks. *Appl. Sci.* **2022**, *12*, 10329. [CrossRef]
22. Yu, S.-N.; Chen, S.-F. Emotion State Identification Based on Heart Rate Variability and Genetic Algorithm. In Proceedings of the 2015 37th Annual International Conference of the IEEE Engineering in Medicine and Biology Society (EMBC), Milan, Italy, 25–29 August 2015; IEEE: Piscataway, NJ, USA, 2015; pp. 538–541.
23. Abuqaddom, I.; Mahafzah, B.A.; Faris, H. Oriented Stochastic Loss Descent Algorithm to Train Very Deep Multi-Layer Neural Networks without Vanishing Gradients. *Knowl.-Based Syst.* **2021**, *230*, 107391. [CrossRef]
24. Ragot, M.; Martin, N.; Em, S.; Pallamin, N.; Diverrez, J.-M. Emotion Recognition Using Physiological Signals: Laboratory vs. Wearable Sensors. In Proceedings of the AHFE 2017 International Conference on Advances in Human Factors and Wearable Technologies, Los Angeles, CA, USA, 17–21 July 2017; pp. 15–22.
25. Bird, J.J.; Manso, L.J.; Ribeiro, E.P.; Ekart, A.; Faria, D.R. A Study on Mental State Classification Using EEG-Based Brain-Machine Interface. In Proceedings of the 2018 International Conference on Intelligent Systems (IS), Funchal, Portugal, 25–27 September 2018; IEEE: Piscataway, NJ, USA, 2018; pp. 795–800.
26. Bird, J.J.; Ekart, A.; Buckingham, C.D.; Faria, D.R. Mental Emotional Sentiment Classification with an Eeg-Based Brain-Machine Interface. In Proceedings of the International Conference on Digital Image and Signal Processing, Oxford, UK, 29–30 April 2019.
27. Ashford, J.; Bird, J.J.; Campelo, F.; Faria, D.R. Classification of EEG Signals Based on Image Representation of Statistical Features. In *Advances in Computational Intelligence Systems: Contributions Presented at the 19th UK Workshop on Computational Intelligence, Portsmouth, UK, 4–6 September 2019*; Springer: Cham, Switzerland, 2020; pp. 449–460.
28. Liu, Z.T.; Xie, Q.; Wu, M.; Cao, W.H.; Li, D.Y.; Li, S.H. Electroencephalogram Emotion Recognition Based on Empirical Mode Decomposition and Optimal Feature Selection. *IEEE Trans. Cogn. Dev. Syst.* **2019**, *11*, 517–526. [CrossRef]
29. Xu, H.; Wang, X.; Li, W.; Wang, H.; Bi, Q. Research on EEG Channel Selection Method for Emotion Recognition. In Proceedings of the IEEE International Conference on Robotics and Biomimetics, ROBIO 2019, Dali, China, 6–8 December 2019; pp. 2528–2535. [CrossRef]
30. Goldberg, D.E.; Holland, J.H. Genetic Algorithms and Machine Learning. *Mach. Learn.* **1988**, *3*, 95–99. [CrossRef]

31. Trevino, V.; Falciani, F. GALGO: An R Package for Multivariate Variable Selection Using Genetic Algorithms. *Bioinformatics* **2006**, *22*, 1154–1156. [CrossRef]
32. Sarker, I.H. Machine Learning: Algorithms, Real-World Applications and Research Directions. *SN Comput. Sci.* **2021**, *2*, 160. [CrossRef]
33. Houssein, E.H.; Hammad, A.; Ali, A.A. Human Emotion Recognition from EEG-Based Brain–Computer Interface Using Machine Learning: A Comprehensive Review. *Neural Comput. Appl.* **2022**, *34*, 12527–12557. [CrossRef]
34. Fausett, L.V. *Fundamentals of Neural Networks: Architectures, Algorithms and Applications*; Pearson Education: Chennai, India, 2006.
35. Irizarry, R.A. *Introduction to Data Science: Data Analysis and Prediction Algorithms with R*; CRC Press: Boca Raton, FL, USA, 2019.
36. Pilnenskiy, N.; Smetannikov, I. Feature Selection Algorithms as One of the Python Data Analytical Tools. *Future Internet* **2020**, *12*, 54. [CrossRef]
37. Fabian, P. Scikit-Learn: Machine Learning in Python. *J. Mach. Learn. Res.* **2011**, *12*, 2825.

Disclaimer/Publisher's Note: The statements, opinions and data contained in all publications are solely those of the individual author(s) and contributor(s) and not of MDPI and/or the editor(s). MDPI and/or the editor(s) disclaim responsibility for any injury to people or property resulting from any ideas, methods, instructions or products referred to in the content.

Article

Affect Analysis in Arabic Text: Further Pre-Training Language Models for Sentiment and Emotion

Wafa Alshehri [1,2,3,*], Nora Al-Twairesh [1,4] and Abdulrahman Alothaim [1,2]

1. STC's Artificial Intelligence Chair, College of Computer and Information Sciences, King Saud University, Riyadh 11451, Saudi Arabia; twairesh@ksu.edu.sa (N.A.-T.); othaim@ksu.edu.sa (A.A.)
2. Department of Information Systems, College of Computer and Information Sciences, King Saud University, Riyadh 11451, Saudi Arabia
3. Department of Computer Sciences, College of Science and Arts, King Khalid University, Almajarda 63931, Saudi Arabia
4. Department of Information Technology, College of Computer and Information Sciences, King Saud University, Riyadh 11451, Saudi Arabia
* Correspondence: waalshehri@kku.edu.sa

Abstract: One of the main tasks in the field of natural language processing (NLP) is the analysis of affective states (sentiment and emotional) based on written text, and attempts have improved dramatically in recent years. However, in studies on the Arabic language, machine learning or deep learning algorithms were utilised to analyse sentiment and emotion more often than current pre-trained language models. Additionally, further pre-training the language model on specific tasks (i.e., within-task and cross-task adaptation) has not yet been investigated for Arabic in general, and for the sentiment and emotion task in particular. In this paper, we adapt a BERT-based Arabic pretrained language model for the sentiment and emotion tasks by further pre-training it on a sentiment and emotion corpus. Hence, we developed five new Arabic models: QST, QSR, QSRT, QE3, and QE6. Five sentiment and two emotion datasets spanning both small- and large-resource settings were used to evaluate the developed models. The adaptation approaches significantly enhanced the performance of seven Arabic sentiment and emotion datasets. The developed models showed excellent improvements over the sentiment and emotion datasets, which ranged from 0.15–4.71%.

Keywords: sentiment analysis; emotion detection; pretrained language models; model adaptation; task-adaptation approach

Citation: Alshehri, W.; Al-Twairesh, N.; Alothaim, A. Affect Analysis in Arabic Text: Further Pre-Training Language Models for Sentiment and Emotion. *Appl. Sci.* **2023**, *13*, 5609. https://doi.org/10.3390/app13095609

Academic Editors: Pawel Dybala, Rafal Rzepka and Michal Ptaszynski

Received: 23 March 2023
Revised: 19 April 2023
Accepted: 28 April 2023
Published: 1 May 2023

Copyright: © 2023 by the authors. Licensee MDPI, Basel, Switzerland. This article is an open access article distributed under the terms and conditions of the Creative Commons Attribution (CC BY) license (https://creativecommons.org/licenses/by/4.0/).

1. Introduction

Natural language processing (NLP) is a field that is concerned with understanding, processing, and analysing natural languages (i.e., human languages). The evolution of the approaches used in NLP tasks is worth noting. Initially, the rule-based approach was dominant in the NLP field, which neglects to consider the contextual meaning of words, and which finds it difficult to cover all the morphologies of the language. With the growth in the availability and accessibility of data, the so-called machine-learning approach emerged. This method has a benefit in terms of accuracy when compared to the rule-based approach. It uses machine learning algorithms, but one of its drawbacks is that it requires complex manual feature engineering. With the emergence of neural networks and, more recently, deep learning, feature engineering has become automatic through the use of word embedding techniques, including Word2Vec [1], Glove [2], FastText [3], and others. Word vectors have been used in the NLP field, and they have achieved state-of-the-art results. Word embeddings are used to map all words into vectors of numbers in the vector space. Language models use pre-trained word embedding as an additional feature to initiliase the first layer of the basic model. The limitations of the word embeddings models are that they cannot handle out-of-vocabulary (OOV) words and meaning or context-dependent representations are lacking.

However, for model training, machine learning and deep learning approaches demand extensive amounts of labelled data, which is time-consuming to annotate and prepare. At present, significant evolution is noticeable in the NLP field, particularly with the emergence of transfer learning. This has reduced the need for massive amounts of training examples. In many NLP applications, most of the recent research that has applied transfer learning techniques has been associated with state-of-the-art results. Transfer learning, according to [4], "The improvement of learning in a new task through the transfer of knowledge from a related task that has already been learned". Transfer learning with pre-trained language models has attracted the interest of the research community in recent years, thus relying on the so-called semi-supervised approach. The language model trains in an unsupervised manner with a significant amount of unlabeled data (corpora), followed by a supervised process of fine-tuning the language model with a labelled dataset that is small and task-specific.

With the advancement of technology and the widespread nature of social networking sites, people have become more expressive of their sentiment and emotion, and others' opinions may also influence them. Many entities have started to consider customer opinions relating to their products or services. The Arabic language is one of the most popular languages in the world and was ranked fourth among the languages used on the internet [5], and it is also the primary language for 22 countries [6]. Therefore, there is a great need for tools and models to analyse Arabic sentiment and emotions on specific topics, phenomena, or trends, which can be benefited from several fields. Affect is the superordinate group; emotions and sentiments are statuses within this group. In other words, affect is a general term which includes both emotions and sentiment states. Sentiment analysis and emotion detection are both NLP tasks that have emerged as hot topics of interest in the NLP research community. According to the Oxford Dictionary [7], a sentiment refers to "a feeling or an opinion, especially one based on emotions", whereas an emotion is "a strong feeling such as love, fear, or anger; the part of a person's character that consists of feelings.", where we can infer that sentiment is a general interpretation of emotion. Sentiment is classified as positive, negative, or neutral, or—using an expanded scale—as very positive, positive, neutral, negative, or very negative. Emotions are often classified according to well-known models, including the Ekman model [8], into happiness, sadness, fear, anger, surprise, and disgust, or using the Plutchik wheel of emotion [9].

Emotion and sentiment analysis in text depends essentially on the language used. This study aims to analyse sentiment and emotion in Arabic, which is one of the most challenging languages. The Arabic language has several varieties, including classical Arabic, modern standard Arabic (MSA), and other dialects. Many challenges are facing the field of text emotion and sentiment analysis in the Arabic language in particular, including the lack of resources, the diversity of dialects that have no standard rules, and the detection of the implicit expression of sentiment or emotion. Furthermore, one root word may be written in more than one form, or one word may have more than one meaning. Additionally, the diacritics change the meaning of the words [5]. The Arabic language differs from other languages due to its morphological richness and complex syntactic synthesis. Therefore, NLP tasks for text emotion and sentiment analysis become more complex in the Arabic language. These challenges, along with others, have delayed progress in this research area, meaning that these tasks have not been adequately investigated and explored in Arabic compared to English. In Arabic, a degree of progress has been recorded in the field of sentiment analysis, where sentiments are typically classified as positive, negative, or neutral. However, progress in emotion detection task is ongoing, and few studies have been conducted that classify emotions deeply (e.g., emotions classification according to Ekman [8], or Plutchik [9]). The evolution that occurred in this area, especially the exploitation of transfer learning and advanced pre-trained language models, led to overcoming many of this field challenges, as well as substantial performance improvements. Arabic research papers predominantly employ machine learning or deep learning algorithms, as opposed to pre-trained language models.

Recent efforts in these fields have focused on adapting pre-trained language models to specific domains and tasks using domain-specific or task-specific unlabeled corpora, by the continuation of the pre-training of language models on this task or domain. Using either a domain-adaptation approach [10–12] or a task-adaptation approach [13,14], model adaptation has led to significant performance enhancements in the English language. As far as we know, model adaptation approaches, especially additional pre-training of the language model on a specific domain, have only been used in two Arabic language studies [15,16]. However, classifying sentiment and emotion was not the focus of these studies. Additionally, further pre-training the language model on a specific task (i.e., within-task and cross-task adaptation) has not been investigated for Arabic in general and for sentiment and emotion tasks in particular. This study aims to tackle these problems and fill these gaps by developing models with the overall aim of advancing the current state of sentiment and emotion classification tasks for Arabic. The pre-trained language model QARiB [17], which has achieved state-of-the-art results in several NLP tasks, including sentiment analysis and emotion detection, was used in this study. QARiB is further pre-trained using sentiment and emotion-specific datasets, assuming that the small task-specific datasets given during the pre-training process are sufficient to improve model performance in that task. The developed model was then evaluated by fine-tuning it on seven sentiment and emotion datasets. In particular, the contributions of this study are:

- Develop five new Arabic language models: QST, QSR, QSRT, QE3, and QE6, which are the first task-specific adapted language models based on QARiB, for Arabic sentiment analysis and emotion detection tasks. The developed models significantly enhanced the performance of seven Arabic sentiment and emotion datasets, and the research community can use these models for sentiment and emotion tasks;
- Conduct comprehensive experiments to investigate the impact of the within-task and cross-task adaptation approaches on the performance of sentiment and emotion classification;
- Analyse the influence of the genre of training datasets (i.e., tweets and reviews) utilised for model adaption on the performance of sentiment classification;
- Make the newly adapted models available to the public (https://huggingface.co/NLP-EXP, accessed on 1 March 2023).

The remainder of the paper is organised as follows: Section 2 offers a concise literature review on the classification of sentiment and emotion in Arabic text. In Section 3, the approach proposed for developing the models, pre-training datasets, and all necessary pre-processing steps and tokenisation is described. In Section 4, the experimental setup is described, including the evaluation datasets, the baseline model, the fine-tuning architecture, and the hyperparameter choices for fine-tuning our models. The results of the experiment are presented and discussed in Section 5. Section 6 concludes the paper besides outlining a few other future directions.

2. Literature Review

This section reviews the literature on the classification of sentiment and emotion in Arabic text using various approaches, including lexicon-based, machine learning, deep learning, and fine-tuning Transformer-based language models.

2.1. Sentiment Analysis in Arabic

Recently, the task of sentiment analysis has attracted attention in the Arabic NLP community. Before the emergence of transfer learning, a single trained model was used for a single task. Building a model from scratch is costly in terms of training time, memory, materials, and the data required to train the model. In particular, deep learning models require large amounts of labelled data to train models and generate satisfactory results. In fact, generating usable datasets requires time and effort due to the limited availability of data not only in general but also in the Arabic language. Therefore, the advent of transfer learning enabled many of these problems to be overcome. It became possible to use a single model for multiple tasks instead of building a specific model for each task. Many models using transfer learning have achieved state-of-the-art results across different NLP tasks. In several languages, pre-trained language models have yielded great results in sentiment analysis [18–20]. To achieve similarly high outcomes in Arabic, several works employed different Arabic Transformer-based pre-trained language models for sentiment analysis.

The Transformer architecture, which was introduced in [21], is entirely dependent on the attention mechanism, which includes the encoder and decoder parts. The BERT (bi-directional encoder representations from Transformers) language model [18] is based on the Transformer architecture and is composed of a set of Transformer encoders layered on top of each other. Two objective functions were used to train BERT. The first being the masked language modelling (MLM) objectives, which use the special <MASK> token to randomly mask samples of input tokens in order to predict the word given their context. BERT was also trained on the next sentence prediction (NSP) objective. Given two sentences, the model predicts whether or not they follow each other.

AraBERT, as proposed by Antoun et al. [22], was the first Transformer-based pre-trained language model in Arabic based on the BERT model. AraBERT was pre-trained on a ~24 GB Arabic corpora and fine-tuned for three NLP tasks, one of which was sentiment analysis. The authors fine-tuned and evaluated AraBERT for sentiment classification on five sentiment datasets, and the model achieved state-of-the-art results compared to mBERT [18] and hULMonA [23]. Several efforts [24,25] have employed the AraBERT model for sentiment classification.

Further, numerous approaches have achieved high performance in this field by fine-tuning the Arabic Transformer-based pre-trained models for sentiment analysis, such as mBERT [18], AraELECTRA [26], ARBERT and MARBERT [27], XLM-R [28], QARiB [17], CAMeLBERT [29], GigaBERT [16], DziriBERT [30], AraXLNet [31], Arabic-ALBERT (https://ai.ku.edu.tr/arabic-albert/, accessed on 1 February 2022), and ArabicBERT [32]. In [33], Elmadany et al. introduced ORCA, a publicly accessible benchmark for evaluating Arabic language understanding tasks. ORCA covers a variety of Arabic varieties and a range of challenging tasks using 60 datasets across seven Natural Language Understanding (NLU) task clusters. The task clusters include (1) sentence classification, (2) topic classification, (3) structured prediction, (4) semantic similarity, (5) natural language inference, (6) question-answering, and (7) word sense disambiguation. The authors used ORCA to compare 18 multilingual and Arabic pre-trained language models and created a public leaderboard with a unified evaluation metric (ORCA score) to support future research. ORCA includes sentiment analysis as one of its tasks; 19 available datasets were used to construct this task. The best performance on this task was attained by the AraElectra language model, which achieved 80.86%. As for the dialect tasks in ORCA, MARBERTv2 [27] and QARiB [17] get the highest ORCA scores, respectively.

At present, research in these fields has been directed towards adapting pre-trained language models to specific domains and tasks using domain-specific or task-specific unlabelled corpora. Model adaptation techniques have given rise to large performance gains in the English language. For example, using a domain-adaptation approach in [10–12], using the task-adaptation approach in [13,14]. In the biomedical domain, for example, [10] developed BioBERT, a BERT model that was further pre-trained on biomedical corpora, and evaluated on various biomedical text mining tasks, such as question answering (QA), named entity recognition (NER), and relation extraction (RE). Further, clinical BERT was introduced in [11] by continuing the pre-training of BERT and BioBERT models using clinical notes. Models were trained for 150k steps and fine-tuned on five NER and natural language inference (NLI) datasets. The results indicate that pre-training language models on biomedical and clinical corpora facilitate the comprehension of complex medical texts. In order to undertake a variety of NLP tasks in the field of finance, Araci et al. [12] developed FinBERT, a BERT-based language model. The model is additionally pre-trained using the financial corpus TRC2-financial, which contains approximately 400k sentences. The model was evaluated using two sentiment datasets: Financial PhraseBank and FiQA

Sentiment. Experiments indicate that FinBERT obtained state-of-the-art results on both datasets, improving accuracy by 15%.

In addition, [13] attempted to adapt the RoBERTa language model [34] using in-domain, within-task, and cross-task model adaptation approaches. For domain-adaptation, the authors additionally pre-trained RoBERTa for 12.5k steps in the biomedical (BioMed), computer science (CS), news, and reviews domains. For all domains, they evaluate each language model using two text classification datasets: CHEMPROT and RCT for BioMed, ACL-ARC and SCIERC for CS, HYPERPARTISAN and AGNEWS for News, and HELPFULNESS and IMDB for Reviews. The domain adaptation approach exhibits performance enhancements over the RoBERTa model on all datasets, with the exception of the AGNEWS dataset. For task adaptation, the RoBERTa was additionally pre-trained on each of the aforementioned datasets for 100 epochs, followed by an evaluation on the same dataset to determine the efficacy of the task-adaptation approach. Compared to the domain-adaptation approach, the task-adaptation approach utilises fewer, but more task-relevant data and is cheaper to implement. The outcomes of this approach were comparable to those of domain adaptation, and improvements over RoBERTa were demonstrated for all datasets. Lastly, the authors conducted an experiment utilising an approach for cross-task transfer. For instance, the RoBERTa is further pre-trained using the HYPERPARTISAN dataset before being evaluated and fine-tuned using the AGNEWS dataset. While the task-adaptation approach has been shown to be effective, the cross-task approach has been shown to have negative effects. A study conducted by [14] also demonstrated the effectiveness of domain and task adaptation approaches. Where BERT was additionally pre-trained using seven text classification datasets, including IMDB, Yelp P., Yelp F., TREC, Yahoo! Answers, AG's News, and DBPedia, which cover three domains: sentiment, topic, and question. The adapted models were subsequently evaluated using the aforementioned datasets, as additional pre-training is contributing to improving the performance of BERT for a particular task.

The domain-adaptation approach was used in Arabic NLP research by [15], they developed a language model in the COVID-19 domain, by further pre-training AraBERT and mBERT using around 1 million tweets. The models evaluated on the ARACOVID19-MFH dataset cover different NLP tasks, such as fake news detection, opinion mining, hate speech, etc. Another work by [16], was introducing a domain-specific language model pre-trained on large-scale news corpora. They utilised roughly 13 million news articles from the Gigaword corpus to further pre-train the XML-RoBERTa model. The model evaluated four NLP tasks: named entity recognition (NER), part-of-speech (POS), relation extraction (RE), and argument role labelling (ARL). Other works that utilised domain adaptive approaches for Arabic sentiment analysis are found in [35–37]. The results of these works show there is a significant performance improvement that could result from domain-specific adaptation.

2.2. Emotion Detection in Arabic

Notably, limited studies have involved the detection of emotion in Arabic text. Most prior studies have used traditional methods, such as the lexicon-based approach [38]. In addition, machine learning or deep learning algorithms and the AIT dataset (affect in tweets for SemEval-2018 competition Task 1) have been utilised in the majority of Arabic emotion detection studies [39–42], due to the scarcity of Arabic resources for this task.

Limited studies have fine-tined Arabic Transformer-based models for emotion detection. One of the studies on emotion and sentiment showcased the AraNet toolkit by Abdul-Mageed et al. [43], where mBERT was fine-tuned on many of the available datasets for different tasks, including sentiment and emotion analysis. For the sentiment task, mBERT was fine-tuned and tested using 15 sentiment datasets, collectively containing approximately 126,766 examples. For the emotion task, mBERT was fine-tuned using two datasets, LAMA-DINA and LAMA-DIST, collectively containing approximately 189,903 tweets. To the best of our knowledge, this was the first research to use an Arabic Transformer-based model (i.e., mBERT) for emotion detection. Notably, AraNet achieved state-of-the-art performance on these tasks.

Using the AraNet-Emo dataset [43], the developer of the ARBERT and MARBERT models [27] fine-tuned these models for emotion classification. In comparison to AraNet [43], XLM-R [28], and AraBERT [22], MARBERT obtained state-of-the-art results with an F1-score of 75.83%. Additionally, the QARiB language model [17] was fine-tuned for emotion detection using the AIT emotion classification (E-c) task dataset. The model attained state-of-the-art performance and outperformed the AraBERT [22], mBERT [18] and ArabicBERT [32] with a macro-F1 score of 46.8%. Elfaik et al. in [44] used AraBERT for extracting features and an attentional LSTM-BiLSTM model for emotion classification. Utilising the AIT dataset, exhaustive experiments were conducted, in which the proposed approach performs better than many versions of the mBERT and AraBERT models. The authors of [45] suggested an ensemble deep-learning method for detecting emotion in Arabic Tweets. The AIT dataset [39] was evaluated using three deep learning models individually (Bi-LSTM, Bi-Directional gated recurrent unit Bi-GRU and MARBERT), and compared to the developed ensemble model. The developed ensemble model significantly outperforms the individual models, as shown by the increase in macro F1 score varying from 5.3% to 23.3%.

Using the AIT dataset [39], Al-Twairesh [46] conducted an experimental study on the development of language models. Extensive experiments were carried out to examine the effectiveness of several language models (including the traditional TF-IDF, different versions of AraVec [47], AraBERT [22], and ArabicBERT [32] models, and multi-DialectBert [48]) on the emotion detection task. The results demonstrate that the ArabicBERT-Large model showed the best performance. One of the most recent works was proposed by Mahmoud et al. [49]. Researchers released the "Arabic Egyptian COVID-19 Twitter Dataset (ArECTD)", one of the largest Arabic emotion detection datasets. The dataset included roughly 78k tweets that were classified into ten emotion classes: "sadness", "fear", "sarcasm", "sympathy", "anger", "surprise", "love", "joy", "hope", and "none." The dataset was evaluated using two Arabic language models, AraBERT and MARBERT, achieving accuracy values of 70.01% and 72.5%, respectively.

To summarise, new pre-trained language models, have shown significant developments in sentiment and emotion classification. However, gaps exist due to insufficient research activity in this area for the Arabic language. In Arabic, most research papers in NLP have focused on using machine learning or deep learning algorithms to address sentiment and emotion classification problems. The evolution that has occurred in the NLP area, especially in the exploitation of transfer learning and advanced pre-trained language models (Transformer-based models), has not been significantly investigated in the Arabic language. However, compared to English, in Arabic studies, machine learning or deep learning algorithms were utilised to analyse sentiment and emotion more often than current pre-trained language models. In particular, the Arabic emotion detection task has limited studies compared to the Arabic sentiment analysis task. In addition, most of the studies have used a limited number of datasets (i.e., experiments using one or two datasets), or small datasets; for example, in emotion detection studies, most have used the AIT dataset. The reason could be the limited number of resources available for Arabic sentiment and emotion classification tasks. Furthermore, model adaptation approaches have given rise to large performance gains in the English language, whether using domain-adaptation approaches [10–12] or task-adaptation approaches [13,14]. Adapting the pre-trained language model to a specific domain or task means continuing the pre-training of language models on this task or domain using an unlabelled domain-specific or task-specific dataset. To the best of our knowledge, model adaptation approaches, in particular, further pre-training the language model on a specific domain have only been undertaken in two studies for the Arabic language [15,16], whereas classifying sentiment and emotion were not the focus of these studies. Additionally, further pre-training the language model on a specific task (i.e., within-task and cross-task adaptation) has not yet been investigated for Arabic in general, and for the sentiment and emotion task in particular. Nevertheless, amid the emerging advancements in Arabic sentiment and emotion classification, further study and experimentation are still needed to address the existing gaps and enhance performance in this field.

3. Methodology

The current study aims to develop pre-trained language models to boost the current state of Arabic sentiment analysis and emotion detection tasks. This section provides more details about the adaptation approach utilised in this work and the developed pre-trained language models. Furthermore, the pre-taring data collection, as well as all required pre-processing steps and tokenisation, are described. Finally, the section concludes with a presentation of the model's evaluation metrics.

3.1. Task-Adaptation Approach

Recent research has demonstrated that further unsupervised pre-training of the language model on the task-specific dataset, followed by fine-tuning on the supervised target task dataset, can yield substantially better performance than directly supervised target task fine-tuning [14]. Further pre-training enables continued training of the pre-trained language model on domain-specific or task-specific corpora instead of building or pre-training the model from scratch, which is more time-consuming and has a high computational cost. For these reasons and to achieve these research goals, we use the task-adaptation approach to enhance and obtain a better result for Arabic sentiment and emotion classification tasks. Due to the specialised language used in the emotion and sentiment context, general-purpose models are inadequate. The focus of this work is on NLP transfer learning or model adaptation methodologies that appear to be a promising solution to this challenge. Moreover, due to the scarcity of Arabic resources for sentiment and emotion tasks, we believe that pre-trained language models can help to solve these problems since they require fewer annotated samples and can be further pre-trained using task-specific (sentiment and emotion) corpora. Since the distribution of the text for the target task differed from general corpora, we used QARiB pre-trained language model [17], to develop models that could learn the semantic relations in the target task's text.

The QARiB model (QCRI Arabic and Dialectal BERT) [17] uses a BERT-based architecture [18], both of which are Transformers-based architectures. QARiB [17] is trained using only the MLM objective with a masking probability of 15%, and the NSP objective was excluded. QARiB has five versions that differ in the size of training datasets, the mixture of formal (MSA) and informal (dialect) Arabic text and using a Farasa [50] tokeniser or not. Arabic Gigaword Fourth Edition [51], Abulkhair Arabic Corpus [52], Open Subtitles [53], and a collection of Arabic tweets constitute the training dataset. A Byte-Pair-Encoding (BPE) tokeniser [54] was employed for dataset tokenisation with a vocabulary size of 64k. The architecture of all of these models is the same as the BERT-base model, with 12 encoder layers, a hidden size of 768, and 12 multi-head attention heads. A BERT-base-QaRiB model version was utilised in our study, and we further pre-trained the model using task-specific datasets. QARiB has achieved excellent results on a variety of Arabic NLP tasks, including sentiment analysis, named entity recognition, dialect identification, emotion classification, and offensive detection, despite employing almost the same structure across tasks. Comparing the model against various Arabic pre-trained language models, including AraBERT [22], mBERT [18], and ArabicBERT [32], showed state-of-the-art results. Adapting QARiB for sentiment and emotion tasks could potentially aid or benefit numerous NLP studies on sentiment and emotion and boost task outcomes. This can be accomplished by investigating whether the newly adapted language models can provide better outcomes than the QARiB language model on these tasks. To determine if this adaptation would be beneficial for sentiment and emotion classification tasks, two model adaptation approaches were implemented as follows: first, within-task adaptation; second, cross-task adaptation. More details about this process are given in the following subsections.

3.1.1. Pre-Training Datasets Collection

The initial phase of this method, as shown in Figure 1, is to gather pre-training task-specific datasets. The further pre-training of the language model on particular tasks needs task-specific unlabelled data. Building or generating a new dataset requires much time and effort. For this reason, we gathered further pre-training datasets from existing and available Arabic sentiment and emotion datasets. We built three sentiment datasets and one emotion dataset by augmenting different available datasets. The datasets were constructed as follows:

1. Sentiment Tweets Dataset (STD): a collection of fourteen sentiment datasets, sourced from Twitter. The statistics for these datasets are reported in Table 1. It contains the following datasets: Arabic Jordanian General Tweets (AJGT) [55], Arabic Speech Act and Sentiment (ArSAS) [56], Arabic Sentiment Twitter Dataset for the Levantine Dialect (ArSenTD-LEV) [57], Arabic-Dialect [58], BBN-Dataset [59], Syrian-Dataset [59], The Arabic Tweets Sentiment Analysis Dataset (ATSAD) [60], The Arabic Sentiment Tweets Dataset (ASTD) [61], SemEval-2017 [62], ArSarcasm [63], The Tweets and Emojis Arabic Dataset for Sentiment Analysis (TEAD) [64], Affect in Tweets (AIT) [39], Multi-Domain Arabic Resources for Sentiment Analysis (MARSA) [65], and AraSenti-Tweet [66]; Table 1 illustrates a summary of the STD dataset statistics which contains over 682,000 tweets, approximately 445,395 positive tweets and approximately 238,355 negative tweets. We only kept the positive and negative classes in our experiment. Other classes such as neutral, mixed, or objective were eliminated. To obtain more specific or relevant sentiment data, we utilised the manually annotated sentiment datasets. In total, 11 out of the 14 datasets were manually annotated. The last column in Table 1 shows the average sequence length for each dataset, which varies from 10 to 23. It can be noticed that the STD has a sequence length of 13 words on average.

2. Sentiment Reviews Dataset (SRD): we augmented four reviews sentiment datasets to build this dataset. The review datasets were utilised as follows: The Opinion Corpus for Lebanese Arabic Reviews (OCLAR) [67], Large-scale Arabic Book Review (LABR) [68], Hotel Arabic-Reviews Dataset (HARD) [69], and the Book Reviews in Arabic Dataset (BRAD 1.0) [70]; The rating in reviews datasets is regarded as a human annotation, in which people score the services using stars between 1 and 5. Scores of 4 and 5 are regarded as positive, scores of 1 and 2 are regarded as negative, and a score of 3 is regarded as neutral. Accordingly, we kept the rating of stars 5 and 4 as a positive class and the rating of stars 2 and 1 as a negative class. Additionally, we regarded 3-star ratings to be neutral and eliminated them from our experiment. Table 2 illustrates a summary of the SRD Dataset Statistics, which contains over 751,000 reviews. About 616,700 of the reviews are positive, while 134,773 are negative. The dataset covers several domains, including restaurants, hotels, and books. The last column in Table 2 displays the average sequence length for each dataset, which ranges from 13 to 80. It is notable that the SRD has an average sequence length of 54 words. The nature of tweets differs from that of reviews in terms of sentence length. Tables 1 and 2 illustrate the average sequence lengths of the STD and SRD datasets, respectively. Overall, the SRD has the longest sentence with an average of 54 words, while the STD's average sequence length is 13. We created datasets with two different data types (tweets and reviews) to investigate if using different data types (tweets and reviews) in the further pre-training phase, would affect the final results.

3. Sentiment Tweets–Reviews Dataset (STRD): This dataset combines the STD and SRD datasets and consists of 1,433,657 sentences. Table 3 shows the dataset's statistics. It is notable that the STRD has an average sequence length of 35 words;

4. Emotion Tweets Dataset (ETD): a collection of five emotion datasets which contains the following: Emotional-Tone [71], LAMA-DINA [72], Affect in Tweets (AIT) [39], AraEmoTw [73], and The SemEval-2018 Affect in Tweets Distant Supervision Corpus [39]. We primarily used Ekman's emotional classes in our experiment, which

include joy, sadness, anger, disgust, fear, and surprise. Therefore, tweets annotated with any other classes were excluded. Table 4 illustrates a summary of the ETD Dataset statistics, which comprises roughly 1.2 million tweets. There are around 227,518 tweets for the joy class, 346,696 for sadness, 344,899 for fear, 291,500 for anger, 27,166 for surprise, and 24,431 for disgust. The average sequence length for each dataset is shown in the final row of Table 4, and it ranges from 14 to 32. The ETD has an average sequence length of 19 words.

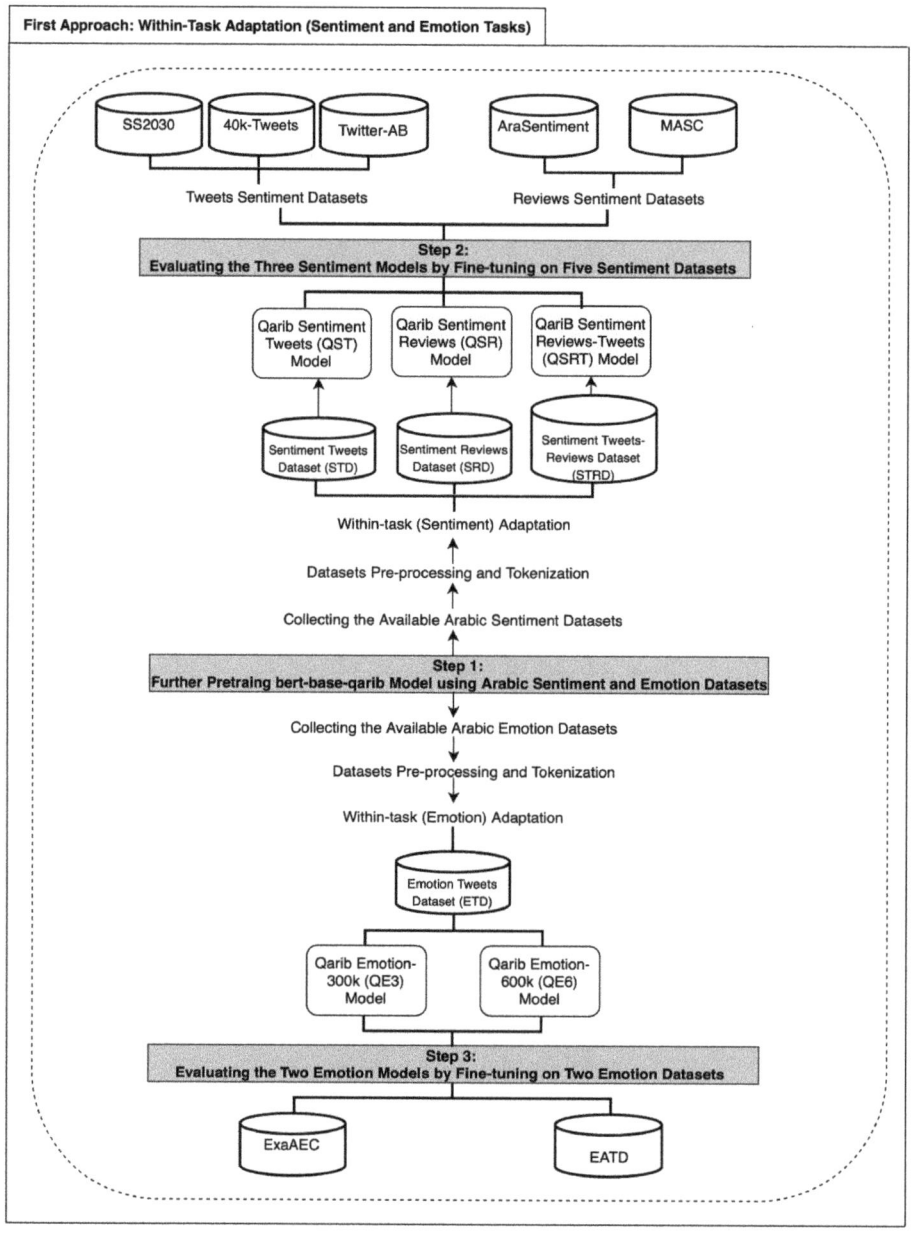

Figure 1. Language Models Development Process Using Within-Task Adaptation Approach.

Table 1. Sentiment Tweets Dataset (STD) Statistics.

Dataset-Name	Size	#Classes	#Pos	#Neg	#Other Classes	#Tweets ≥ 3	Dup	Size-after-Pre Processing	Avg SeqLen
AJGT [55]	1800	2	813	727	-	260	-	1540	10
ArSAS [56]	19,897	4	4323	7325	8113	79	57	11,648	23
ArSenTD-LEV [57]	4000	5	895	1292	885	0	928	2187	23
Arabic-Dialect [58]	52,210	3	5546	15,086	30,033	1118	427	20,632	21
BBN-Dataset [59]	1200	3	481	558	127	34	-	1039	10
Syrian-Dataset [59]	2000	3	447	1349	202	0	2	1796	19
ATSAD [60]	56,795	2	16,106	16,476	-	3980	20,233	32,582	13
ASTD [61]	9693	4	736	1592	7274	91	-	2328	17
ArSarcasm [63]	10,547	3	1634	3473	5340	71	29	5107	17
SemEval2017 [62]	3352	3	728	1108	1470	6	40	1836	17
TEAD [64]	555,923	2	391,530	162,763	-	673	957	554,293	13
Affect in Tweets [39]	1800	7	720	808	262	10	-	1528	17
MARSA [65]	56,782	3	16,647	19,858	18,726	384	1167	36,505	15
AraSenti-Tweet [66]	17,573	4	4789	5940	6461	367	13	10,729	16
Total (STD)	683,750	2	445,395	238,355	78,893	7073	25,406	682,197	13

Table 2. Sentiment Reviews Dataset (SRD) Statistics.

Dataset-Name	Size	#Positive	#Negative	#Neutral	#Tweets ≥ 3	Dup	Size-after-Preprocessing	Avg Len of Seq
OCLAR [67]	3916	1595	275	418	1587	41	1870	13
LABR [68]	63,257	39,069	7568	12,201	2328	2091	46,637	66
HARD [69]	409,562	263,453	51,864	80,326	8570	5349	315,317	22
BRAD [70]	510,599	312,583	75,066	106,785	14,429	1736	387,649	80
Total (SRD)	751,473	616,700	134,773	199,730	26,914	9217	751,473	54

Table 3. Sentiment Tweets-Reviews Dataset (STRD) Statistics.

Dataset-Name	Size	#Positive	#Negative	Duplicate	Avg Len of Seq	Size-after-Preprocessing
STD	682,197	445,395	238,355	-	13	-
SRD	751,473	616,700	134,773	-	54	-
Total (STRD)	1,433,670	1,062,095	373,128	13	35	1,433,657

Table 4. Emotion Tweets Dataset (ETD) Statistics.

Dataset-Name	Emotional-Tone [71]	LAMA-DINA [72]	Affect in Tweets [39]	AraEmoTw [73]	SemEval-2018 AIT DISC [39]	Total (ETD)
Size	10,065	8502	5600	226,774	1,019,435	1,262,210
#Joy	1169	1265	1389	38,591	185,104	227,518
#Sad	1222	964	792	23,871	319,847	346,696
#Fear	1140	1376	797	44,036	297,550	344,899
#Anger	1423	902	1390	70,851	216,934	291,500
#Surprise	992	1141	-	25,033	-	27,166
#Disgust	-	986	-	23,445	-	24,431
#Other-Classes	3832	1777	-	-	-	5609
#Tweets ≥ 3	279	88	37	421	-	825
#Duplicate	8	3	1195	526	-	1737
Size-after-preprocessing	5946	6634	4368	225,827	1,019,435	1,262,205
Avg Len of Seq	14	15	17	32	16	19

3.1.2. Pre-Training Datasets Pre-Processing

The datasets collected in the previous sections were already pre-processed by their authors. Therefore, we performed light pre-processing to prepare the datasets for the further pre-training task. Tables 1–4 show the statistics of the STD, SRD, STRD, and ETD datasets. The following pre-processing steps were performed:

- The URLs, user mentions, and hashtags present in any of the collected sentences were replaced with the tokens [رابط] and [مستخدم] and [هاشتاق];
- Sentences with three words or fewer were removed;
- Null and duplicated were eliminated.

3.1.3. Pre-Training Datasets Tokenisation

The original QARiB model [17] utilises byte-pair-encoding (BPE) [54] tokenisation with a set of vocabulary generated from several Arabic corpora. Following [17], we utilised a BPE tokeniser with a 64k tokens vocabulary to tokenise the dataset in our experiment. As stated in [10], employing a custom vocabulary (e.g., domain-specific vocabulary) prevents benefiting from the pre-training from the BERT checkpoint. Moreover, in [74], it was shown that pre-training with a custom vocabulary has outcomes that coincide with the outcomes that result from pre-training with a general vocabulary. Due to these reasons, our models have been further pre-trained with the QARiB model's BPE vocabulary.

3.1.4. Within-Task Adaptation Approach

The main phase of the approach presented in Figure 1 is to continue pre-training QARiB with sentiment and emotion datasets that we collected and preprocessed in the previous sections. The proposed approach (i.e., within-task adaptation) can be explained as: The QARiB language model is further pre-trained using training data for a target task. The model is trained on an unlabelled task-specific dataset (e.g., sentiment dataset) and evaluated by performing fine-tuning on the labelled dataset from the same task (e.g., sentiment dataset). This can be expressed as:

$$QARiB_T \rightarrow PretrainingDataset_T \rightarrow FinetuningDataset_T \qquad (1)$$

"T" refers to the target task, which might be either sentiment or emotion. The overall process of the within-task adaptation approach of the model's development is illustrated in Figure 1.

The adaption of our models followed the settings for training the QARiB language model since this model was trained using only the masked language modelling objective. In this experiment, we further pre-trained five versions of the QARiB language model from the BERT-base-QaRiB checkpoint. Three of them were trained with sentiment-specific datasets including STD, SRD, and STRD. The objective was to investigate the impact of continuous pre-training on sentiment task performance. Furthermore, two models were trained using ETD, which is an emotion-specific dataset. The goal is to determine whether further pre-training may improve the results of emotion classification. The four datasets were divided into train and test sets at a ratio of 80% and 20%, respectively. We utilised the training script run_mlm.py provided by huggingface (https://github.com/huggingface/transformers.git, accessed on 1 February 2022), and we ran it on a single GPU provided by Google Colab Pro+. Every 5000 steps, a model checkpoint was saved. The models were trained using the QARiB model's default hyperparameters, as stated below:

- A batch size of 64;
- A maximum sequence length of 64 words;
- A learning rate of 8×10^{-5}.

The next subsections offer a detailed explanation of each model we developed for each task (sentiment and emotion).

Sentiment Models

For the sentiment task, we further pre-train three different sentiment models using the three sentiment datasets: STD, SRD, and STRD. The first model is the QARiB-Sentiment-Tweets (QST) Model, which is a QARiB language model that was further pre-trained for 300 k training steps and roughly 28 epochs using the STD (shown in Table 1). The second model is the QARiB-sentiment-reviews (QSR) Model, which is a QARiB language model that was further pre-trained for 300 k training steps and roughly 26 epochs using the SRD (shown in Table 2). Despite the fact that the QST and QSR models were trained for the same number of training steps (300 k steps) and roughly the same data size (see Tables 1 and 2), Table 5 shows that the QSR has lower validation loss and better perplexity results than the QST. The reason could be that the review data has a longer text and is less noisy than the tweet dataset. As a result, we decided to continue training from the last checkpoint of the QSR model and trained the third model, which we call the QARiB-sentiment-reviews-tweets (QSRT) model, for an additional 300 k training steps (600 k in total with QSR training steps). The QSRT model was trained for roughly 52 epochs using STRD. Table 5 shows that the QSRT model outperformed the QST and QSR models in terms of training results. This was expected, given the increased amount of training data and the variety of data types (i.e., tweets and reviews). Table 5 provides additional details about the model's training results.

Table 5. The Adapted Model's Training Results.

Model	Training Steps	Epochs	Train_loss	Val_loss	Perplexity
QST Model	300 k	28.46	0.4839	2.704	14.9399
QSR Model	300 k	26.08	0.7821	2.6746	14.5071
QSRT Model	600 k	52.16	0.0209	2.5515	12.8269
QE3 Model	300 k	15.52	0.467	2.7762	16.0576
QE6 Model	600 k	31.04	0.435	2.7778	16.0838

Emotion Models

For the emotion task, we further pre-train two different emotion models using the emotion dataset ETD. The first model is the QARiB-emotion-300k (QE3) model, which is a QARiB language model that was further pre-trained for 300 k training steps and roughly 28 epochs using the ETD (shown in Table 4). The second emotion model, known as the QARiB-Emotion-600k (QE6) Model, was trained by continuing the training from the last checkpoint of the QE3 model using the same dataset ETD. The QE6 model trained for an additional 300 k steps (600 k in total including the QE3 training steps) and around 31 epochs. Table 5 shows that there was no improvement in the training results of the QE6 models when compared to the QE3 models. In fact, there was an increase in QE6 validation loss and perplexity. According to Table 5, the QSRT model has the lowest validation loss and perplexity score. Furthermore, we can see that the sentiment models outperform the emotion models in terms of validation loss and perplexity score. Afterwards, each of the three sentiment models was evaluated by fine-tuning using various sentiment datasets. Additionally, the emotion models were tested by fine-tuning them using various emotion datasets to examine the effectiveness of the within-task adaptation approach.

3.1.5. Cross-Task Adaptation Approach

In this phase, QARiB is further pre-trained on a task-specific unlabeled dataset (e.g., sentiment data), and fine-tuned on the labeled dataset from the other task (e.g., emotion dataset) and vice versa. This can be expressed as:

$$QARiB_T \rightarrow PretrainingDataset_T \rightarrow FinetuningDataset_S \qquad (2)$$

where $S \neq T$ are the source and target task and can be either sentiment or emotion. The overall process of the cross-task adaptation approach of QARiB is illustrated in Figure 2.

One of the objectives of this experiment is to discover if there is a relationship between sentiment and emotion tasks. To do this, we utilised this approach to determine if the model trained on sentiment data may benefit and enhance the emotion outcomes. In the other direction, we wanted to investigate if the model developed using emotion data may improve and enhance sentiment results. To accomplish this, we used the same pre-trained sentiment models (QST, QSR, and QSRT) as well as emotion models (QE3, and QE6) (described in Section 3.1.4). The sentiment models were fine-tuned using different emotion datasets, while the emotion models were fine-tuned using various sentiment datasets as illustrated in Figure 2.

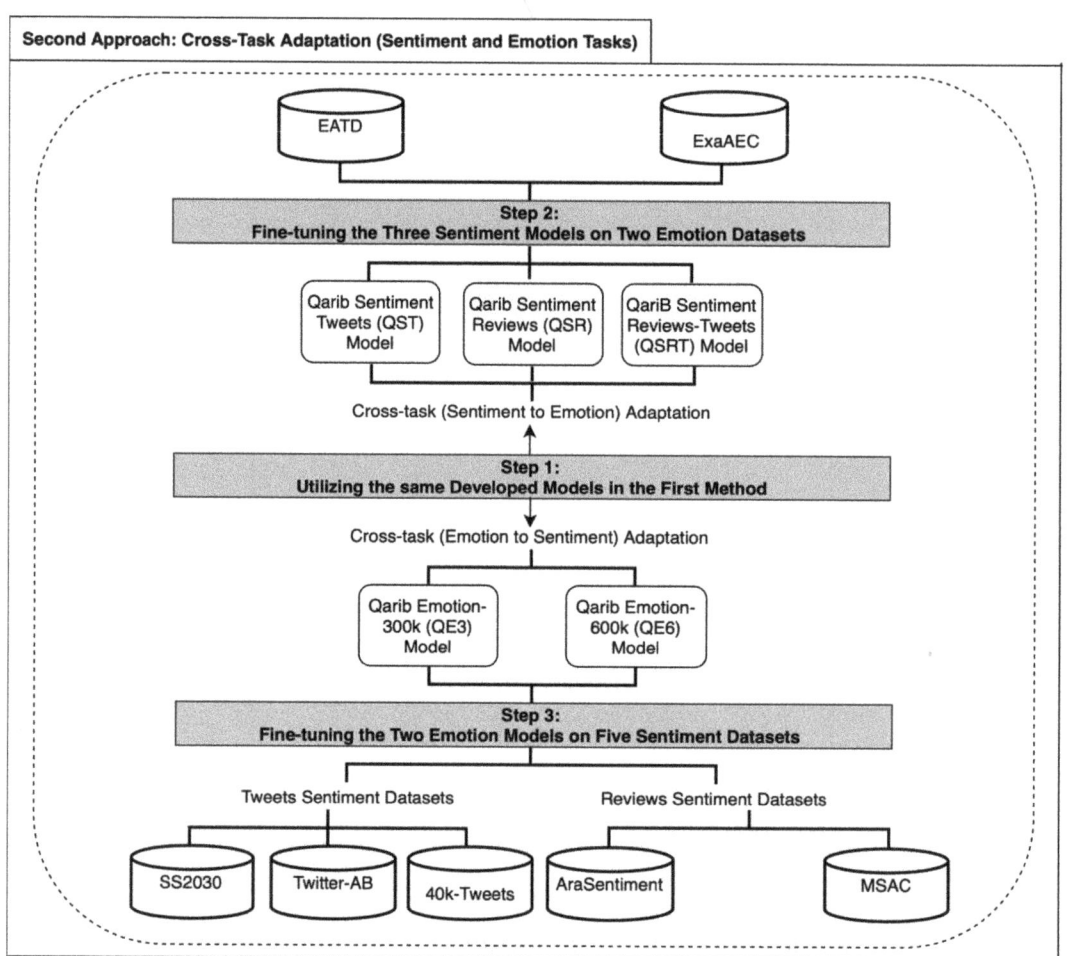

Figure 2. Language Models Development Process Using the Cross-Task Adaptation Approach.

3.2. Performance Evaluation Metrics

In this work, we present the experimental results to ensure the efficiency of our models using the following metrics:

3.2.1. Precision

Precision is defined as the proportion of true positives to all positives.

$$Precision = TP/(TP + FP) \qquad (3)$$

3.2.2. Recall

Recall is the proportion of correctly identified examples from a specified class to all examples of that class.

$$Recall = TP/(TP + FN) \qquad (4)$$

3.2.3. Accuracy

Accuracy is defined as the proportion of the total number of correct prediction examples to the total number of predictions.

$$Accuracy = (TP + TN)/(TP + TN + FP + FN) \tag{5}$$

3.2.4. F1-Score

The F1-Score provides one value that combines precision and recall, also known as the harmonic mean. One of the most commonly used metrics, computed as:

$$F1 = 2 \times (precision \times Recall)/(precision + Recall) \tag{6}$$

where:

- Positive examples correctly predicted are denoted by TP (true positive)
- Negative examples correctly predicted are denoted by TN (true negative)
- Incorrect positive predictions are denoted by FP (false positive)
- The wrong negative predictions are denoted by FN (false negative).

We report the macro average (also known as the unweighted mean) for each precision, recall, and F1-score. Using the macro average, each class is given equal importance, even if the classes are imbalanced. The results are discussed and compared specifically based on their macro-F1 score.

4. Experiments

We evaluated our adapted models on two Arabic NLP downstream tasks: sentiment analysis and emotion detection. In order to investigate whether further pre-training of the QARiB model [17] using task-specific unlabeled data could continue to improve the performance of the QARiB model [17] on sentiment and emotion tasks, the experiments aimed at addressing the following research questions for this study.

1. Does the type of training data (tweets or reviews) used in QARiB's language model further pre-training stage affect the end-task results?

To provide an answer to RQ1, the sentiment fine-tuning datasets that we used came from two distinct domains (Twitter and reviews). We intended to study the training and fine-tuning using various data types and evaluate model performance on each dataset from different points of view. For illustration, the QST model was trained using just tweets. However, we wanted to study the extent to which it performed well with review datasets (e.g., ArSentiment, and MASC). In contrast, we also wanted to evaluate the performance of the QSR model trained using only reviews on the tweet datasets (e.g., SS2030, 40k-Tweets, Twitter-AB).

2. What sentiment classification performance can be achieved if the QARiB language model is further pre-trained on a sentiment-specific dataset?

To address RQ2, we investigated whether sentiment models such as QST, QSR, and QSRT, which were further pre-trained on unlabeled sentiment datasets, could improve performance when fine-tuned on various labelled sentiment datasets. In this experiment, five sentiment datasets were used to fine-tune sentiment models.

3. What emotion classification performance can be achieved if the QARiB language model is further pre-trained on an emotion-specific dataset?

To find the answer to RQ3, we studied how well emotion models, such as QE3, and QE6 models, that were trained on emotion-unlabeled datasets, performed when fine-tuned on a variety of emotion-labelled datasets. We fine-tuned the emotion models using two-emotion datasets in an attempt to enhance the classification results. Moreover, we fine-tuned the BERT-base-QaRiB model as a baseline model on all seven sentiment and emotion datasets and compared the results.

4. Is there a relationship between sentiment and emotion representation? (i.e., can further pre-training QARiB with a sentiment dataset boost emotion classification results and vice versa?)

To provide an answer to RQ4 and see if there is a relationship between the sentiment and emotion tasks, we fine-tuned sentiment models QST, QSR, and QSRT on the two emotion datasets to examine whether the model trained on sentiment data could improve or increase the performance of emotion classification. Second, we fine-tuned the QE3 and QE6 models on the five sentiment datasets to see whether the model trained using emotion data could improve or enhance the results of the sentiment classification performance.

This section describes the experiment's setup, including the evaluation datasets, the baseline model compared to our models, the fine-tuning architecture, and the hyperparameter choices for fine-tuning our models.

4.1. Fine-Tuning Datasets

The datasets used for the evaluation process were chosen from the available Arabic sentiment and emotion dataset. For fine-tuning our models, we used five sentiment datasets and two emotion datasets. For all fine-tuning experiments, we applied the standard train/development/test set split of 80/10/10. Below is a description of the datasets utilised:

4.1.1. Sentiment Datasets

In order to cover different domains or sources, we chose the five sentiment datasets from different domains, including Twitter and reviews. The SS2030 [75], 40k-Tweets [76], and Twitter-AB [77] datasets were sourced from Twitter. In addition, ArSentiment [78], and MASC [79] were reviews datasets.

- SS2030 dataset [75]: sentiment dataset that has been gathered from Twitter includes 4252 tweets focusing on a variety of social issues in Saudi Arabia. The data set was manually annotated, and it consists of two classes (2436 positive, 1816 negative).
- Twitter-AB [77]: This dataset consists of 2000 tweets that were gathered from Twitter and have been classified into 1k positive, and 1k negative. The dataset was manually labelled and included both MSA and the Jordanian dialect, encompassing diverse topics related to politics and the arts.
- 40k-Tweets [76]: There are 40,000 tweets in this dataset, 20,000 of which are positive and 20,000 of which are negative. These tweets are written in both MSA and an Egyptian dialect. Furthermore, the gathered tweets are manually labelled and span a wide range of topics such as politics, sports, health, and social problems.
- The ArSentiment [78] is a large and multi-domain reviews dataset consisting of over 45k reviews on the 5-rating scale, for movies, hotels, restaurants, and products. We used a rating scale to assign labels to data, 1 and 2 stars have been considered negative, 3 stars have been considered neutral, and 4 and 5 stars have been considered positive.
- Multi-domain Arabic Sentiment Corpus (MASC) [79]: a review dataset that was scraped from a variety of websites including Google Play, Twitter, Facebook, and Qaym. The dataset, which included several different domains, was manually annotated into two classes: positive, and negative.

We selected datasets of varying sizes, some of which contained 40,000 sentences such as 40k-Tweets, and ArSentiment. Others, such as SS2030, Twitter-AB, and MASC had sizes of less than 7000 sentences. The selection of sentiment datasets with diverse domains and sizes was motivated by a desire to examine the impact of adaptation approaches from multiple perspectives. The statistics and classes distribution of the sentiment datasets are shown in Table 6. In addition, Table 7 provides the number of train, development, and test samples for each dataset.

Table 6. Sentiment Datasets Statistics.

Dataset	Size	#Classes	#Positive	#Negative	#Neutral	Source
SS2030	4252	2	2436	1816	-	
Twitter-AB	1961	2	999	962	-	Twitter
40k-Tweets	39,993	2	19,998	19,995	-	
ArSentiment	45,498	3	33,003	9336	3159	Reviews
MASC	6733	2	4476	2257	-	

Table 7. Train/Dev/Test Samples for Sentiment Datasets.

Dataset	Train Samples	Dev Samples	Test Samples
SS2030	3401	426	425
Twitter-AB	1568	197	196
40k-Tweets	31,994	4000	3999
ArSentiment	36,398	4550	4550
MASC	5386	674	673

4.1.2. Emotion Datasets

We evaluated our models on emotion Arabic tweet dataset (EATD) [80] and ExaAEC dataset [81]. In comparison to sentiment datasets, the labelled emotion datasets for the Arabic language are small and scarce. All these datasets are derived from Twitter. Table 8 illustrates the distribution of classes for each dataset. The number of train, development, and test samples for each dataset is presented in Table 9.

Table 8. Emotion Datasets Statistics.

Dataset-Name	EATD	ExaAEC
Size	2021	4738
#Classes	4	6
#Joy	629	472
#Sad	414	1909
#Fear	359	195
#Anger	619	191
#Surprise	-	795
#Disgust	-	1176

Table 9. Train/Dev/Test Samples for Emotion Datasets.

Dataset	Train Samples	Dev Samples	Test Samples
EATD	1616	203	202
ExaAEC	3790	474	474

- EATD [80]: an Arabic emotion dataset gathered from Twitter. The dataset was classified into four classes including anger, disgust, joy, and sadness. The annotation of the dataset was automatically for over 22k tweets based on emojis and manually for a subset of 2021 tweets. The manually annotated dataset has been utilised in our experiments.
- ExaAEC [81]: a multi-label Arabic emotion dataset consisting of approximately 20,000 tweets categorized as "neutral", "joy", "love", "anticipation", "acceptance", "surprise", "sadness", "fear", "anger", and "disgust." Each tweet in this dataset was manually annotated with one or two emotions. Given that the dataset contains tweets with multiple labels, we select a subset containing only tweets with a single label and according to the Ekman model, as follows: 'sadness' 1909, 'disgust' 1176, 'surprise' 795, 'joy' 472, 'fear' 195, 'anger' 191, for a total of approximately 4738 tweets.

4.2. Fine-Tuning Architecture

Fine-tuning the BERT model is "simple and direct", as indicated by [18], and only requires the addition of one more layer after the last BERT layer and training for a small number of iterations. The input sequence used to fine-tune the language model, in this case, is represented by the tokens [CLS] and [SEP] appended to the beginning and end of the sentence, respectively. The [CLS] token is used for all classification-related tasks. As a result, our models can be utilised for a variety of downstream text classification tasks with only minor architecture changes needed. Specifically, we fine-tuned our models for sentiment and emotion classification in Arabic text using the same fine-tuning strategy as BERT [18]. Trainer is a class within the Transformers library that can be utilised to fine-tune a variety of pre-trained Transformers-based models using a specific dataset. For the purpose of instantiating our sequence classification models, we utilised the AutoModelForSequenceClassification class. Due to the fact that our models were not pre-trained on the process of classifying sentences, the head of the model that had been pre-trained was removed, and in its place, a new head more suited to each task was added. The new head's weights were initially selected at random. This indicates that during model fine-tuning, just the weights of the new layers will be updated. In other words, during fine-tuning, all of the layers in our models will be frozen. For classification tasks, we added a fully connected feed-forward layer to the model and used the standard SoftMax activation function for prediction. It is worth noting that we fine-tuned our models independently for each task and dataset, using the same fine-tuning architecture. For a specific number of epochs, we fine-tuned our models on the training set. After that, the model checkpoint with the lowest validation loss was chosen automatically. We then used this checkpoint to do an evaluation of the test set.

4.3. Fine-Tuning Hyper-Parameters

Evaluating or fine-tuning the pre-trained language model is time-consuming, and manually experimenting with various hyperparameters might take days. Hyperparameter optimisation libraries, like Ray-tune [82], allow for the automatic selection of optimal values for model hyperparameters. This library is compatible with a wide variety of machine learning frameworks, including PyTorch and TensorFlow. This library was used in the experiments we conducted for this work. We ran ten trials for each dataset, and the hyperparameters were randomly chosen by the tool. After the hyperparameter search was completed, we obtained the best hyperparameters, which were used to fine-tune our final model. It should be noted that, due to computational and time constraints, we did not run the search for more than ten trials. In fact, for some datasets, such as 40k-Tweets and ArSentiment, the training time ranges from 6 to 10 hours and may exceed that time depending on the hyperparameters chosen by RayTune.

4.4. Baseline Models

As a baseline, to estimate how well our models performed, we compared them to the BERT-base-QaRiB model's [17] performance on the same tasks. We used a currently available BERT-base-QaRiB model and performed supervised fine-tuning, as described in Section 4.2, of the model's parameters for each dataset. Moreover, the results of this study were compared to the benchmark results provided by the datasets' original papers [75–80]. Except for the ExaAEC dataset, of which only a subset was used in this work, which is incompatible with the version used in [81].

5. Results and Discussion

5.1. Exp-I: Experiment to Investigate the Influence of the Within-Task Adaptation Approach on Sentiment and Emotion Classification Performance

The results of fine-tuning the BERT-base-QaRiB, QST, QSR, and QSRT models on the SS2030, Twitter-AB, 40k-Tweets, ArSentiment, and MASC datasets are shown in Table 10. Table 11 presents the results obtained by fine-tuning the BERT-base-QaRiB, QE3, and QE6 models using the EATD and ExaAEC datasets. The results are discussed, compared, and analysed specifically based on the macro-F1 score for the partition of the test set. In addition, the results of each dataset of the base studies [75–80] are presented in Tables 10 and 11. The results that showed an enhancement above the results obtained by the baseline model (i.e., the BERT-base-QaRiB model) are typically highlighted in bold. The results that are highlighted in bold and underlined are the best results that have been achieved for each dataset according to the model used. In total, 26 separate experiments were carried out utilising various sentiment and emotion datasets.

Table 10. Results of fine-tuning sentiment models on sentiment datasets (All-metrics).

Source	Datasets	Models	Precision	Recall	Accuracy	Macro-F1
Twitter	SS2030	SVM [75]	90.1%	-	89.83%	89.7%
		BERT-base-QaRiB	90.87%	90.56%	91.06%	90.70%
		QST	92.97%	92.89%	93.18%	**92.93%**
		QSR	91.35%	91.70%	91.76%	**91.51%**
		QSRT	92.05%	92.47%	92.47%	**92.24%**
	Twitter-AB	SVM [77]	-	-	87.2%	-
		BERT-base-QaRiB	94.47%	94.32%	94.39%	94.37%
		QST	96.93%	96.93%	96.94%	**96.93%**
		QSR	96.41%	96.49%	96.43%	**96.43%**
		QSRT	97.43%	97.47%	97.45%	**97.45%**
	40k-Tweets	LSTM [76]	86.94%	88.9%	88.05%	87.24%
		BERT-base-QaRiB	90.56%	90.51%	90.50%	90.49%
		QST	91.38%	91.37%	91.37%	**91.37%**
		QSR	91.34%	91.27%	91.27%	**91.27%**
		QSRT	90.66%	90.65%	90.65%	**90.65%**
Reviews	ArSentiment	SVM [78]	-	-	59.9%	-
		BERT-base-QaRiB	81.21%	72.52%	90.62%	75.93%
		QST	79.04%	75.34%	90.48%	**76.83%**
		QSR	81.61%	76.53%	91.45%	**78.53%**
		QSRT	81.20%	75.83%	91.25%	**78.15%**
	MASC	LLR [79]	-	-	-	97.8%
		BERT-base-QaRiB	95.65%	93.81%	95.10%	94.61%
		QST	97.04%	97.20%	97.33%	**97.12%**
		QSR	95.95%	95.37%	95.99%	**95.65%**
		QSRT	96.14%	96.53%	96.58%	**96.33%**

Table 11. Results of fine-tuning emotion models on emotion datasets (All-metrics).

Datasets	Models	Precision	Recall	Accuracy	Macro-F1
EATD	SVM [80]	69.67%	69.04%	-	68.52%
	BERT-base-QaRiB	85.31%	86.47%	85.64%	85.46%
	QE3	92.79%	89.63%	90.59%	**90.10%**
	QE6	88.20%	89.12%	88.61%	**88.31%**
ExaAEC	BERT-base-QaRiB	65.87%	62.84%	74.26%	63.81%
	QE3	65.53%	63.72%	75.11%	**64.21%**
	QE6	63.80%	64.90%	74.68%	**64.16%**

In Table 10, the results that outperformed the BERT-base-QaRiB model results are highlighted in bold. All sentiment models, including QST, QSR, and QSRT, outperformed the BERT-base-QaRiB model on all sentiment datasets. In addition, when comparing the QST model to the QSR model across all of the experiments (given in Table 10), we observed that the sentiment datasets that were sourced from Twitter, including SS2030, Twitter-AB, and 40k-Tweets, the QST model outperformed the QSR model. Based on this, we may infer that data distributions of tasks within the same source or domain could be similar. This also indicates that further pre-training of the model using task-specific datasets from the same genre or domain of the fine-tuning datasets yields better results than utilising datasets from a different genre or domain.

Compared to the BERT-base-QaRiB model, Table 10 reveals a performance gain of 2.22% for the QST model and 0.80% for the QSR model on the SS2030 dataset. In addition, the QST and QSR models showed improvements in the Twitter-AB dataset by 2.56% and 2.06%, respectively. In fact, the 40k-Tweets dataset performance increased only by 0.88% using the QST model and by 0.77% using the QSR model. The explanation might be that the 40k-Tweets dataset is a multi-domain dataset including several domains, such as politics and arts, and these domains were not included or covered extensively during the training of the model. Comparing the improvement in the performance of our models QST and QSR on the SS2030 and Twitter-AB datasets to the 40k-Tweets dataset may suggest that the models perform better on small datasets as opposed to large datasets. Compared to the BERT-base-QaRiB model, Table 10 reveals an improvement in the performance of 0.90% for the QST model and 2.60% for the QSR model on the ArSentiment dataset. Meanwhile, performance on the MASC dataset improved by 2.51% using the QST model and by 1.04% using the QSR model.

The QSRT model outperformed BERT-base-QaRiB on the SS2030, Twitter-AB, 40k-Tweets, ArSentiment, and MASC datasets by 1.54%, 3.08, 0.15%, 2.22%, and 1.72%, respectively. Our best sentiment model for SS2030, 40k-Tweets and MASC datasets was the QST model, which achieved 92.93%, 91.37%, and 97.12% F1-scores, respectively. Moreover, the QSR model obtained the highest F1 score on the ArSentiment dataset by achieving 78.53%. The QSRT model was the best sentiment model on the Twitter-AB dataset with macro-F1 of 97.45%. This may suggest that there is no need to perform further pre-training of a model, with a large amount of training data. Instead, training with task-specific datasets that share the same domain as the fine-tuning datasets could result in higher performance.

Table 10 demonstrates that the developed models significantly outperform the results of the original studies for the SS2030, Twitter-AB, 40k-Tweets, and ArSentiment datasets in terms of accuracy by 3.35%, 10.25% 3.32%, and 31.55%, respectively. In terms of the F1 score, it improved by 0.04% on the MASC dataset. The results reported in Table 10 show that the within-task adaptation approach has a beneficial impact on the final results of the sentiment analysis task. In other words, further pre-training of the QARiB language model with unlabeled sentiment datasets and fine-tuning using labelled sentiment datasets improved or enhanced the final results of the sentiment analysis task. In addition, further pre-training using sentiment-specific datasets with the same source or domain as the fine-tuning datasets leads to better enhancement in the sentiment classification results.

In terms of the results of emotion detection, it is challenging for the model that the emotion datasets used, such as EATD and ExaAEC, have multiple classes (i.e., four and six emotion classes). Nevertheless, compared to BERT-base-QaRiB, our QE3 and QE6 models performed better on all emotion datasets, as shown in Table 11 in bold. Compared to the BERT-base-QaRiB model, the QE3 model performed 4.64% better on the EATD dataset. Using the same dataset and the QE6 model, a 2.84% improvement in performance was observed. Results for the ExaAEC dataset were enhanced by 0.40% with the QE3 model and by 0.35% with the QE6 model. The reason could be that the ExaAEC dataset contains six emotion classes in addition to being an unbalanced dataset, as demonstrated in Table 8.

It can be observed that the QE3 model was the best emotion model across all emotion datasets, including the EATD, and ExaAEC datasets, obtaining macro-F1 scores of 90.10%, and 64.21%, respectively. Furthermore, Table 11 shows that the QE3 model outperforms the SVM model in terms of the F1 score by 21.58% for the EATD dataset. These results indicate that QE3 outperformed QE6 on the two emotion datasets. In addition, as shown in Table 5, the training results of the QE6 models were not superior to those of the QE3 models. In fact, QE6 validation loss and perplexity increased. Together, these findings provide an important insight, namely that pre-training the model for longer training steps is not necessary to achieve optimal performance.

Finally, the results reported in Table 11 show that the within-task adaptation approach has a positive impact on the final results of the emotion detection task. Accordingly, pre-training the QARiB language model with unlabeled emotion datasets and fine-tuning it with labelled emotion datasets improves emotion detection task performance. In addition, further pre-training of the model for longer training steps is unnecessary to get the highest performance.

5.2. Exp-II: Experiment to Investigate the Influence of the Cross-Task Adaptation Approach on Sentiment and Emotion Classification Performance

Table 12 summarises the results of fine-tuning the BERT-base-QaRiB, QE3, and QE6 models using the sentiment datasets SS2030, Twitter-AB, 40k-Tweets, ArSentiment, and MASC. Table 13 shows the results of fine-tuning the BERT-base-QaRiB, QST, QSR, and QSRT models using EATD and ExaAEC datasets. The results are discussed, compared, and analysed specifically based on the macro-F1 score for the partition of the test set. In addition, the results of each dataset of the original studies [75–80] are presented in Tables 12 and 13. The results that showed an improvement over those obtained by the baseline model (i.e., BERT-base-QaRiB model) are highlighted in bold. The results that are in bold and underlined represent the highest results that have been obtained for each dataset and according to which model. In total, 16 experiments were carried out using various sentiment and emotion datasets.

Table 12. Results of fine-tuning emotion models on sentiment datasets (All-metrics).

Source	Datasets	Models	Precision	Recall	Accuracy	Macro-F1
Twitter	SS2030	SVM [75]	90.1%	-	89.83%	89.7%
		BERT-base-QaRiB	90.87%	90.56%	91.06%	90.70%
		QE3	93.21%	92.62%	93.18%	**92.89%**
		QE6	91.09%	91.50%	91.53%	**91.27%**
	Twitter-AB	SVM [77]	-	-	87.2%	-
		BERT-base-QaRiB	94.47%	94.32%	94.39%	94.37%
		QE3	96.41%	96.49%	96.43%	**96.43%**
		QE6	96.06%	95.83%	95.92%	**95.90%**
	40k-Tweets	LSTM [76]	86.94%	88.9%	88.05%	87.24%
		BERT-base-QaRiB	90.56%	90.51%	90.50%	90.49%
		QE3	91.22%	91.19%	91.20%	**91.20%**
		QE6	91.40%	91.40%	91.40%	**91.40%**
Reviews	ArSentiment	SVM [78]	-	-	59.9%	-
		BERT-base-QaRiB	81.21%	72.52%	90.62%	75.93%
		QE3	80.57%	77.83%	91.10%	**79.12%**
		QE6	80.25%	77.50%	91.32%	**78.65%**
	MASC	LLR [79]	-	-	-	97.8%
		BERT-base-QaRiB	95.65%	93.81%	95.10%	94.61%
		QE3	96.03%	95.95%	96.29%	**95.99%**
		QE6	95.45%	95.60%	95.84%	**95.52%**

Table 13. Results of fine-tuning sentiment models on emotion datasets (all-metrics).

Datasets	Models	Precision	Recall	Accuracy	Macro-F1
EATD	SVM [80]	69.67%	69.04%	-	68.52%
	BERT-base-QaRiB	85.31%	86.47%	85.64%	85.46%
	QST	91.58%	89.92%	91.09%	**90.18%**
	QSR	88.18%	88.89%	88.61%	**88.27%**
	QSRT	89.74%	90.26%	90.10%	**89.78%**
ExaAEC	BERT-base-QaRiB	65.87%	62.84%	74.26%	63.81%
	QST	67.98%	63.58%	76.79%	**65.05%**
	QSR	67.19%	63.83%	75.53%	**64.98%**
	QSRT	71.33%	64.25%	75.74%	**64.38%**

Overall, as shown in Tables 12 and 13, all sentiment models, including QST, QSR, and QSRT, and emotion models, including QE3, and QE6, outperformed the BERT-base-QaRiB model for all sentiment and emotion datasets. In the tables, results that exceeded the BERT-base-QaRiB model results are highlighted in bold. Table 12 reveals that the QE3 model outperformed BERT-base-QaRiB on the SS2030, Twitter-AB, 40k-Tweets, ArSentiment, and MASC datasets by 2.18%, 2.06%, 0.70%, 3.18%, and 1.38%, respectively. In addition, the QE6 model outperformed BERT-base-QaRiB on the SS2030, Twitter-AB, 40k-Tweets, ArSentiment, and MASC datasets by 0.57%, 1.53%, 0.90%, 2.71%, and 0.91%, respectively.

When comparing emotion models, including QE3 and QE6, across five sentiment datasets, QE3 outperforms QE6 on the SS2030, Twitter-AB, ArSentiment, and MASC datasets, obtaining macro-F1 by 92.89%, 96.43%, 79.12%, and 95.99%, respectively. On the 40k-Tweets dataset, the QE6 model outperforms the QE3 and obtains macro-F1 by 91.40%. These results may provide insight into the fact that pre-training the model for longer training steps does not necessarily give optimal performance. In addition, the further pre-training of the model using sentiment data and fine-tuning it with an emotion dataset can improve the final emotion classification results.

In Table 13, compared to the BERT-base-QaRiB model, the QST model improved performance on the EATD dataset by 4.71%, making it the best model on this dataset with a macro-F1 of 90.18%. On the same dataset and using the QSR model, a 2.80% improvement in performance was observed. On the same dataset, the QSRT model showed a performance improvement of 4.32%, which was better compared to the QSR models. Using the ExaAEC dataset, QST model outperformed all other models with an improvement of 1.24% and a macro-F1 of 65.05%. While using the QSR and QSRT models, performance improvements of 1.17% and 0.57% were achieved.

Comparing the results of the QST and QSR models on emotion datasets, including EATD and ExaAEC on all emotion datasets, we noticed that the QST model outperformed the QSR model. This was expected because the QST model was further trained using tweet data, and the emotion datasets were also taken from Twitter. These findings indicate that we might get better results if we further pre-train the model using a task-specific dataset from the same genre for fine-tuning the datasets. On emotion datasets, the QSRT model performed somewhat worse than the QST and QSR models on the ExaAEC dataset, although the QSRT model was trained with a larger dataset and a different data genre (Twitter and reviews). This may indicate that a large quantity of data may not be necessary for training the model and that a dataset of the same genre as the dataset used for fine-tuning may be more efficient. In general, Table 12 demonstrates that the developed models significantly outperform the results of the base studies for the SS2030, Twitter-AB, 40k-Tweets, and ArSentiment datasets, except the MASC dataset, in terms of accuracy by 3.35%, 9.23%, 3.35%, and 31.42%, respectively. Furthermore, Table 13 shows that the QST model outperforms the SVM model in terms of f1-score by 21.66% for the EATD dataset.

In conclusion, the results presented in Tables 12 and 13 illustrate the effectiveness of the cross-task adaptation approach on the final results of the sentiment and emotion classification tasks. These results suggest that the data distribution between sentiment and emotion may be converging. We can see how each task can influence and improve the results of the other. This may give an important insight into how convergent tasks with converged data distribution might enhance each other's performance. For instance, the Arabic emotion detection task has more limited resources than sentiment. Therefore, this study may help researchers tackling emotion detection to obtain better results by utilising sentiment resources. Additionally, when comparing the results of the two task-adaptation approaches (cross-task and within-task), it can be shown that the cross-task adaptation results sometimes outperform the within-task approach. On the emotion datasets EATD and ExaAEC, for instance, the sentiment model QST outperformed the emotion models QE3 and QE6. Additionally, the emotion models QE3 and QE6 outperformed the sentiment models on the sentiment datasets 40k-Tweets and ArSentiment.

6. Conclusions

The experiments described in the previous sections examine the effect of two adaptation approaches: within-task and cross-task adaptation. In total, five new models were developed using the previous approach: the QST, QSR, QSRT, QE3, and QE6 models. Different evaluation experiments were conducted by fine-tuning each model for two downstream tasks, sentiment analysis and emotion detection. Using five sentiment datasets, including SS2030, Twitter-AB, 40k-Tweets, ArSentiment, and MASC, in addition to two emotion datasets, EATD and ExaAEC, 42 experiments were carried out in total. The sentiment and emotion datasets covered both small- and large-resource settings. The experiments reveal the following: first, the within-task and cross-task adaptation approaches have influenced the final results and boosted performance for all tasks (i.e., sentiment and emotion). Second, our newly developed QST, QSR, QSRT, QE3, and QE6 models outperformed the BERT-base-QaRiB model on all sentiment and emotion datasets. Third, the training using task-specific datasets that share the same domain as the fine-tuning datasets results in higher performance. Fourth, additional pre-training of the model for longer training steps is unnecessary to get the highest performance. Finally, cross-task adaptation shows that sentiment and emotion data may converge, and each task might enhance the results of the other.

This study showed that pre-training the QARiB language model on small-scale sentiment or emotion data improves model understanding of this domain data and yields considerable improvements. Because of the scarcity of emotion datasets, one of the limitations of this research is that the model was only evaluated on two small emotion datasets. In general, findings reveal interesting areas for future research. The findings indicate that these approaches (i.e., within-task and cross-task adaptation) can improve the performance of QARiB. Consequently, any pre-trained Arabic language model can be utilised with the approaches that we have investigated. While Arabic language models like AraBERT and MARBERT already perform effectively well on sentiment and emotion tasks, they may benefit significantly from further task-specific pre-training. In addition, we believe that pre-training on larger task-specific data could further enhance performance. Finally, the developed language models are publicly available to be used by the NLP community for research purposes, and we hope this work helps researchers interested in the domain of Arabic sentiment and emotion analysis.

Author Contributions: Conceptualisation, W.A., N.A.-T. and A.A.; data curation, W.A.; formal analysis, W.A.; funding acquisition, A.A.; methodology, W.A., N.A.-T. and A.A.; software, W.A.; supervision, N.A.-T. and A.A.; validation, W.A.; writing—original draft, W.A.; writing—review and editing, N.A.-T. and A.A. All authors have read and agreed to the published version of the manuscript.

Funding: This research was funded by the Deanship of Scientific Research, King Saud University.

Institutional Review Board Statement: Not applicable.

Informed Consent Statement: Not applicable.

Data Availability Statement: The datasets have been taken from this link https://arbml.github.io/masader/ (accessed on 7 September 2022).

Acknowledgments: The authors are grateful to the Deanship of Scientific Research, King Saud University for funding through the Vice Deanship of Scientific Research Chairs.

Conflicts of Interest: The authors declare no conflict of interest.

References

1. Mikolov, T.; Chen, K.; Corrado, G.; Dean, J. Efficient estimation of word representations in vector space. *arXiv* **2013**, arXiv:1301.3781.
2. Pennington, J.; Socher, R.; Manning, C.D. Glove: Global vectors for word representation. In Proceedings of the 2014 Conference on Empirical Methods in Natural Language Processing (EMNLP), Doha, Qatar, 25–29 October 2014; pp. 1532–1543.
3. Bojanowski, P.; Grave, E.; Joulin, A.; Mikolov, T. Enriching word vectors with subword information. *Trans. Assoc. Comput. Linguist.* **2017**, *5*, 135–146. [CrossRef]

4. Torrey, L.; Shavlik, J. Transfer learning. In *Handbook of Research on Machine Learning Applications and Trends: Algorithms, Methods, and Techniques*; IGI Global: Hershey, PA, USA, 2010; pp. 242–264.
5. Oueslati, O.; Cambria, E.; Ben HajHmida, M.; Ounelli, H. A review of sentiment analysis research in Arabic language. *Future Gener. Comput. Syst.* **2020**, *112*, 408–430. [CrossRef]
6. Abdullah, M.; Hadzikadicy, M.; Shaikhz, S. SEDAT: Sentiment and emotion detection in Arabic text using CNN-LSTM deep learning. In Proceedings of the 2018 17th IEEE International Conference on Machine Learning and Applications (ICMLA), Orlando, FL, USA, 17–20 December 2018; pp. 835–840.
7. Stevenson, A. *Oxford Dictionary of English*; Oxford University Press: New York, NY, USA, 2010.
8. Ekman, P.; Friesen, W.V.; O'sullivan, M.; Chan, A.; Diacoyanni-Tarlatzis, I.; Heider, K.; Krause, R.; LeCompte, W.A.; Pitcairn, T.; Ricci-Bitti, P.E. Universals and cultural differences in the judgments of facial expressions of emotion. *J. Pers. Soc. Psychol.* **1987**, *53*, 712. [CrossRef]
9. Plutchik, R. A general psychoevolutionary theory of emotion. In *Theories of Emotion*; Elsevier: Amsterdam, The Netherlands, 1980; pp. 3–33.
10. Lee, J.; Yoon, W.; Kim, S.; Kim, D.; Kim, S.; So, C.H.; Kang, J. BioBERT: A pre-trained biomedical language representation model for biomedical text mining. *Bioinformatics* **2020**, *36*, 1234–1240. [CrossRef] [PubMed]
11. Alsentzer, E.; Murphy, J.; Boag, W.; Weng, W.-H.; Jindi, D.; Naumann, T.; McDermott, M. Publicly Available Clinical BERT Embeddings. In Proceedings of the 2nd Clinical Natural Language Processing Workshop, Minneapolis, MN, USA, 5 June 2019; pp. 72–78.
12. Araci, D. FinBERT: Financial Sentiment Analysis with Pre-trained Language Models. *arXiv* **2019**, arXiv:1908.10063.
13. Gururangan, S.; Marasović, A.; Swayamdipta, S.; Lo, K.; Beltagy, I.; Downey, D.; Smith, N.A. Don't Stop Pretraining: Adapt Language Models to Domains and Tasks. In Proceedings of the 58th Annual Meeting of the Association for Computational Linguistics, Online, 5–10 July 2020.
14. Sun, C.; Qiu, X.; Xu, Y.; Huang, X. How to fine-tune bert for text classification? In *China National Conference on Chinese Computational Linguistics*; Springer: Berlin/Heidelberg, Germany, 2019; pp. 194–206.
15. Ameur, M.S.H.; Aliane, H. AraCOVID19-MFH: Arabic COVID-19 Multi-label Fake News & Hate Speech Detection Dataset. *Procedia Comput. Sci.* **2021**, *189*, 232–241.
16. Lan, W.; Chen, Y.; Xu, W.; Ritter, A. An Empirical Study of Pre-trained Transformers for Arabic Information Extraction. In Proceedings of the 2020 Conference on Empirical Methods in Natural Language Processing (EMNLP), Online, 19–20 November 2020; pp. 4727–4734.
17. Abdelali, A.; Hassan, S.; Mubarak, H.; Darwish, K.; Samih, Y. Pre-training bert on arabic tweets: Practical considerations. *arXiv* **2021**, arXiv:2102.10684.
18. Kenton, J.D.M.-W.C.; Toutanova, L.K. BERT: Pre-training of Deep Bidirectional Transformers for Language Understanding. In Proceedings of the NAACL-HLT, Minneapolis, MN, USA, 2–7 June 2019; pp. 4171–4186.
19. Dadas, S.; Perełkiewicz, M.; Poświata, R. Pre-training polish transformer-based language models at scale. In Proceedings of the Artificial Intelligence and Soft Computing: 19th International Conference, ICAISC 2020, Zakopane, Poland, 12–14 October 2020; Part II 19. Springer: Berlin/Heidelberg, Germany, 2020; pp. 301–314.
20. Polignano, M.; Basile, P.; de Gemmis, M.; Semeraro, G.; Basile, V. AlBERTo: Italian BERT Language Understanding Model for NLP Challenging Tasks Based on Tweets. In Proceedings of the CLiC-it, Bari, Italy, 13–15 November 2019.
21. Vaswani, A.; Shazeer, N.; Parmar, N.; Uszkoreit, J.; Jones, L.; Gomez, A.N.; Kaiser, .; Polosukhin, I. Attention is all you need. In Proceedings of the Advances in Neural Information Processing Systems, Long Beach, CA, USA, 4–9 December 2017; pp. 5998–6008.
22. Antoun, W.; Baly, F.; Hajj, H. AraBERT: Transformer-based Model for Arabic Language Understanding. In Proceedings of the LREC 2020 Workshop Language Resources and Evaluation Conference, Marseille, France, 11–16 May 2020; p. 9.
23. ElJundi, O.; Antoun, W.; El Droubi, N.; Hajj, H.; El-Hajj, W.; Shaban, K. hulmona: The universal language model in arabic. In Proceedings of the Fourth Arabic Natural Language Processing Workshop, Florence, Italy, 1 August 2019; pp. 68–77.
24. Obied, Z.; Solyman, A.; Ullah, A.; Fat'hAlalim, A.; Alsayed, A. BERT Multilingual and Capsule Network for Arabic Sentiment Analysis. In Proceedings of the 2020 International Conference On Computer, Control, Electrical, And Electronics Engineering (ICCCEEE), Khartoum, Sudan, 26 February–1 March 2021; pp. 1–6.
25. Wadhawan, A. AraBERT and Farasa Segmentation Based Approach For Sarcasm and Sentiment Detection in Arabic Tweets. In Proceedings of the Sixth Arabic Natural Language Processing Workshop, Kyiv, Ukraine, 19 April 2021; pp. 395–400.
26. Antoun, W.; Baly, F.; Hajj, H. AraELECTRA: Pre-Training Text Discriminators for Arabic Language Understanding. In Proceedings of the Sixth Arabic Natural Language Processing Workshop, Kyiv, Ukraine, 19 April 2021; pp. 191–195.
27. Abdul-Mageed, M.; Elmadany, A. ARBERT & MARBERT: Deep Bidirectional Transformers for Arabic. In Proceedings of the 59th Annual Meeting of the Association for Computational Linguistics and the 11th International Joint Conference on Natural Language Processing (Volume 1: Long Papers), Online, 1–6 August 2021; pp. 7088–7105.
28. Conneau, A.; Khandelwal, K.; Goyal, N.; Chaudhary, V.; Wenzek, G.; Guzmán, F.; Grave, É.; Ott, M.; Zettlemoyer, L.; Stoyanov, V. Unsupervised Cross-lingual Representation Learning at Scale. In Proceedings of the 58th Annual Meeting of the Association for Computational Linguistics, Online, 5–10 July 2020; pp. 8440–8451.

29. Inoue, G.; Alhafni, B.; Baimukan, N.; Bouamor, H.; Habash, N. The interplay of variant, size, and task type in Arabic pre-trained language models. In Proceedings of the Sixth Arabic Natural Language Processing Workshop, Kyiv, Ukraine, 19 April 2021; pp. 32–104.
30. Abdaoui, A.; Berrimi, M.; Oussalah, M.; Moussaoui, A. Dziribert: A pre-trained language model for the algerian dialect. *arXiv* **2021**, arXiv:2109.12346.
31. Alduailej, A.; Alothaim, A. AraXLNet: Pre-trained language model for sentiment analysis of Arabic. *J. Big Data* **2022**, *9*, 1–21. [CrossRef]
32. Safaya, A.; Abdullatif, M.; Yuret, D. Kuisail at semeval-2020 task 12: Bert-cnn for offensive speech identification in social media. In Proceedings of the Fourteenth Workshop on Semantic Evaluation, Barcelona, Spain, 12–13 December 2020; pp. 2054–2059.
33. Elmadany, A.; Nagoudi, E.M.B.; Abdul-Mageed, M. ORCA: A Challenging Benchmark for Arabic Language Understanding. *arXiv* **2022**, arXiv:2212.10758.
34. Liu, Y.; Ott, M.; Goyal, N.; Du, J.; Joshi, M.; Chen, D.; Levy, O.; Lewis, M.; Zettlemoyer, L.; Stoyanov, V. Roberta: A robustly optimized bert pretraining approach. *arXiv* **2019**, arXiv:1907.11692.
35. Khaddaj, A.; Hajj, H.; El-Hajj, W. Improved generalization of arabic text classifiers. In Proceedings of the Fourth Arabic Natural Language Processing Workshop, Florence, Italy, 1 August 2019; pp. 167–174.
36. El Mekki, A.; El Mahdaouy, A.; Berrada, I.; Khoumsi, A. Domain adaptation for Arabic cross-domain and cross-dialect sentiment analysis from contextualized word embedding. In Proceedings of the 2021 Conference of the North American Chapter of the Association for Computational Linguistics: Human Language Technologies, Online, 6–11 June 2021; pp. 2824–2837.
37. Alqahtani, Y.; Al-Twairesh, N.; Alsanad, A. A Comparative Study of Effective Domain Adaptation Approaches for Arabic Sentiment Classification. *Appl. Sci.* **2023**, *13*, 1387. [CrossRef]
38. Badaro, G.; Jundi, H.; Hajj, H.; El-Hajj, W.; Habash, N. Arsel: A large scale arabic sentiment and emotion lexicon. In Proceedings of the OSACT 3: The 3rd Workshop on Open-Source Arabic Corpora and Processing Tools, Miyazaki, Japan, 8 May 2018; p. 26.
39. Mohammad, S.; Bravo-Marquez, F.; Salameh, M.; Kiritchenko, S. Semeval-2018 task 1: Affect in tweets. In Proceedings of the 12th International Workshop on Semantic Evaluation, New Orleans, LA, USA, 5–6 June 2018; pp. 1–17.
40. Badaro, G.; El Jundi, O.; Khaddaj, A.; Maarouf, A.; Kain, R.; Hajj, H.; El-Hajj, W. Ema at semeval-2018 task 1: Emotion mining for arabic. In Proceedings of the 12th International Workshop on Semantic Evaluation, New Orleans, LA, USA, 5–6 June 2018; pp. 236–244.
41. Aljwari, F. Emotion Detection in Arabic Text Using Machine Learning Methods. *IJISCS-Int. J. Inf. Syst. Comput. Sci.* **2022**, *6*, 175–185.
42. Khalil, E.A.H.; El Houby, E.M.F.; Mohamed, H.K. Deep learning for emotion analysis in Arabic tweets. *J. Big Data* **2021**, *8*, 1–15. [CrossRef]
43. Abdul-Mageed, M.; Zhang, C.; Hashemi, A. AraNet: A Deep Learning Toolkit for Arabic Social Media. In Proceedings of the 4th Workshop on Open-Source Arabic Corpora and Processing Tools, with a Shared Task on Offensive Language Detection, Marseille, France, 12 May 2020; pp. 16–23.
44. Elfaik, H. Combining context-aware embeddings and an attentional deep learning model for Arabic affect analysis on twitter. *IEEE Access* **2021**, *9*, 111214–111230. [CrossRef]
45. Mansy, A.; Rady, S.; Gharib, T. An ensemble deep learning approach for emotion detection in arabic tweets. *Int. J. Adv. Comput. Sci. Appl.* **2022**, *13*, 01304112. [CrossRef]
46. Al-Twairesh, N. The evolution of language models applied to emotion analysis of Arabic tweets. *Information* **2021**, *12*, 84. [CrossRef]
47. Soliman, A.B.; Eissa, K.; El-Beltagy, S.R. Aravec: A set of arabic word embedding models for use in arabic nlp. *Procedia Comput. Sci.* **2017**, *117*, 256–265. [CrossRef]
48. Talafha, B.; Ali, M.; Za'ter, M.E.; Seelawi, H.; Tuffaha, I.; Samir, M.; Farhan, W.; Al-Natsheh, H. Multi-dialect Arabic BERT for Country-level Dialect Identification. In Proceedings of the Fifth Arabic Natural Language Processing Workshop, Barcelona, Spain, 12 December 2020; pp. 111–118.
49. Mahmoud, A.E.-S.; Lazem, S.; Abougabal, M. *Benchmarking a Large Twitter Dataset for Arabic Emotion Analysis*; Research Square: Durham, NC, USA, 2022.
50. Abdelali, A.; Darwish, K.; Durrani, N.; Mubarak, H. Farasa: A fast and furious segmenter for arabic. In Proceedings of the 2016 Conference of the North American Chapter of the Association for Computational Linguistics: Demonstrations, San Diego, CA, USA, 12–17 June 2016; pp. 11–16.
51. Parker, R.; Graff, D.; Chen, K.; Kong, J.; Maeda, K. "Arabic Gigaword." LDC Catalog No. LDC2009T30. 2009. Available online: https://catalog.ldc.upenn.edu/LDC2009T30 (accessed on 1 March 2022).
52. El-Khair, I.A. 1.5 billion words arabic corpus. *arXiv* **2016**, arXiv:1611.04033.
53. Lison, P.; Tiedemann, J. Opensubtitles2016: Extracting large parallel corpora from movie and tv subtitles. In Proceedings of the Tenth International Conference on Language Resources and Evaluation (LREC 2016), Portorož, Slovenia, 23–28 May 2016.
54. Sennrich, R.; Haddow, B.; Birch, A. Neural Machine Translation of Rare Words with Subword Units. In Proceedings of the 54th Annual Meeting of the Association for Computational Linguistics (Volume 1: Long Papers), Berlin, Germany, 7–12 August 2016; pp. 1715–1725.

55. Alomari, K.M.; ElSherif, H.M.; Shaalan, K. Arabic tweets sentimental analysis using machine learning. In *International Conference on Industrial, Engineering and Other Applications of Applied Intelligent Systems*; Springer: Berlin/Heidelberg, Germany, 2017; pp. 602–610.
56. Elmadany, A.; Mubarak, H.; Magdy, W. Arsas: An arabic speech-act and sentiment corpus of tweets. *OSACT* **2018**, *3*, 20.
57. Baly, R.; Khaddaj, A.; Hajj, H.; El-Hajj, W.; Shaban, K.B. ArSentD-LEV: A Multi-Topic Corpus for Target-based Sentiment Analysis in Arabic Levantine Tweets. In Proceedings of the OSACT 3: The 3rd Workshop on Open-Source Arabic Corpora and Processing Tools, Miyazaki, Japan, 8 May 2018; p. 37.
58. Boujou, E.; Chataoui, H.; El Mekki, A.; Benjelloun, S.; Chairi, I.; Berrada, I. An open access NLP dataset for Arabic dialects: Data collection, labeling, and model construction. *arXiv* **2021**, arXiv:2102.11000.
59. Salameh, M.; Mohammad, S.; Kiritchenko, S. Sentiment after translation: A case-study on arabic social media posts. In Proceedings of the 2015 Conference of the North American Chapter of the Association for Computational Linguistics: Human Language Technologies, Denver, CO, USA, 31 May–5 June 2015; pp. 767–777.
60. Kwaik, K.A.; Chatzikyriakidis, S.; Dobnik, S.; Saad, M.; Johansson, R. An Arabic tweets sentiment analysis dataset (ATSAD) using distant supervision and self training. In Proceedings of the 4th Workshop on Open-Source Arabic Corpora and Processing Tools, with a Shared Task on Offensive Language Detection, Marseille, France, 12 May 2020; pp. 1–8.
61. Nabil, M.; Aly, M.; Atiya, A. Astd: Arabic sentiment tweets dataset. In Proceedings of the 2015 Conference on Empirical Methods in Natural Language Processing, Lisbon, Portugal, 17–21 September 2015; pp. 2515–2519.
62. Rosenthal, S.; Farra, N.; Nakov, P. SemEval-2017 task 4: Sentiment analysis in Twitter. In Proceedings of the 11th International Workshop on Semantic Evaluation (SemEval-2017), Vancouver, BC, Canada, 3–4 August 2017; pp. 502–518.
63. Farha, I.A.; Magdy, W. From arabic sentiment analysis to sarcasm detection: The arsarcasm dataset. In Proceedings of the 4th Workshop on Open-Source Arabic Corpora and Processing Tools, with a Shared Task on Offensive Language Detection, Marseille, France, 12 May 2020; pp. 32–39.
64. Abdellaoui, H.; Zrigui, M. Using tweets and emojis to build tead: An Arabic dataset for sentiment analysis. *Comput. Sist.* **2018**, *22*, 777–786. [CrossRef]
65. Alowisheq, A.; Al-Twairesh, N.; Altuwaijri, M.; Almoammar, A.; Alsuwailem, A.; Albuhairi, T.; Alahaideb, W.; Alhumoud, S. MARSA: Multi-domain Arabic resources for sentiment analysis. *IEEE Access* **2021**, *9*, 142718–142728. [CrossRef]
66. Al-Twairesh, N.; Al-Khalifa, H.; Al-Salman, A.; Al-Ohali, Y. Arasenti-tweet: A corpus for arabic sentiment analysis of saudi tweets. *Procedia Comput. Sci.* **2017**, *117*, 63–72. [CrossRef]
67. Al Omari, M.; Al-Hajj, M.; Hammami, N.; Sabra, A. Sentiment classifier: Logistic regression for arabic services' reviews in lebanon. In Proceedings of the 2019 International Conference on Computer and Information Sciences (ICCIS), Sakaka, Saudi Arabia, 3–4 April 2019; pp. 1–5.
68. Aly, M.; Atiya, A. Labr: A large scale arabic book reviews dataset. In Proceedings of the 51st Annual Meeting of the Association for Computational Linguistics (Volume 2: Short Papers), Sofia, Bulgaria, 22–27 May 2013; pp. 494–498.
69. Elnagar, A.; Khalifa, Y.S.; Einea, A. Hotel Arabic-reviews dataset construction for sentiment analysis applications. In *Intelligent Natural Language Processing: Trends and Applications*; Springer: Berlin/Heidelberg, Germany, 2018; pp. 35–52.
70. Elnagar, A.; Lulu, L.; Einea, O. An annotated huge dataset for standard and colloquial arabic reviews for subjective sentiment analysis. *Procedia Comput. Sci.* **2018**, *142*, 182–189. [CrossRef]
71. Al-Khatib, A.; El-Beltagy, S.R. Emotional tone detection in arabic tweets. In *International Conference on Computational Linguistics and Intelligent Text Processing*; Springer: Berlin/Heidelberg, Germany, 2017; pp. 105–114.
72. Alhuzali, H.; Abdul-Mageed, M.; Ungar, L. Enabling Deep Learning of Emotion With First-Person Seed Expressions. In Proceedings of the Second Workshop on Computational Modeling of People's Opinions, Personality, and Emotions in Social Media, New Orleans, LA, USA, 6 June 2018; pp. 25–35. [CrossRef]
73. Alqahtani, G. Multimodal Emotion Detection of Social Networks Data Using Deep Learning. Master's Thesis, University of Patras, Patras, Greece, 2022.
74. Beltagy, I.; Lo, K.; Cohan, A. SciBERT: A pretrained language model for scientific text. *arXiv* **2019**, arXiv:1903.10676.
75. Alyami, S.N.; Olatunji, S.O. Application of Support Vector Machine for Arabic Sentiment Classification Using Twitter-Based Dataset. *J. Inf. Knowl. Manag.* **2020**, *19*, 1–13. [CrossRef]
76. Mohammed, A.; Kora, R. Deep learning approaches for Arabic sentiment analysis. *Soc. Netw. Anal. Min.* **2019**, *9*, 52. [CrossRef]
77. Abdulla, N.A.; Ahmed, N.A.; Shehab, M.A.; Al-Ayyoub, M. Arabic sentiment analysis: Lexicon-based and corpus-based. In Proceedings of the 2013 IEEE Jordan Conference on Applied Electrical Engineering and Computing Technologies (AEECT), Amman, Jordan, 3–5 December 2013; pp. 1–6.
78. ElSahar, H.; El-Beltagy, S.R. Building large arabic multi-domain resources for sentiment analysis. In *International Conference on Intelligent Text Processing and Computational Linguistics*; Springer: Berlin/Heidelberg, Germany, 2015; pp. 23–34.
79. Al-Moslmi, T.; Albared, M.; Al-Shabi, A.; Omar, N.; Abdullah, S. Arabic senti-lexicon: Constructing publicly available language resources for Arabic sentiment analysis. *J. Inf. Sci.* **2018**, *44*, 345–362. [CrossRef]
80. Hussien, W.A.; Tashtoush, Y.M.; Al-Ayyoub, M.; Al-Kabi, M.N. Are emoticons good enough to train emotion classifiers of arabic tweets? In Proceedings of the 2016 7th International Conference on Computer Science and Information Technology (CSIT), Amman, Jordan, 13–14 July 2016; pp. 1–6.

81. Sarbazi-Azad, S.; Akbari, A.; Khazeni, M. ExaAEC: A New Multi-label Emotion Classification Corpus in Arabic Tweets. In Proceedings of the 2021 11th International Conference on Computer Engineering and Knowledge (ICCKE), Mashhad, Iran, 28–29 October 2021; pp. 465–470.
82. Liaw, R.; Liang, E.; Nishihara, R.; Moritz, P.; Gonzalez, J.E.; Stoica, I. Tune: A research platform for distributed model selection and training. *arXiv* **2018**, arXiv:1807.05118.

Disclaimer/Publisher's Note: The statements, opinions and data contained in all publications are solely those of the individual author(s) and contributor(s) and not of MDPI and/or the editor(s). MDPI and/or the editor(s) disclaim responsibility for any injury to people or property resulting from any ideas, methods, instructions or products referred to in the content.

Article

Personality Types and Traits—Examining and Leveraging the Relationship between Different Personality Models for Mutual Prediction

Dušan Radisavljević [1,*], Rafal Rzepka [2] and Kenji Araki [2]

[1] Graduate School of Information Science and Technology, Hokkaido University, Nishi 9, Kita 14, Kita-ku, Sapporo 060-0814, Japan
[2] Faculty of Information Science and Technology, Hokkaido University, Nishi 9, Kita 14, Kita-ku, Sapporo 060-0814, Japan; rzepka@ist.hokudai.ac.jp (R.R.); araki@ist.hokudai.ac.jp (K.A.)
* Correspondence: radisavljevic@ist.hokudai.ac.jp; Tel.: +81-11-706-7389

Abstract: The popularity of social media services has led to an increase of personality-relevant data in online spaces. While the majority of people who use these services tend to express their personality through measures offered by the Myers–Briggs Type Indicator (MBTI), another personality model known as the Big Five has been a dominant paradigm in academic works that deal with personality research. In this paper, we seek to bridge the gap between the MBTI, Big Five and another personality model known as the Enneagram of Personality, with the goal of increasing the amount of resources for the Big Five model. We further explore the relationship that was previously reported between the MBTI types and certain Big Five traits as well as test for the presence of a similar relationship between Enneagram and Big Five measures. We propose a new method relying on psycholingusitc features selected based on their relationship with the MBTI model. This approach showed the best performance through our experiments and led to an increase of up to 3% in automatic personality recognition for Big Five traits on the per-trait level. Our detailed experimentation offers further insight into the nature of personality and into how well it translates between different personality models.

Keywords: personality; automatic personality recognition; psychometrics; psycholinguistic features; machine learning; personality computing

Citation: Radisavljević, D.; Rzepka, R.; Araki, K. Personality Types and Traits—Examining and Leveraging the Relationship between Different Personality Models for Mutual Prediction. *Appl. Sci.* **2023**, *13*, 4506. https://doi.org/10.3390/app13074506

Academic Editor: Yu-Dong Zhang

Received: 23 February 2023
Revised: 28 March 2023
Accepted: 30 March 2023
Published: 2 April 2023

Copyright: © 2023 by the authors. Licensee MDPI, Basel, Switzerland. This article is an open access article distributed under the terms and conditions of the Creative Commons Attribution (CC BY) license (https://creativecommons.org/licenses/by/4.0/).

1. Introduction

Personality is a popular concept that emerged in the early 20th century to explain the differences present on an individual level between people. Academics describe it as a set of distinct emotional, cognitive and behavioural patterns that differ from individual to individual but remain relatively consistent over time and context. As such, personality is essential to determining one's life and identity. It influences various life choices [1] and how one is perceived by others [2,3] and also has a significant impact on how one experiences the world around them [4].

The pivotal role of personality in human interaction has inspired various researchers to formulate theories and models that enhance our understanding of the concept [5]. While the theories tend to be more general in nature and cover the conceptual aspects of personality through how it works, *personality models* are what is often used in practice to describe one's personality and its effects on an individual.

The process of assessing ones personality through these models historically consisted of a psychological expert administering a test during an interview, with the test choice being dictated by the choice of model. Though reliable and effective, this approach requires expert knowledge and is usually time consuming. However, the combination of new machine-learning techniques and increase of online communication has led to interest from scholars into the possibility of automating the tasks of prediction, interpretation and generation of dimensions that personality models use.

Although the dimensions and the number of them differ on a model-to-model basis, it is possible to separate personality models into two groups based on the kind of value they assign along the said dimensions. These groups are:

- *Trait-based personality models*, which use traits or, in other words, assign a continuous value along the dimension.
- *Type-based personality models*, which rely on types to describe a personality. Types can also be viewed as categories or a form of a discrete value selected from a dimension's domain.

Examples of a type-based model would be the likes of *Myers–Briggs Type Indicator* (also known as the *MBTI*) and *Enneagram of Personality* (often referred to as simply *Enneagram*), while the *Big Five* model would be an example of a trait-based one.

1.1. Motivation

New computational algorithms and the development of modern technologies have led to the formation of *personality computing* —a research field at the intersection of personality psychology and artificial intelligence [6]. In addition to advancements in machine learning, another contributing factor to the rapid development of the field has been the increase in popularity of social media services, as this has encouraged people to share their interests publicly in online spaces [1].

However, while the amount of personality-related information increases daily, the need for more relevant personality-labelled data remains one of the most cited issues in the field [7]. This paradoxical phenomenon can be explained by the difference in personality model preference between academia and the non-psychological population that is more prevalent on social media. More specifically, while the Big Five model has seen extensive use in personality research, the MBTI has been the more popular choice for describing personality in online spaces.

The trait-based approach of the Big Five model as well as its empirical support, cross-cultural applicability and reliability [8–10] have caused it to be a popular choice in scientific circles, with a majority of research on personality computing centred around it. Several studies have additionally contributed to this preference by confirming its validity [11,12], while the MBTI has often been criticised for lacking this evidence [13–16]. If we consider the data that is publicly available through https://books.google.com/ngrams (accessed on 28 March 2023), we can note that, although both the Big Five and MBTI personality models have seen an increase in popularity within the last couple of decades, the "Big Five" *n*-gram appears significantly more frequently when it comes to book titles (shown in Figure 1).

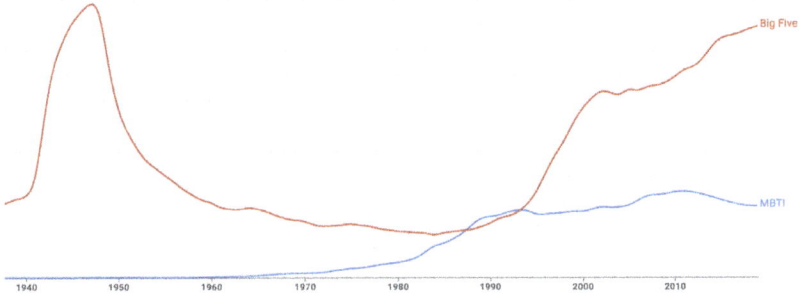

Figure 1. The frequency of personality models in book titles, showcasing interest in scientific circles for different personality models. The graph originates from https://books.google.com/ngrams (accessed on 28 March 2023). We note that the Big Five model is usually referred to as simply the "Big Five" in book titles.

On the other hand, if we use the frequency with which people search for a particular personality model using the Google (https://www.google.com/ (accessed on 28 March 2023)) search engine as an indicator of interest, we note that the preference seems to be different from that in academia. When observing the data available through the Google Trends API (https://trends.google.com/home (accessed on 28 March 2023)), we notice that the MBTI has drawn much more interest than the Big Five, especially in the last several years (depicted in Figure 2). This can be attributed to the fact that the way in which MBTI assigns personality (using a four-letter acronym) is easier to interpret and report for the non-psychological population, causing it to be more prevalent on social media platforms and, thus, attract greater attention.

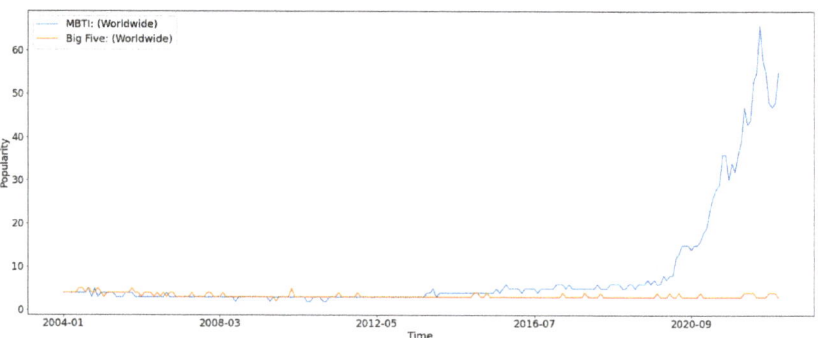

Figure 2. The frequency of personality models appearing as Google search terms, showcasing the interests of the general public. Data source: https://trends.google.com/trends/ (accessed on 28 March 2023).

Despite the differences between the two previously mentioned personality models, several studies [13,17,18] have pointed towards a statistically significant correlation between the MBTI types and certain traits belonging to the Big Five model. The existence of this relationship between the two raises the question of whether it can somehow be leveraged in overcoming each of their individual shortcomings. In our study, we aim to bridge the gap between these two personality models and, as a result, provide a significant increase in resources for the more scientifically accredited Big Five model.

1.2. Contribution

Our work further examines the relationship between the MBTI and the Big Five personality models. In addition, we look into the possible existence of a relationship between the Big Five and the Enneagram model, as well as the nature thereof. We conduct detailed experiments involving different sets of features and employ various regression algorithms, providing insight into their effectiveness on the task of personality prediction. Throughout this work, we adopt a comprehensive reporting style in order to assure replicability and better comparability with previous works, as the lack of comparison between studies has been a prevalent issue in personality computing [19]. The main goal of our experiments is to highlight the relationship between different personality models and further our understanding of personality. We can summarize our contributions as follows:

1. The series of detailed experiments that we conduct provides insight into the effectiveness of different features and regression algorithms for the task of personality prediction. Additionally, the choice of algorithms allows for greater interpretability of the results while maintaining a simplistic approach.
2. We propose a simple framework based on psycholinguistic features that leverage the relationship between different personality models. Our method led to an increase of 1.2 to 3.3% in the Pearson r correlation coefficient between predicted values and

the gold-standard labels when compared to the baseline approach on a dimension-to-dimension level.
3. Using psycholinguistic features helps further explore the relationship between language and how type-based personality shapes its use in online spaces. While the relationship between the Big Five and language use has been thoroughly studied due to the lexical background of the Big Five model [20], similar studies for the type-based models are limited to the best of our knowledge.

1.3. Paper Overview

The structure of this paper is as follows. Section 2 serves as a gentle introduction to the field by providing a brief history of personality research while also contextualizing the importance of our research. In Section 3, we introduce several closely related works. Section 4 describes our approach and provides a description of our experimental settings. In Section 5, we go over the results of our experiments, while in Section 6, we further discusses their implications as well as limitations. Finally, Section 7 summarizes our findings and concludes this paper.

2. Background on Personality Research

2.1. Brief History of Personality

Personality is a relatively new concept, dating back to the early 20th century. Despite this, it is speculated that the efforts to classify people based on their communication, behavioural and thinking patterns long predate written sources. The earliest known literary work that touches upon the topic of individual differences is a book called "Characters" [21] by the Greek philosopher Theophrastus. Dating back to the fourth century BC, it includes 30 short descriptions of different moral types, known as characters, that can be interpreted as prototypes of the modern personality types. However, it should be noted that some translators of the work have since noted that the word "trait", rather than "character", would be better suited as certain characteristics overlap between the descriptions.

Another early work that has proven important for the development of modern personality theories is that of physicians Hippocrates and Galen of Pergamon, which was later documented in Galen's book "De Temperamentis" [22]. Despite appearing roughly five centuries after "Characters", their work on the Four Humours theory has arguably had a greater influence on modern personality psychology and philosophy.

Hippocrates was first one to suggest that an imbalance in humour, or vital bodily fluids (from the Latin *humor*—meaning fluid), can influence behaviour. He described each humour as a combination of values along the two different pillars—dry/wet and hot/cold. For example, blood was considered to be hot and wet, while black bile was considered to be a combination of the cold and dry pillars. Following their work, Galen speculated the existence of a moderate value between the two pillars, combining the values along them to describe a total of nine different temperaments, four of which he considered to be primary [23]. These temperaments—namely, *sanguine*, *choleric*, *melancholic* and *phlegmatic*, have left an impact on both the English language and various works that involve personality.

The theory of the four temperaments formulated by Hippocrates and Galen has since been thoroughly explored by philosophers and psychologists in their theories that attempted to explain reasons behind individual differences. Prominent philosopher Immanuel Kant further explored the theory in their book "Anthropology From a Pragmatic Point of View" [24], arguing that, rather than nine, there are, in fact, only four temperaments. He described these temperaments as independent from one another, formulating a comprehensive list of traits that can be used in describing them. This approach can be seen as related to how the modern type-based personality models function.

In addition to inspiring Kant's work, the four temperaments theory has also drawn interest from Wilhelm Wundt—the man who is widely considered to be the father of experimental psychology. Wundt proposed that a two dimensional approach was sufficient to accurately describe personality. He introduced the dimensions of emotional *intensity* (strong–weak) and activity *changeability* (changeable–unchangeable) and expressed the four temperaments along the axes of these dimensions. Some authors have since pointed out that these interpretations of the four temperaments made by Kant and Wundt have an "uncanny" resemblance to *Neuroticism* and *Extroversion*—two dimensions that belong to the Big Five personality model [25].

The seminal contributions made by Immanuel Kant and Wilhelm Wundt can be seen as highly influential in the field of personality research [26], as they have helped to develop theories of several prominent researchers. The impact of their work is observed in personality theories proposed by Gordon Allport and Carl Gustav Jung, who, in turn, helped shape the modern Big Five and MBTI models, respectively. The concepts of introversion and extroversion proposed by Jung can be traced back to their study of Kant's work on morality. While in the case of Gordon Allport, the concept of a "cardinal trait" that is present in their work and refers to a single dominant trait that shapes the personality can be tied to the Kantian idea of self.

2.2. Personality Evaluation Tools

Through the history of personality psychology, several prominent theories and models have been used to describe personality [5,27–29]. However, in this subsection, we focus primarily on introducing the three personality models mentioned in the introductory section—the Big Five, Myers–Briggs Type Indicator and the Enneagram of Personality, all of which have been used in our experiments.

2.2.1. Big Five Model

The Big Five model or the Five Factor Model [30] has often been described as the "dominant paradigm" in the field of personality research and as "one of the most influential models" in psychology [31]. Its origins can be traced to a list of 4500 terms relating to personality traits that were introduced by Gordon Allport and Henry Odbert [32]. Using factor analysis, this list was initially reduced to 16 traits, only to be narrowed down to the final five from which the model received its name. As such, it is a result of contributions from many different authors, with roots in the English lexicon [20]. The five traits, or factors, that make up the Big Five model are most frequently labelled as:

1. *Openness*—measure of curiosity;
2. *Conscientiousness*—measure of efficiency;
3. *Extroversion*—measure of energy;
4. *Agreeableness*—measure of compassion;
5. *Neuroticism*—measure of sensitivity.

When using the Big Five model to measure personality, a continuous value is assigned to person along each of these traits (Figure 3) with the exact scale of these numbers largely depending on the test used [7]. An example of this would be a person scoring 85/100 along the *Openness* trait, which indicates they are less likely to be cautious when exploring new things.

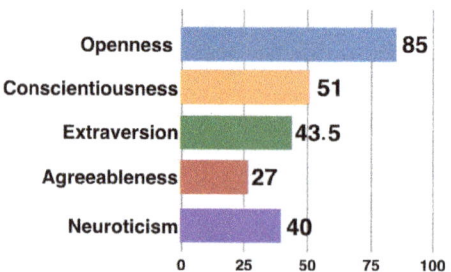

Figure 3. Graphical depiction of a sample Big Five test result.

2.2.2. Myers–Briggs Type Indicator

The *Myers–Briggs Type Indicator* [33], or *MBTI* for short, is another very popular personality evaluation tool. The Myers–Briggs Company (https://www.themyersbriggs.com/en-US/Connect-with-us/Blog/2018/October/MBTI-Facts--Common-Criticisms (accessed on 28 March 2023) , which is in charge of distributing the evaluation test, has stated that millions of people use the test annually in over 100 countries.

Unlike the Big Five model, MBTI focuses on personality types, rather than traits, and as such is based on the theoretical works of Carl Gustav Jung [34], who is often credited as the progenitor of personality types. Building on the three dichotomies that Jung originally proposed, namely Introverted/Extroverted, Sensing/Intuitive and Thinking/Feeling, Isabel Briggs-Myers and Katherine Cook Briggs later introduced the fourth dichotomy labelled as Judging/Perceiving, which finalized the initial MBTI model [35]. These dichotomies can be explained as measures in the following way:

1. *Extroverted/Introverted*—Describes how an individual gains energy. Often abbreviated as *E-I*;
2. *Sensing/Intuitive*—Describes how an individual gains information. Often abbreviated as *S-N*;
3. *Thinking/Feeling*—Describes how an individual makes decisions. Often abbreviated as *T-F*;
4. *Judging/Perceiving*—Describes how an individual observes the world around them. Often abbreviated as *J-P*.

Personality measured by the MBTI model is based on the idea that every individual has one pronounced value from the four dichotomies mentioned (Figure 4). For example, the type *INTJ* would refer to an individual who is *Introverted*, *iNtuitive*, *Thinking* and *Judging*. Thus, the MBTI offers a total of 16 different combinations that describe 16 unique personality types [36].

Figure 4. Example MBTI test results. The blue background indicates the dominant value of the dichotomy.

2.2.3. Enneagram of Personality

The *Enneagram of Personality* is another prominent personality model that uses types to measure individual differences [37]. The origins of the MBTI and the Big Five model can be traced to some of the early philosophical theories on individuality [26]; however, the exact origins of the Enneagram are disputed, with the Armenian philosopher George Gurdjieff often being credited with introducing it to the Western world [38].

The Enneagram is usually depicted as a circle with nine equidistant points that are connected with intersecting lines—the figure from which the model received its name. Personality types offered by the Enneagram are referred simply by a number from *One* to

Nine (e.g., Type *Eight*), with these types often being associated with different virtues, vices or ego fixations (depicted in Figure 5).

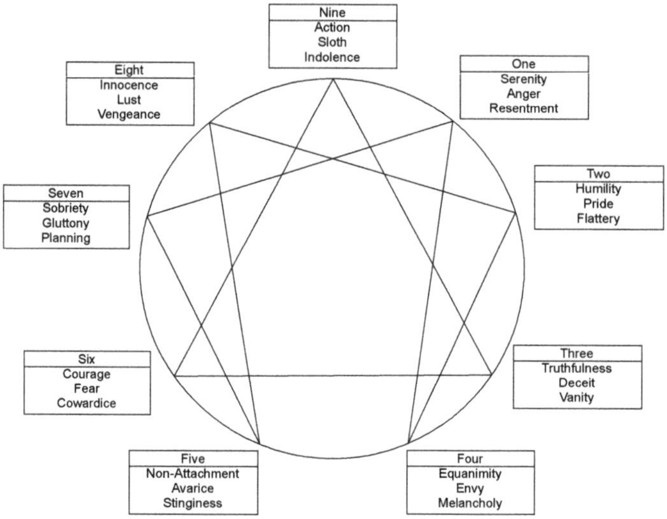

```
                    Nine
                   Action
                    Sloth
                  Indolence
      Eight                        One
    Innocence                   Serenity
      Lust                        Anger
    Vengeance                   Resentment

                                      Two
                                    Humility
     Seven                            Pride
    Sobriety                         Flattery
    Gluttony
    Planning


                                      Three
      Six                          Truthfulness
    Courage                           Deceit
     Fear                             Vanity
    Cowardice

             Five            Four
        Non-Attachment    Equanimity
           Avarice           Envy
          Stinginess      Melancholy
```

Figure 5. Enneagram types with their virtues, passions and ego-fixations as described by the Bolivian philosopher Óscar Ichazo.

2.3. Personality Computing

In recent years, several studies have pointed towards the existence of a relationship between personality and essential aspects of life, such as career selection and success [39–41], political participation and affiliation [42–44], religion [45], investment in social roles [46] and quality of life [47]. Due to the importance of personality and the rapid development of new technologies, Vinciarelli et al. [1] coined the term *personality computing* to describe the practice of applying machine-learning approaches to tasks involving personality [6].

The same authors described the main tasks of personality computing to be: (1) *Automatic Personality Recognition* (the identification of one's true personality from verbal or behavioural evidence), (2) *Automatic Personality Perception* (the recognition of personality that others might assign to an individual based on their behaviour) and (3) *Automatic Personality Synthesis* (the generation of artificial personalities). They concluded that any study that seeks to understand, predict or synthesise human behaviour could greatly benefit from personality computing approaches, closely relating the research field to affective computing [48] and its endeavors to build emotionally intelligent machines.

Although the main focus of our research is the translation from one personality model into another, it can best be described as a subtask of the automatic personality recognition task.

3. Related Research

Early research works involving personality centred around smaller sources of data, such as essays [49–51], emails [52] or blogs [53,54]. An important turning point for the personality computing field was the release of the *MyPersonality* dataset, which utilized information from the social media platform *Facebook* (https://www.facebook.com/ (accessed on 28 March 2023)). The dataset originated from the research of Kosinski et al. [55] and consisted of about 15.5 million Facebook statuses and some 7.5 million user profiles. As such, it represented the first publicly available large-scale dataset that included labels for the Big Five model. While *MyPersonality* has since been removed from the internet due to privacy concerns, with only a small portion of it being left accessible [56], it has inspired many future researchers in examining social media platforms as a potential source of data (Table 1) [6,57,58].

A very influential work in personality computing is that of Mairesse et al. [51]. Their work used the EAR [59] and Essays [49] datasets to test the effectiveness of different features using classification, regression and ranking models. When using a smaller set of data, they reported that simpler algorithms, such as Naive Bayes and regression trees offered better performance; however, ranking models achieved better scores for a larger dataset. While their approach closely related to our endeavours in analysing the effectiveness of different features, they focused on a single personality model. On the other hand, our work analyses the effectiveness of different features and how well they capture the relationship between multiple personality models, rather than focusing only on the Big Five model and language usage.

An interesting study comparing multiple personality models is that of Celli and Lepri [60], who compared the effectiveness of predicting labels for the MBTI and the Big Five model on a dataset originating from the social media platform Twitter (https://twitter.com/ (accessed on 28 March 2023)). They treated the problem as nine separate binary classification tasks, five of which were for classifying Big Five traits and four for the MBTI types. For the model architecture, they used a combination of n-gram, LIWC dimensions and metadata features for Support Vector Machine and another meta-classifier based on the work of Thornton et al. [61]. While their approach offered novel insight into the difference in effectiveness of the automatic personality recognition task for two different models, the relationship between models themselves was left unexplored.

Several notable contributions have been made on the automatic personality recognition task using deep-learning methods. Sun et al. [62] used a concatenation of bidirectional LSTMs and a convolutional neural network in order to predict personality traits from two Big Five personality datasets—the Essays dataset [49] and another one coming from the YouTube platform [63]. Kazameini et al. [64] also used the Essays dataset in their experiments; however, they adopted a multi-step approach that used a combination of Mairesse features [51] and BERT token representations [65] for prediction of the Big Five personality using a Support Vector Machine algorithm.

On the other hand, Kerz et al. [66] used a two-step approach that relied on BERT and BLSTM to predict Big Five personality traits from the Essays dataset, and MBTI types from the MBTI Kaggle dataset [67]. While there have been many more studies that used deep-learning methods and showed promising results [58,68,69], most of them tend to focus on a single personality model, most commonly the Big Five.

In addition, the vast majority of studies that utilize deep-learning methods treat the problem of Big Five trait prediction as a classification problem, rather than a regression one. The only exception is the previously mentioned work of Kerz et al. [66], which focused on multiple personality models. However, due to Big Five traits and MBTI types stemming from two different sources of data, it is impossible to obtain deeper insight into the relationship between two personality models.

Our approach is heavily inspired by the work of Gjurković et al. [7], who introduced a dataset containing labels for the MBTI, Big Five and the Enneagram model stemming from a single social media platform. They were the first to explore the possibility of using labels from one personality model as features in order to increase the scores of the automatic personality recognition task for another one. Their approach relied on a combination of MBTI/Enneagram predictions and a set of n-gram features to predict Big Five traits. In our work, we seek to extend their case study by taking a more detailed approach, similar to Mairesse et al. [51].

As with the works previously introduced in this section, we focus on the automatic personality recognition task. However, to do so, we seek to leverage the relationship present between multiple personality models and the way it is reflected through different features. The approach we take is detailed in nature for two reasons. The first is to assure comparability with the approach that we used as a baseline, avoiding some of the common issues in the personality computing research field [6,19]. The second reason is that, due to the complex nature of personality, gradual introduction and experimentation with different features is the best way to accurately single out the effects they have on the task of personality recognition [51].

Table 1. A non-exhaustive overview of various personality computing works that utilize social media platforms as a source of data. Different to our work, most of the listed research tends to focus on a single personality model, most often the Big Five model.

Social Media Platform	Authors	Modality	Personality Model
Facebook	Schwartz et al. [70]	Text	Big Five
	Farnadi et al. [71]	Text	Big Five
	Verhoeven et al. [72]	Text	Big Five
	Celli et al. [73]	Text	Big Five
	Park et al. [74]	Text	Big Five
	Youyou et al. [75]	Text	Big Five
	Segalin et al. [76]	Images	Big Five
	Tandera et al. [77]	Multimodal	Big Five
	Kulkarni et al. [78]	Text	Big Five
	Ramos et al. [79]	Text	Big Five
	Xue et al. [80]	Text	Big Five
	Marengo et al. [81]	Text	Big Five
Flickr	Cristani et al. [82]	Images	Big Five
Instagram	Osterholz et al. [83]	Images	Big Five
Reddit	Gjurković and Šnajder [84]	Text	MBTI
	Wu et al. [85]	Text	MBTI
	Gjurković et al. [7]	Text	Big Five, Enne. and MBTI
	Radisavljević et al. [86]	Text	Big Five and MBTI
Sina Weibo	Zhou et al. [87]	Text	Big Five (Extroversion)
TikTok	Meng and Leung [88]	Multimodal	Big Five
Twitter	Plank and Hovy [89]	Text	MBTI
	Verhoeven et al. [90]	Text	MBTI
	Tighe and Cheng [91]	Text	Big Five
	Celli and Lepri [60]	Text	Big Five and MBTI
	Balakrishnan et al. [92]	Text	Big Five and Dark Triad [1]
	Cahyani and Faishal [93]	Text	Big Five
Youtube	Biel and Gatica–Perez [63]	Video	Big Five
	Bassignana et al. [94]	Text	MBTI

[1] Dark Triad is a group of three traits associated with negative behaviour. These traits are largely independent from the traits measured by the Big Five model.

4. Materials and Methods

When looking at tasks that use regression approaches, there are several different methods that can be utilized for improving results over the baseline. Some of the more common examples involve different regularization methods or data manipulation in the form of data augmentation or even data cleanup.

However, as the primary focus of our work is to improve our understanding of personality and the way it is reflected across different personality models, we decided to use two methods that are more linear in nature—(1) feature selection and (2) model selection. The exact design choices behind these methods will be further discussed in Sections 4.2 and 4.3, respectively. To better contextualize the choice we made, we start this section by briefly introducing the dataset used, as well as the approach that serves as a baseline for our experiments.

4.1. Pandora Dataset and the Baseline Approach

The data used in our experiments stems from the social media platform Reddit (https://www.reddit.com/ (accessed on 28 March 2023)), and is the contribution of Gjurković et al. [7]. As such, it represents a direct extension of their previous work that introduced the MBTI9k [84], another dataset containing MBTI labels. We decided that the PANDORA (Personality ANd Demographics Of Reddit Authors) dataset was suitable for our experiments due to the following reasons:

1. Most of the work done in personality computing that focuses on using data for different personality models utilizes datasets coming from separate sources, with some examples being the works of Mehta et al. [95] and Kerz et al. [66]. On the other hand, in the work of some authors, while the data originates from the same source, it contains no overlap between users labelled with different personality models (e.g., the work of Celli and Lepri [60]). To the best of our knowledge, the PANDORA dataset is the **only dataset that contains personality-relevant information for multiple personality models, with an overlap between user groups labelled with each of these models**.
2. The topical diversity of Reddit opens up the possibility of looking into the effects that interests and hobbies have on personality prediction. As a social media platform, Reddit is divided into a series of different "subreddits"—or smaller message boards. These message boards are often centred around a single topic or interest that individuals participating on them tend to share. **It is possible that information on these topical interests can be leveraged to improve the results of the personality prediction task**.

The PANDORA dataset consists of 10,288 users that have labels for either the MBTI, Big Five or Enneagram personality models, with some users being labelled for more than one personality model. Additionally, some users also have their demographic information, such as gender or age stated. The dataset also includes 17,640,062 comments written by users in the period between the January 2015 and May 2019. Table 2 includes an overview of the dataset, providing insight into the exact number of users and comments that contain labels for either MBTI, Big Five or Enneagram personality models, as well as information on the amount of overlap between these groups.

Table 2. The number of users and comments labelled with each of the personality models in the PANDORA dataset. The data in this table were adapted from the work of Gjurković et al. [7] CC-BY-NC.

Personality Model	Number of Users	Number of Comments
Big Five	1608	3,006,566
Enneagram	794	1,458,816
MBTI	9084	15,597,237
Big Five and Enneagram	64	235,883
Big Five and MBTI	393	1,086,324
Enneagram and MBTI	793	1,457,625
All three models	63	234,692
Total	10,288	17,640,062

Observing the the way data are distributed, we note that none of the personality traits appear to follow a normal distribution (Figure 6). While this is not particularly unusual, it should be noted that most previous work reported a tendency towards normal distribution for the Big Five traits (e.g., the works of Mairesse et al. [51] or work of Uysal and Pohlmeier [96]). In the PANDORA dataset, however, most of the personality traits follow a skewed distribution, with the only exception being the *Neuroticism* trait for which the labels seem to follow a bimodal distribution.

This phenomenon can be attributed to several different reasons, such as selection bias [84] or the propensity of openly stating personality traits being dictated by certain personality traits, e.g., high *Openness*. An additional possibility is that certain subreddits covering certain topics or interests tend to be more prevalent in the dataset, and thus the number of individuals with particular personality traits associated with such interests and topics tend to be more numerous. To test this possibility, we conducted an experiment into the effect that subreddit participation has on personality predictions, and the details of this study are described in Section 4.2.3.

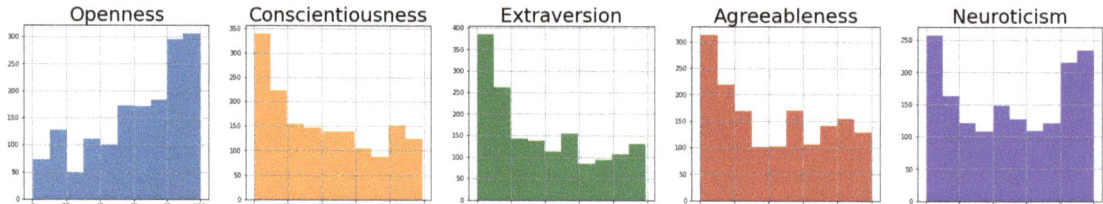

Figure 6. Probability distribution for each of the Big Five personality trait labels present in the PANDORA dataset. While the previous research has suggested Big Five traits usually follow a normal distribution, labels present in the PANDORA dataset [7] seem to follow a more skewed one.

While the data distribution of personality traits present in the dataset seems a bit unusual, it should be noted that the correlations between the Big Five traits and the MBTI types reported in data are largely in agreement with research that has previously examined relationship between these two sets of dimensions [13,18]. The only exception would be the *Openness* personality trait that in the case of PANDORA shows an unusually low correlation with the *S-N* type, despite previous works in the field reporting an agreement between these two dimensions that is higher than chance (Figure 7).

The approach used as a baseline in our work was described by the authors of PANDORA as a domain-adaptation task [97] of transferring the MBTI and Enneagram labels into the more scientifically relevant Big Five ones [7]. To this end, they started by training four logistic regression models [98]—one for each of the MBTI types, and an additional one for the types present in the Enneagram. In order to train these models, they utilized a subset of users that included the MBTI/Enneagram labels only—in other words, a subset that had no overlap with users labelled with the Big Five personality model. The set of users with labels for both the MBTI/Enneagram models and the Big Five were later used as part of a validation set.

The labels obtained from these five regression models were then used to predict MBTI/Enneagram values for the set of users without assigned labels for these type-based models. The predictions were then either used independently or combined with other features (e.g., gender, POS tags, stylistic features and named entities in text) and n-grams into a single feature set in order to predict the Big Five personality traits. The experiments were conducted using two different algorithms: (1) a linear regression model with a L2 regularization norm [99] (also known as Ridge regression) and (2) a neural network with the L2 regularization norm BERT [65] for textual encoding. The end results indicated better performance for the linear regression model, with about 15% higher results compared with the deep-learning model.

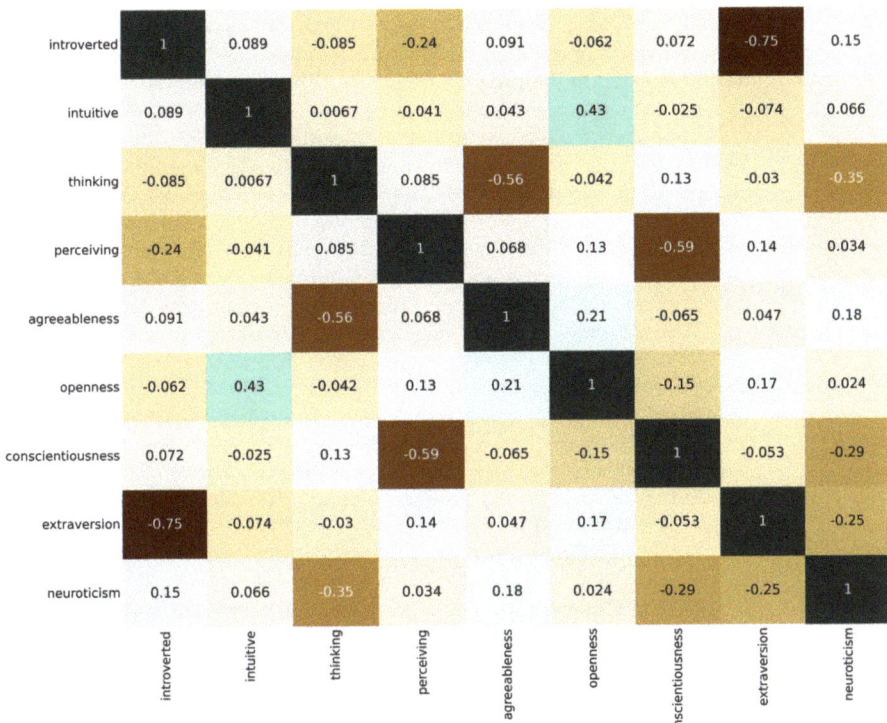

Figure 7. Heatmap indicating linear correlation between labels for the Big Five traits and MBTI types present in the PANDORA dataset [7].

4.2. Feature Selection Approach

The first step of our experiments focused on finding the optimal set of features for the task of automatic personality recognition. These features were then combined with the predictions of the type-based personality labels in an effort to leverage a relationship between them and the Big Five model. We theorize that the following three feature sources can benefit the results and lead to possible improvements over the baseline approach:

1. *Class predictions for the Big Five personality traits*—a set of features obtained from predicting Big Five labels as classes rather than values (e.g., "High Extroversion" instead of 74% Extroversion) by applying a technique known as *binning*. These predictions were then used in combination with other features to predict the Big Five personality traits.
2. *Language-based features originating from Linguistic Inquiry and Word Count (LIWC)*—a set of psycholinguistic features produced as a result of statistical analysis conducted by LIWC.
3. *Information about user participation and engagement on the social media platform Reddit*—a feature set constructed from frequency with which users post on the most and least popular message boards present in the dataset.

4.2.1. Big Five Classification Predictions as Features

Our first hypothesis is based on the idea that the difference in the domain between personality models can ultimately impede the prediction results. While Enneagram types are represented using a whole number on a scale from 1 to 9, the MBTI types can be described using a binary value of either 0 or 1 in order to aptly represent each of the four dichotomies. On the other hand, Big Five personality traits are labelled using a positive number within the range of 0 to 100, with some labels even being represented as a single precision decimal number. It is possible to minimize these differences by introducing an additional step to the prediction process, which would treat the task as a classification problem rather than a regression one.

To convert the Big Five labels from continuous values into discrete ones, we applied binning as described by Segalin et al. [76], with slight modifications. In their work, the authors used two different techniques in order to separate Big Five traits into binary classes—(1) utilizing the mean value of the particular personality trait and (2) by using the first and third quartiles of the distribution as class delimiters and discarding all values between the them. As the data distributions of the PANDORA dataset and the data used by Segalin et al. [76] differ, we introduced slight adjustments to the approaches they used.

When it comes to technique (1), instead of relying on the mean value, we decided to use the median point of the personality trait distribution as a separator between classes. The reasoning behind this is that the median values tend to be more resilient to skewed data distributions, thus, making it a better fit for the PANDORA dataset (Figure 6). For technique (2), we decided against discarding any non-extreme value, and instead binned the Big Five traits into three classes rather than two. In doing so, we prevented any loss of information since the Big Five personality labels present in the PANDORA dataset are relatively smaller in size when compared to MBTI (Table 2).

Despite the recent success of different deep-learning approaches in predicting Big Five traits as classes [64,100,101], we decided to use the same regression algorithm as in the case of predicting MBTI and Enneagram types, so to allow for better comparability. The features used for this task include n-grams and MBTI/Enneagram predictions as described in Gjurković et al. [7]. These predictions were then later used for the regression model with Figure 8 illustrating the steps taken in predicting the continuous values for the Big Five personality traits.

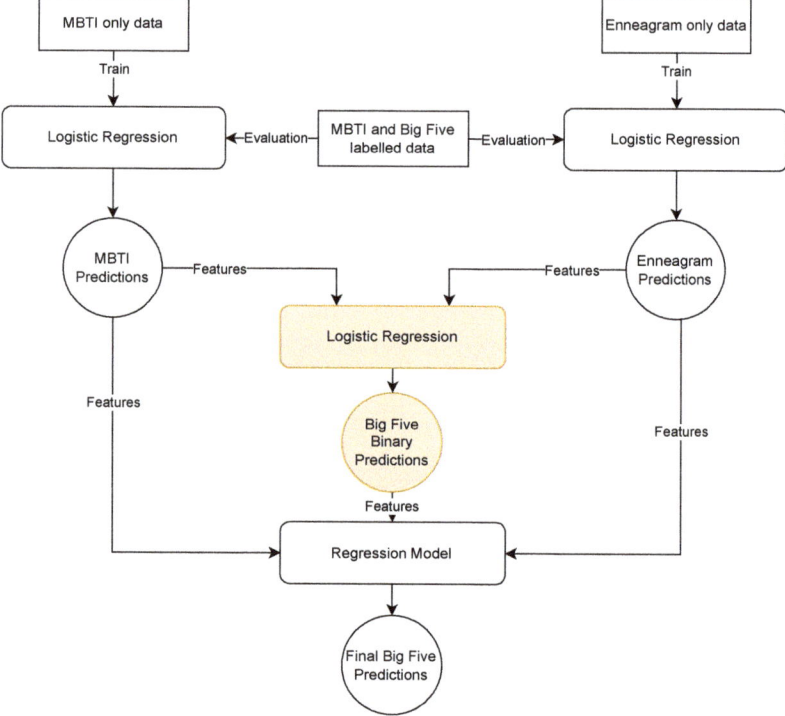

Figure 8. Illustration of the model stack after introducing the Big Five classification predictions.

Orange highlights the added logistic regression model that predicts Big Five traits as one of either two or three classes, depending on if the median or quartile values are used as separators between classes. The newly added model acts as a weak learner in the model stack.

4.2.2. Language-Based Features

Boyd and Pennebaker [102] stated language to be one of the most important indicators of personality. As such, many linguistic and psycholinguistic features have seen extensive use in the field of personality computing [103]. These language-based features often tend to be researched from the aspect of their relationship to the Big Five model (with some examples being the works of Mairesse et al. [51] and Holtgraves [104]). However, these features have rarely been used to connect the relationship between multiple personality models, with only few works attempting to do so [66,86]. This can be explained by the fact that, while the Big Five model is a direct result of a statistical analysis of the English lexicon [20], the MBTI and Enneagram do not share a similar lexical background.

To further examine the relationship between these personality models and language, we rely on the Linguistic Inquiry and Word Count, also known as LIWC (a full list of dimensions and their overview can be found in the https://www.liwc.app/help/psychometrics-manuals (accessed on 28 March 2023) [105], which has been a popular tool for analysis of the ways in which language is used. LIWC utilizes over 100 internal dictionaries that test for the presence of various linguistic features, capturing the social and psychological states people express through language.

Each of these dictionaries consists of words, word stems, emoticons and other text features that help to better identify the psychological category of interest from the textual data. For example, the "affiliation" dictionary comes with some 350 entries among which are words, such as "community", "together" and other verbal constructs indicating a person's desire to connect with others. Using these dictionaries, LIWC compares the words present in the provided text with the list of words contained in these internal dictionaries, thus, calculating the percentage of match for each of its dimensions.

It is important to note that, while a high number of features returned by LIWC can be considered psycholinguistic, as they are used to capture emotional and psychological states and processes (e.g., "posemo" for positive emotions or "anxiety" for anxiety-related words), not all of them fall into this category. Certain LIWC features, such as word count or function words, can be viewed as purely linguistic when used in isolation. However, if these features are paired with others in a set, they can be also considered psycholinguistic. For example, a higher word count when paired with anxiety words can be indicative of a certain emotional state. As we focus on using multiple LIWC features in combination for our experiments, we refer to them as psycholinguistic, thus, accenting their role as indicators of one's psychological state.

Due to the high quantity of LIWC features being present, we later performed a feature selection approach based on their relationship with the type-based personality models. By doing this, we sought to optimize the approach to our experiments and to avoid potential noise in the set of psycholinguistic features.

4.2.3. Subreddit Participation as Features

Similarly to shaping our behavioural and communication patterns in face to face interactions, personality can also be reflected in our interactions in online spaces [106]. As such, we found it interesting to examine the possibility of participation on particular subreddits influencing the results of personality prediction. As subreddits are often grouped around a single interest point, this experiment can be seen as an examination into the effect that personality has on interests in particular topics.

When observing all of the subreddits individually, connecting interests to particular personalities seems to be a difficult task. This is due to the large number of Reddit participants that frequent each subreddit but have not disclosed any personality-relevant data. In an effort to avoid this problem, we focused primarily on measuring the frequency of participation on different subreddits through measuring the number of users for each subreddit and the amount of messages posted on them through time.

In order to obtain detailed Reddit information, we used the *PushShift Reddit dataset* [107]. Through this, we collected participation statistics for different subreddits in the span of time between the chronologically first and last comments present in the PANDORA dataset. After that, we selected the 50 most popular subreddits for the construction of a feature vector through using information, such as the amount of comments posted on the subreddit and the number of participating users for the observed time period. Subsequently, these feature vectors were then normalized in an effort to effectively use them for the linear regression models.

4.3. Model Selection Approach

After examining the effects that each of the feature sets have on personality prediction, we conducted experiments applying several different algorithms for this task. For the sake of maintaining the comparability with the baseline approach, we tested several different linear regression models. Additionally, we conducted experiments using a deep-learning model (KerasReressor) and an ensemble-learning approach (XGBoost).

4.3.1. Lasso Regression

While the baseline approach relied on the linear regression model implementing the L2 regularization norm, due to the large number n-gram features present in the feature set, it is possible that a model implementing L1 regularization can yield better results [108]. While L2 regularization introduces the squared magnitude of coefficients as a penalty function, the L1 uses the absolute value—making it more robust to outliers. As a result, the L1 regularization can impact and potentially eliminate some less important features from the numerous n-grams used in the feature set. A regression model that uses the L1 regularization norm is also known as Lasso regression (**L**east **A**bsolute **S**hrinkage and **S**election **O**perator).

The difference between these two models can be mathematically formulated in the following way. If we have m features and n observations in our data, with $x_{i,j}$, we can mark the j-th feature of the i-th observation. Next, if we use w to represent the weight of our function, for the i-th feature of the observation that we are interested in predicting, (which we mark with y), the basic regression formula can be written as Equation (1). For the L2 regularization with the regularization parameter $\alpha \epsilon [0,1]$ that we multiply with the sum of squared weights w, the regression formula takes the form of Equation (2). Finally, if we take the absolute value of weights instead of squaring them, we obtain the L1 regularization as shown in Equation (3).

$$\sum_{i=1}^{n}(y_i - \sum_{j=1}^{m} x_{i,j}w_j)^2 \tag{1}$$

$$\sum_{i=1}^{n}(y_i - \sum_{j=1}^{m} x_{i,j}w_j)^2 + \alpha \sum_{j=1}^{m} w_j^2 \tag{2}$$

$$\sum_{i=1}^{n}(y_i - \sum_{j=1}^{m} x_{i,j}w_j)^2 + \alpha \sum_{j=1}^{m} |w_j| \tag{3}$$

4.3.2. Elastic-Net

While the elimination of less important features can prove to be beneficial, it is possible that taking a more moderate approach can lead to even better results. While L1 regularization tends to be more strict by removing features, L2 only minimizes their effect. A balanced combination of the two regularization norms can prove to be beneficial in improving predictions, as it combines the best aspects of both the Ridge and Lasso regression models. The algorithm that relies on both the L1 and L2 norm is known as Elastic-Net and can be mathematically formulated in the following way:

$$\sum_{i=1}^{n}(y_i - \sum_{j=1}^{m} x_{i,j} w_j)^2 + \alpha_1 \sum_{j=1}^{m} w_j^2 + \alpha_2 \sum_{j=1}^{m} |w_j| \qquad (4)$$

Our theory is that the combination of two regularizations can yield better results as it would simultaneously minimize the effect of the outliers on the prediction while also preserving those features that could potentially capture the intricate nature and finer differences between personality traits.

4.3.3. Huber Regressor

Lasso, Ridge and Elastic-Net all rely on the ordinary least squares formula for their loss function. One problem with this is that outliers often have too much influence on the predictions. This is true for both the models that implement the L1 and the L2 regularization norms, despite the fact that the L1 norm utilized the median as the central value for the sake of minimizing this effect. There are several different regression approaches that offer the complete elimination of outliers, with one example being RANSAC (RANdom SAmple Consensus).

However, due to the size of the data, as well as the nature of the task, we propose that it is best to minimize the effect of outliers rather than to completely eliminate them. For this reason, we decided to experiment with the Huber regressor, which is available through Pythons Sklearn package (https://scikit-learn.org/stable/modules/generated/sklearn.linear_model.HuberRegressor.html (accessed on 28 March 2023)). The Huber regressor, much like the Ridge regression model, implements L2 regularization.

However, it does so by using M-estimators [109] rather than mean of the distribution as its central value, thus, making it more resistant to outliers. We suggest that, due to this property, it will result in slightly better predictions compared with the baseline approach. The loss function of the Huber regressor can be mathematically formulated in the following way:

$$\min_{w,\sigma} \sum_{i=1}^{n} \left(\sigma + H_\epsilon \left(\frac{x_{i,j} w - y_i}{\sigma} \right) \sigma \right) + \alpha \sum_{j=1}^{m} w_j^2 \qquad (5)$$

where $H_\epsilon(z)$ takes the values of:

$$H_\epsilon(z) = \begin{cases} z^2, & \text{if } |z| < \epsilon, \\ 2\epsilon|z| - \epsilon^2, & \text{otherwise} \end{cases} \qquad (6)$$

4.3.4. Epsilon-Support Vector Regression—SVR

Similarly to Huber regressor, the Epsilon-Support Vector Regression, or SVR, has shown good resistance to outliers. Based on a classification algorithm of Support Vector Machines (SVM), SVR uses a kernel trick to perform regression in higher dimensions. As a result, SVR tends to generalize well without its computational complexity depending on dimensionality of the problem [110]. This generalization is mainly the result of SVR using an ϵ-insensitive region (also known as an ϵ-tube), that is often used to better approximate functions that have continuous values. With this property in mind, as well as the fact that SVR is known to perform well on smaller sets of data, our theory is that the overall prediction scores can be improved by applying the SVR algorithm in our experiments.

4.3.5. Keras Regressor

Deep-learning methods have recently shown promising results in the field of automatic personality recognition [58,62,80]. While there have been several different architectures for deep-learning models that achieved promising results, we decided to focus on KerasRegressor—a part of the Keras library.

Keras (https://keras.io/ (accessed on 28 March 2023)) is a high-level library for deep learning in the Python programming language that allowing for the easy and efficient construction of neural networks. As a part, KerasRegressor represents a deep-learning model trained to predict continuous values, such as stock prices and weather conditions. In our work, we experimented with several different architectures for the KerasRegressor model, subsequently selecting the best performing one.

The model consists of four fully connected layers, with the input shaped to match the data. We use a truncated normal kernel initializer and ReLU activation function with Adam functioning as an optimizer. The model is compiled using the root mean squared error as a loss function and trained over 30 epochs with a batch size of 32 to accommodate for the size of the data.

4.3.6. Boosting Algorithms

Boosting algorithms are a useful option when working with weak estimators. Boosting hierarchically builds a model, attempting to minimize the error over time. The three most popular ensemble learning models that implement gradient boosting algorithms are XGBoost (https://xgboost.readthedocs.io/en/stable/ (accessed on 28 March 2023)), LightGBM (https://lightgbm.readthedocs.io/en/v3.3.2/ (accessed on 28 March 2023)) and Catboost (https://catboost.ai/ (accessed on 28 March 2023)). While the first two algorithms utilize asymmetrical trees—with XGBoost growing vertically and LightGBM horizontally, Catboost relies on symmetrical trees.

These algorithms have performed well on many different prediction tasks so far. For our task, we selected XGBoost due to it relying on asymmetrical trees that expand level-wise rather than leaf-wise, as well as the splitting method it uses. Additionally, recent works in personality computing have reported promising results when using XGBoost for prediction of the MBTI personality types [111]. However, it should be noted that boosting algorithms are not advised for smaller sets of data or in cases where the features outnumber the data samples as this can lead to overfitting.

These factors can potentially pose an issue in the case of the PANDORA dataset, as the amount of Big Five labels is relatively small. In order to minimize the risk of overfitting, we used 100 early stopping epochs [112] as well as performed five-fold cross validation during the training process.

Since XGBoost uses many different hyperparameters, it is difficult to tell which combination would lead to most optimal results. For this purpose, we relied on the Optuna (https://github.com/optuna/optuna (accessed on 28 March 2023)) package for searching the hyperspace, in an effort to find the best possible combination of parameters for our experiments. Table 3 lists the parameters and their values as calculated by the optimization package used in our work.

Table 3. Hyperparameters used by the XGBoost model for our experiments.

Parameter Name	Parameter Value
Number of estimators	10,000
Learning rate	0.002
Maximum depth of a tree	3
L1 regularization term	5.25
L2 regularization term	34.85
Subsample of columns	0.1
Subsample of training instances	0.7
Gamma	0

4.4. Ethical Approach to Personality Research

Before going over the results of our experiments, it is important to address the issue of ethics in personality computing. While the field itself has been rapidly developing [6], one of the most frequently cited reasons for the lack of easily accessible personality-relevant data has been privacy concerns. In fact, it is due to these reasons that the work of Kosinski et al. [55] on the *MyPersonality* dataset had to be removed from the internet.

A report by Fang et al. [19] stated that only about 10% of research papers reflect on the ethics and fairness of research into personality. Due to this, we suggest that this poses an issue of utmost importance for the research field. The improper handling of private data can lead to personal information being used in unintended and harmful ways, such as profiling and targeting individuals with particular services or advertisements.

In order to assure ethical research in our study, we made sure to comply with the guidelines specified by the Reddit social media platform. Additionally, our study complies with the set of rules specified by the authors of the PANDORA dataset (https://psy.takelab.fer.hr/datasets/all/pandora/ (accessed on 28 March 2023)) [7]. Due to this, we made sure to remove data from any user whose information can no longer be publicly accessed through Reddit. Additionally, we report all findings of our research on the aggregate level only, assuring the protection of privacy for the participants.

5. Results

In Section 4, we detailed several approaches that rely on different features and algorithms. Due to the large quantity of these approaches and for the sake of providing a detailed and structured comparison between the results of our experiments, we separate this chapter into two subsections, with Figure 9 providing a general overview of the flow of our experiments. In the first subsection, we focus on listing the results achieved through the feature selection approach, while the second part reports the results for each of the different regression algorithms applied. Additionally, we provide a more streamlined and concise overview of our findings in (Section 6).

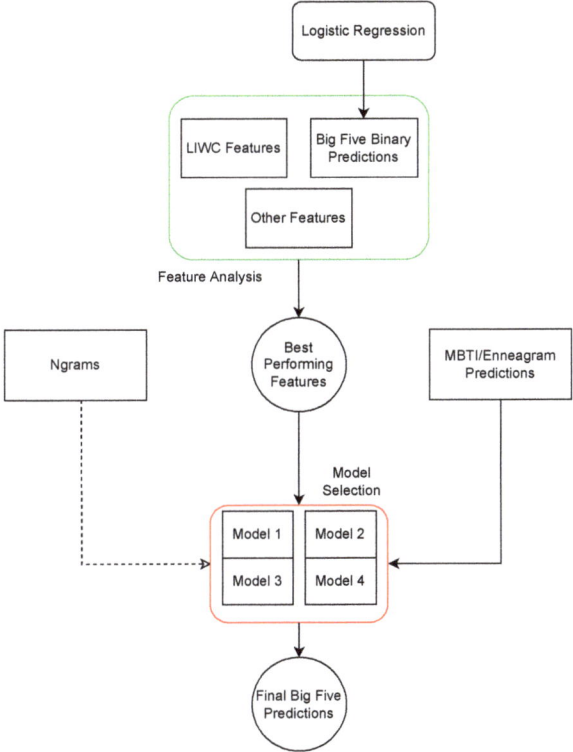

Figure 9. Simplified illustration of the pipeline behind our method. The green rectangle depicts feature selection approaches, while the red one highlights the model selection step.

Before detailing the results of our experiments, we briefly go over the results of the baseline approach as well as the evaluation criteria. Gjurković et al. [7] were the first to test the hypothesis of using MBTI and Enneagram predictions to successfully predict Big Five labels. In their experiments, they used several different feature sets, such as n-grams and MBTI/Enneagram predictions that were the result of logistic regression models. These features were later combined with regression models to acquire the predicted Big Five labels. In Table 4, we present the correlations between the predicted MBTI types and Big Five traits with the ground truth labels present in the PANDORA dataset as reported by Gjurković et al. [7].

Their best performing model was a L2 regularized linear regression model that used a combination of n-grams and predictions of the MBTI/Enneagram labels as features, yielding the best results for nearly all of the Big Five personality traits. The only exception was the *Openness* trait, which demonstrated better performances when the same regression model was used, only without the MBTI/Enneagram predictions in the feature set. The metric used to evaluate the performance of these models is known as the Pearson correlation coefficient [113]. The results of their experiments are reported in Table 5 for the sake of providing a reference when comparing the results with those achieved through experiments.

While Gjurković et al. chose the Pearson correlation coefficient for their evaluation metric, it is essential to note that various metrics have been used in personality computing to evaluate the performances of different models [6]. When focusing strictly on regression problems, the Pearson r correlation is not an uncommon metric; however, in the cases where the data do not follow a normal distribution, Spearman's rank correlation coefficient might be a slightly better choice. Additionally, it was proposed by Fang et al. [19] to use more than a single metric when evaluating personality-recognition approaches.

For example, using the mean squared error (MSE) in addition to correlation metric would reveal the absolute difference between predictions and how the model follows the trend. In our work, we chose to focus on exploring the effectiveness of different features and algorithms for personality prediction, and how they capture the relationship between multiple models. For the sake of allowing direct comparisons with previous work, we decided to use only a single metric that has been used, which is the Pearson r correlation coefficient.

Table 4. The results of the baseline approach. The Pearson correlation coefficient scores were adapted from the work of Gjurković et al. [7] CC-BY-NC.

	Openness	Conscient.	Extroversion	Agreeableness	Neuroticism
Pearson corr.	0.250	**0.273**	0.387	**0.270**	**0.283**

Bolded numbers represent the best result reported by Gjurković et al. [7].

Table 5. The Pearson correlation coefficient between the gold-standard Big Five labels and the predicted values of MBTI types and Big Five traits. Correlations adapted from the work of Gjurković et al. [7] CC-BY-NC.

	Openness	Conscient.	Extroversion	Agreeableness	Neuroticism
Predicted…					
Introverted	−0.082	0.039	−0.262	−0.003	−0.002
Intuitive	0.127	−0.021	0.049	0.060	0.001
Thinking	−0.001	0.038	−0.039	−0.259	−0.172
Perceiving	0.018	−0.241	0.007	0.034	0.039
Predicted…					
Openness	0.147	−0.082	0.212	0.145	0.070
Conscient.	−0.007	0.237	0.013	−0.112	−0.090
Extroversion	0.098	−0.028	0.272	0.044	0.022
Agreeableness	0.006	−0.079	0.023	0.264	0.176
Neuroticism	−0.048	−0.025	−0.042	0.231	0.162

Underlined numbers indicate significant correlation ($p < 0.05$).

For the sake of readability, when reporting correlation scores for the four MBTI dichotomies, we report a score for only a single value out of the two in a dichotomy. The reason behind this decision can be explained by the fact that a score for the other value would be equal to the same number being multiplied by -1, as each value represents an antipodal point of the other. For example, if the *Introverted* value of the E-I type shows correlation of 0.125, the correlation of the *Extroverted* value would be -0.125.

5.1. Feature Analysis

5.1.1. Big Five Classification Predictions—Median Split

When observing the results of the Big Five predictions achieved through the classification method, we notice that the overall correlation coefficients, in fact, decreased in comparison to the baseline approach (Table 6). While the correlations between the predicted Big Five traits treated as classes and the actual Big Five labels seem to be comparable to the results previously reported in the work of Gjurković et al. [7], the predictions of every single personality trait decreased, with the only exception being the *Openness* trait.

The likely reason for this is due to the predicted values made by the classification model of the *Openness* trait being more statistically independent from other personality traits, when compared to the statistical correlation exhibited by predictions made by the regression model (Table 7).

Table 6. Pearson correlation coefficient between the actual Big Five traits and the predictions achieved through usage of n-grams, MBTI/Enneagram predictions and median-split Big Five predictions as features.

Features	Openness	Conscient.	Extroversion	Agreeableness	Neuroticism
n-grams + Median Split Preds.	**0.260**	0.184	0.336	0.246	0.257
n-grams + Median + Other Preds.	**0.270**	0.225	0.375	0.263	0.255

Bold numbers mark a result that is outperforming the baseline.

Table 7. Correlations between the gold-standard Big Five labels and predictions that use median-delimited Big Five categories as features.

	Openness	Conscient.	Extroversion	Agreeableness	Neuroticism
Median Preds.					
Openness	0.198	−0.053	<u>0.121</u>	0.005	−0.036
Conscient.	−0.078	<u>0.227</u>	−0.039	−0.037	<u>−0.061</u>
Extroversion	<u>0.145</u>	−0.043	<u>0.329</u>	0.012	<u>−0.105</u>
Agreeableness	0.029	0.031	−0.001	<u>0.236</u>	<u>0.171</u>
Neuroticism	−0.008	−0.023	−0.028	<u>0.160</u>	<u>0.246</u>

<u>Underlined</u> numbers indicate significant correlation ($p < 0.05$).

5.1.2. Big Five Classification Predictions—Quartile Split

When treating the prediction of Big Five traits as a three-class classification problem rather than a two-class one, we note that the correlations between these features and actual Big Five labels were worse than when the median value was used as suggested by the results reported in Table 8. While *Openness* remains the only personality trait to see improvements over the baseline approach, this is only when MBTI/Enneagram predictions are included in the feature set (Table 9).

Table 8. Correlations between the gold-standard Big Five labels and predictions that use the quartile-delimited Big Five categories as features.

	Openness	Conscient.	Extroversion	Agreeableness	Neuroticism
Quartile Preds.					
Openness	<u>0.235</u>	−0.025	<u>0.160</u>	−0.011	−0.018
Conscient.	<u>−0.071</u>	<u>0.251</u>	0.010	−0.039	<u>−0.078</u>
Extroversion	<u>0.171</u>	<u>0.066</u>	<u>0.350</u>	0.039	−0.011
Agreeableness	−0.011	<u>−0.086</u>	−0.028	<u>0.284</u>	<u>0.169</u>
Neuroticism	−0.039	<u>−0.107</u>	−0.067	<u>0.152</u>	<u>0.234</u>

<u>Underlined</u> numbers indicate significant correlation ($p < 0.05$).

Table 9. Pearson correlation coefficient between the actual Big Five traits and predictions achieved through the usage of n-grams, MBTI/Enneagram predictions and quartile-split Big Five class predictions as features.

Features	Openness	Conscient.	Extroversion	Agreeableness	Neuroticism
n-grams + Quartile Preds.	0.243	0.222	0.372	0.266	0.246
n-grams + Quartile + Other Preds.	**0.259**	0.258	0.386	0.249	0.265

Bold numbers mark a result that is outperforming the baseline.

5.1.3. Language-Based Features

When conducting analysis into psycholinguistic features, it should be noted that several researchers in the past have found correlations between the Big Five traits and different LIWC dimensions [49,59,104]. However, as the exact list of LIWC dimensions that correlate with each Big Five trait tends to differ on a dataset basis, it is possible that contextual information, in addition to personality, can have a huge influence on language usage that is measured by LIWC.

While it is possible to perform detailed research into how the Big Five personality traits have influenced language usage on the social media platform Reddit, such a study and its results could potentially introduce information leak into the prediction model if used for the regression task. In addition to this, the usage of psycholinguistic information based on its relationship with the Big Five personality traits would possibly minimize the effectiveness of MBTI/Enneagram predictions present in the feature set.

Instead, we focused on a statistical analysis of the MBTI types and how they influence language use on Reddit, as suggested by data present in the PANDORA dataset. Through this approach, we not only open the possibility of this information being leveraged in our prediction model but also provide insight into the linguistic nature of MBTI types. Tables 10 and 11 include information about all the correlations present between LIWC dimensions and MBTI types.

Table 10. Correlation between MBTI types and LIWC features present in the PANDORA dataset. LIWC features that correlate with at least one MBTI type are shown in the table.

	Introverted	Intuitive	Thinking	Perceiving
LIWC Dim.				
achieve	−0.020	0.050 **	0.091 **	−0.057 **
adverb	−0.010	−0.012	−0.193 **	0.030 **
affect	−0.079 **	0.014	−0.208 **	0.025 *
AllPunc	0.024 *	−0.001	0.013	−0.009
anger	−0.034 **	0.002	0.068 **	0.075 **
anx	−0.027 **	0.009	−0.184 **	−0.038 **
Apostro	−0.019	−0.022 *	−0.100 **	0.010
article	0.049 **	0.034 **	0.197 **	0.014

Table 10. Cont.

	Introverted	Intuitive	Thinking	Perceiving
assent	−0.057 **	−0.043 **	−0.111 **	0.051 **
auxverb	−0.030 **	−0.002	−0.024 *	−0.004
bio	−0.062 **	−0.031 **	−0.080 **	−0.011
body	−0.053 **	−0.049 **	−0.038 **	0.010
cause	−0.007	0.048 **	0.153 **	0.019
certain	−0.019	0.059 **	−0.046 **	0.014
cogmech	−0.026 *	0.049 **	−0.068 **	0.007
Colon	0.033 **	−0.023 *	−0.038 **	0.009
Comma	0.022 *	0.045 **	0.007	0.002
conj	−0.053 **	−0.012	−0.137 **	−0.029 **
Dash	0.036 **	−0.016	0.024 *	−0.034 **
death	0.061 **	0.037 **	0.088 **	0.069 **

Note: * when ($p < 0.05$) and ** when ($p < 0.01$).

Table 11. Continuation of Table 10—Correlation between MBTI types and LIWC features present in the PANDORA dataset. LIWC features that correlate with at least one MBTI type are shown in the table.

	Introverted	Intuitive	Thinking	Perceiving
LIWC Dim. (2)				
Dic	−0.047 **	−0.012	−0.121 **	−0.047 **
discrep	0.027 *	0.001	0.068 **	−0.026 *
excl	0.003	−0.006	−0.028 **	0.037 **
Exclam	−0.077 **	−0.033 **	−0.211 **	−0.038 **
family	−0.037 **	−0.067 **	−0.105 **	−0.075 **
feel	−0.061 **	−0.015	−0.265 **	−0.030 **
filler	−0.040 **	−0.064 **	−0.151 **	0.055 **
friend	−0.118 **	−0.037 **	−0.213 **	−0.020
funct	−0.030 **	0.010	−0.103 **	−0.035 **
future	0.023 *	−0.009	0.108 **	−0.005
health	−0.024 *	0.023 *	−0.069 **	−0.059 **
hear	0.013	−0.022 *	−0.126 **	0.071 **
home	0.007	−0.084 **	−0.051 **	−0.097 **
humans	−0.067 **	0.017	−0.037 **	0.001
i	−0.038 **	−0.076 **	−0.236 **	−0.023 *
incl	−0.087 **	0.001	−0.173 **	−0.076 **
ingest	−0.015	−0.058 **	−0.020	−0.042 **
inhib	0.030 **	0.037 **	0.127 **	−0.047 **
insight	−0.019	0.078 **	−0.067 **	0.020
ipron	0.006	0.064 **	−0.057 **	0.026 *
leisure	0.034 **	−0.028 **	−0.033 **	0.051 **
money	0.028 **	−0.014	0.167 **	−0.068 **
motion	−0.036 **	−0.055 **	0.001	−0.095 **
negate	0.035 **	−0.014	0.095 **	0.024 *
negemo	−0.030 **	0.016	−0.017	0.057 **
nonfl	−0.005	0.005	−0.022 *	0.037 **
number	0.054 **	−0.032 **	0.074 **	−0.001
OtherP	0.052 **	−0.004	0.043 **	0.024 *
Parenth	0.047 **	−0.011	0.003	−0.016
past	−0.003	−0.044 **	−0.126 **	−0.025 *
percept	0.001	−0.050 **	−0.224 **	0.022 *
posemo	−0.081 **	0.005	−0.265 **	−0.012
ppron	−0.089 **	−0.064 **	−0.256 **	−0.053 **
preps	0.005	0.045 **	0.002	−0.066 **
present	−0.062 **	−0.012	−0.105 **	0.004
pronoun	−0.070 **	−0.027 **	−0.230 **	−0.033 **
QMark	0.001	−0.019	0.055 **	0.048 **
quant	0.036 **	0.055 **	0.032 **	0.002
Quote	−0.001	0.022 *	0.021 *	0.013
relativ	−0.010	−0.042 **	−0.037 **	−0.075 **
relig	0.038 **	0.041 **	−0.008	0.048 **
sad	0.032 **	0.025 *	−0.121 **	0.001
see	0.043 **	−0.060 **	−0.064 **	0.024 *

Table 11. Cont.

	Introverted	Intuitive	Thinking	Perceiving
SemiC	0.011	0.009	0.021 *	−0.006
sexual	−0.058 **	−0.007	−0.051 **	0.029 **
shehe	−0.072 **	−0.038 **	−0.130 **	−0.036 **
Sixltr	0.014	0.075 **	0.128 **	−0.016
social	−0.127 **	−0.009	−0.145 **	−0.051 **
space	−0.020	−0.008	0.044 **	−0.047 **
swear	−0.049 **	−0.026 *	0.055 **	0.086 **
tentat	0.034 **	0.017	0.001	0.031 **
they	0.011	0.015	0.069 **	−0.045 **
time	0.013	−0.052 **	−0.118 **	−0.059 **
verb	−0.046 **	−0.025 *	−0.127 **	−0.009
WC	0.006	−0.007	0.070 **	−0.023 *
we	−0.066 **	0.030 **	−0.059 **	−0.023 *
work	−0.018	−0.017	0.147 **	−0.068 **
WPS	0.020	0.023 *	0.044 **	0.040 **
you	−0.066 **	0.020	−0.025 *	−0.030 **

Note: * when ($p < 0.05$) and ** when ($p < 0.01$).

The results reported in Tables 10 and 11 highlight that the two MBTI types that tend to correlate with most of the LIWC dimensions are *T-F* with 69 and *J-P* with 54 statistically significant correlations. To establish whether an entire set of correlating LIWC dimensions can contribute to better predicting Big Five traits, we tested all the correlating LIWC dimensions as features, sorting them into four different groups—one for each MBTI type—and a fifth group that includes all 78 LIWC dimensions that were found to correlate with at least one type. The results of combining these features with *n*-grams, as well as MBTI/Enneagram predictions in the set of features, are reported in Table 12.

Table 12. The Pearson correlation coefficient between the actual Big Five traits and the ones predicted using combinations of *n*-grams, MBTI/Enne. predictions and various LIWC dimensions are divided in sets based on which MBTI type they correlate with.

Features	Openness	Conscient.	Extroversion	Agreeableness	Neuroticism
n-grams + ...					
Introverted Correlating LIWC	0.229	0.148	0.321	0.212	0.249
Intuitive Correlating LIWC	0.232	0.159	0.324	0.230	0.238
Thinking Correlating LIWC	0.216	0.150	0.340	0.203	0.241
Perceiving Correlating LIWC	0.228	0.154	0.319	0.216	0.243
All MBTI Types Correlating LIWC	0.214	0.150	0.330	0.206	0.237
n-grams + MBTI/Enne. + ...					
Introverted Correlating LIWC	0.234	**0.274**	0.379	0.258	0.279
Intuitive Correlating LIWC	0.239	**0.283**	0.384	**0.272**	0.266
Thinking Correlating LIWC	0.221	0.273	**0.389**	**0.286**	**0.298**
Perceiving Correlating LIWC	0.235	**0.274**	0.386	0.249	0.282
All MBTI Types Correlating LIWC	0.227	0.271	0.382	0.253	**0.289**

Bold numbers mark a result that is outperforming the baseline approach.

5.1.4. Identifying Useful LIWC Dimensions

While the results of predictions that utilize correlating LIWC dimensions gave promising results for predicting certain traits, it was only when paired with the previously computed MBTI/Enneagram predictions in the feature set that the results improved over baseline. This signals that, despite the LIWC features being an efficient indicator of personality traits, it is only when the relationship between the multiple personality models is leveraged that they become the most effective.

This is especially apparent for the **Conscientiousness** trait, which had some of the worst results without the MBTI/Enneagram predictions in the feature set but ended up outperforming the baseline results when predictions of the type-based personality models were reintroduced into the feature set. It is worth noting, however, that, due to a large amount of LIWC dimensions correlating with several MBTI types, that the potential benefit of certain psycholinguistic features are reduced by the sudden increase in the number of features.

Since this overlap between the LIWC dimensions that correlate with two or more MBTI types ranges from 51.85% shared between the *J-P* and *S-N* types to 86.79% between the *E-I* and *T-F* types, we needed to analytically determine which LIWC dimensions best describe the relationship between MBTI types and the Big Five traits that correlate with them. However, we suggest that several factors can be considered in order to improve the results by helping to select the adequate LIWC dimensions for the feature set.

The first factor is too high of a correlation with the MBTI type. While most of the correlations reported in Tables 10 and 11 tend to be marginally significant, those that have exceeded the absolute value of 0.2 indicate a stronger relationship with the MBTI type, and as such are not a good indicator of the relationship present between MBTI and Big Five models. The second factor is the degree of correlation between MBTI types themselves. If the LIWC category correlates with both of the MBTI types that tend to correlate with each other, that LIWC category should be disregarded from the feature set. Finally, the third factor is the relationship present between the LIWC dimensions themselves. If the LIWC dimensions correlate with one another, only a single one should be selected, as LIWC categories need to be statistically independent from one another.

Using these three factors as criteria, we were left with the following list of LIWC dimensions that correlate with each of the four MBTI types:

1. **Extroverted/Introverted (E-I)** type:
   - shehe—third person singular pronouns (she, her, him...)
   - incl—inclusive words (e.g., with, and...)
   - number—numbers (first, thousand...)
   - present—present tense verbs (is, does, do...)
   - posemo—words associated with positive emotions (love, happy, hope...)
   - pronoun—total pronouns (I, they, it...)

2. **Sensing/Intuitive (S-N)** type:
   - WPS—average words per sentence
   - past—past tense verbs (walked, were...)
   - social—social words (we, thank, care...)
   - ipron—impersonal pronouns (that, what, it...)
   - Colon—number of colons (:)

3. **Thinking/Feeling (T-F)** type:
   - you—second person singular pronouns (u, yourself, you...)
   - article—number of article (a, an, the...)
   - sad—words relating to sadness (:(, cry...)

4. **Judging/Perceiving (J-P)** type:
   - Exclam—number of exclamations (!)
   - i—first person singular pronouns (me, myself, I...)
   - hear—auditory words (hear, sound...)
   - tentat—tentative phrases (if, any, something...)

Combining these LIWC dimensions in combination with their respective MBTI types, we achieved results in predicting the Big Five traits, which are reported in Table 13.

The results reported in Table 13 indicate that choosing LIWC dimensions with the method we described can further increase the results when predicting Big Five traits, especially in cases when the MBTI type and the Big Five trait have been found to statistically correlate with one other. This is visible for all the MBTI types as the prediction results for *Openness* increased when using LIWC dimensions that were selected for the *S-N* type, *Agreeableness* increased when using LIWC features for *T-F*, and so on. However, it should be noted that, despite the prediction scores for the *Neuroticism* trait increasing when using LIWC dimensions selected for the *E-I* and *S-N* types, we propose that this can be attributed either to a possible relationship between Enneagram and *Neuroticism* or due to the nature of the data, as *Neuroticism* is the only trait to follow a bimodal distribution in this dataset (Figure 6).

Unlike the MBTI types, Enneagram types have shown no presence of a statistically significant correlation with any of the LIWC dimensions.

Table 13. Pearson correlation coefficient between the gold-standard Big Five labels and predictions achieved through using combinations of *n*-grams, MBTI/Enneagram predictions and selected LIWC dimensions for each of the MBTI traits (from top to bottom: (1) *E-I*, (2) *S-N*, (3) *T-F* and (4) *J-P*).

Features	Openness	Conscient.	Extroversion	Agreeableness	Neuroticism
n-grams + MBTI/Enne. Preds. + ...					
shehe, incl, number, present, posemo, pronoun	**0.256**	0.270	**0.407**	0.263	**0.296**
WPS, past, social, ipron, Colon	**0.265**	0.275	**0.392**	0.273	**0.290**
you, article, sad	0.250	0.272	**0.381**	0.289	0.283
Exclam, i, hear, tentat	0.246	**0.283**	**0.384**	0.269	0.272

Bold numbers mark a result that is outperforming the baseline approach.

5.1.5. Effect of Enneagram Predictions on the Big Five Predictions

While MBTI and Big Five personality models have been previously compared in the works of several authors [13,18,66,86], the relationship between the Enneagram and Big Five traits has not been thoroughly explored. This can be largely attributed to Enneagram often being underutilized in both academia and consulting—the two areas where the Big Five model and MBTI have enjoyed success, respectively. However, taking a closer look into the possible relationship between these models can help to better answer the question as to how Enneagram predictions can help the process of predicting Big Five personality traits.

The results reported in Table 14 indicate that, for certain Big Five traits, the results change drastically when the Enneagram predictions are removed from the feature set. This is visible when comparing them to the results reported in Table 13 as well as the results reported for the baseline approach. With this in mind, we make the following observations:

1. The model's performance when predicting the *Neuroticism* trait without Enneagram predictions in the feature set decreases in comparison to all the feature sets that previously included Enneagram predictions.
2. The model's performance when predicting both *Conscientiousness* and *Agreeableness* increases in almost every case when Enneagram predictions are removed from the feature set. The only exception to this is when the following LIWC dimensions appear in the feature set: shehe, incl, number, present, posemo and pronoun.
3. Predictions of the *Openness* trait either stay the same or only slightly fluctuate when Enneagram predictions are removed from the feature set, indicating that predicting this trait benefits only slightly from the Enneagram predictions.

Table 14. The Pearson correlation coefficient between the gold-standard Big Five labels and predictions achieved through using *n*-grams, MBTI predictions and selected LIWC dimensions in the feature set with the Enneagram predictions omitted. LIWC dimensions selected for each MBTI type follow the same order described in Table 13 (e.g., from top to bottom: (1) *E-I*, (2) *S-N*, (3) *T-F* and (4) *J-P*)).

Features	Openness	Conscient.	Extroversion	Agreeableness	Neuroticism
n-grams + ... Baseline without Enne. Preds.	0.250	**0.281**	0.374	**0.276**	0.258
n-grams + MBTI Preds. + ...					
shehe, incl, number, present, posemo, pronoun	0.242	**0.277**	0.380	**0.266**	0.268
WPS, past, social, ipron, Colon	0.253	**0.285**	0.378	**0.278**	0.267
you, article, sad	0.250	**0.281**	0.370	**0.299**	0.256
Exclam, i, hear, tentat	0.248	**0.293**	0.371	**0.274**	0.248

Bold numbers indicate results that outperform results on the same model that use Enneagram predictions in their feature set.

The first of these three observations points towards the possible relationship existing between the Enneagram types and Big Five's *Neuroticism* trait. This can be confirmed when observing the correlations between the Enneagram types and the Big Five traits that Gjurković et al. [7] reported in their work, which we list in Table 15 for reference.

Despite these results, it is still difficult to conclude whether the relationship between Enneagram and *Neuroticism* is result of the data's nature, or a case of language usage that is associated with *Neuroticism* being shared with many of the Enneagram types. This is largely due to a lack of literature comparing Enneagram types to Big Five traits, especially from the perspective of language usage. As the PANDORA dataset contains a rather small number of Enneagram labels, as well as Big Five ones, it would be difficult to conduct an in-depth analysis into the topic from this dataset alone. However, we hope that these findings can help to motivate future research into the relationship between Enneagram types, Big Five traits and language usage shared between them, as we propose that it would be greatly beneficial for personality computing tasks conducted in the future.

Table 15. The Pearson correlation coefficient of the gold-standard Big Five labels with the predicted values of Enneagram types as reported by Gjurković et al. [7] CC-BY-NC.

Features	Openness	Conscient.	Extroversion	Agreeableness	Neuroticism
Pred. Type					
Enneagram Type 1	0.002	0.032	−0.028	0.047	0.025
Enneagram Type 2	−0.011	<u>0.108</u>	0.030	<u>0.135</u>	<u>0.046</u>
Enneagram Type 3	<u>0.085</u>	0.014	<u>0.071</u>	−0.064	<u>−0.069</u>
Enneagram Type 4	0.041	0.017	0.033	<u>0.166</u>	<u>0.159</u>
Enneagram Type 5	<u>0.067</u>	−0.035	<u>−0.060</u>	<u>−0.121</u>	<u>−0.076</u>
Enneagram Type 6	−0.051	0.004	−0.035	0.046	<u>0.113</u>
Enneagram Type 7	−0.043	−0.019	<u>0.078</u>	<u>−0.085</u>	<u>−0.088</u>
Enneagram Type 8	0.022	−0.044	<u>0.063</u>	<u>−0.129</u>	<u>−0.075</u>
Enneagram Type 9	−0.034	−0.016	<u>−0.102</u>	0.041	−0.005

<u>Underlined</u> numbers indicate significant correlation ($p < 0.05$).

5.1.6. Subreddit Participation

Analysing data from the PushShift dataset, we found that, in the period between the chronologically first and last comments present in the PANDORA dataset there has been activity on 879,826 different subreddits. Out of all these subreddits, the 50 most popular ones have mostly been those with a more general topics, such as r/AskReddit and r/worldnews. However, it is worth noting that, in the set of most popular subreddits, several personality related ones were included, e.g., r/mbti and several subreddits dedicated to specific MBTI types, such as r/INTP and r/ENFP. On the other hand, the PANDORA dataset included information on some 46,214 different subreddits, a considerably smaller number.

After forming feature vectors based on either the number of unique users participating in subreddits within the time window matching that of the PANDORA dataset or on the number of total comments, we found that these two feature vectors are nearly identical. This is due to the fact that total number of comments and number of unique users participating in subbreddits showed a high Pearson correlation of 0.83. Consequently, we decided to only focus on the feature vector that is formed by using the total number of comments as a measurement of subreddit popularity. The results of predictions when using these features in the feature set are shown in Table 16.

Table 16. The Pearson correlation coefficient between the actual Big Five traits and predictions achieved through usage of n-grams, MBTI/Enneagram predictions and subreddit participation in the feature set.

Features	Openness	Conscient.	Extroversion	Agreeableness	Neuroticism
n-grams + Subreddits	0.208	0.160	0.331	0.171	0.224
n-grams + Subreddits + MBTI/Enne.	0.225	**0.274**	0.387	**0.274**	0.252

Bold numbers mark a result that is outperforming the baseline approach.

While subreddit participation is visibly less effective when predicting *Openness* and *Neuroticism*, it caused a slight increase in the results when predicting *Conscientiousness* and *Agreeableness* with the success rates of predicting *Extroversion* remaining the same. We suggest that this is caused by the fact that vast numbers of subreddits present in the feature vector tend to be more general in nature, rather than topic-specific. This has contributed to them attracting different people that all, most likely, do not share much in the way of personality traits. However, as the relationship between interests and MBTI types has not been thoroughly studied in the existing literature, we decided not to further investigate the usefulness of this feature set, instead choosing to leave it for future works.

5.2. Model Selection

Features introduced in the previous subsection were all evaluated on the same L2 regularized regression model, which was also used as part of the method that achieved the baseline results. While some features led to improvements, they also, in turn, introduced additional complexity in the feature space. To additionally improve the results, we conducted experiments with several different regression models capable of weighing features in order to bring the most out of the them for the task of predicting Big Five traits.

For the sake of conciseness, as well as for easier comparison between the results, we chose to report the results for all the different models within a single table (Table 17). For features used as input to these models, we decided to select the best performing set, which was a combination of n-grams, MBTI/Enneagram predictions and certain LIWC features, which were selected in the way described in the subsection detailing the methods for selecting the best language based features (Section 5.1.4).

The first section of results in Table 17 outlines the results achieved by using the Ridge regression model, which are same as those previously reported in Table 13. When comparing these results to other sections of the table, we note that certain models, such as SVR, Huber regressor and Elastic-Net, led to improvements in predicting most of the Big Five traits, while Lasso regression, KerasRegressor and XGBoost demonstrated poor performance on the task overall. Out of the better performing models, Elastic-Net stands out as it scored the best on three out of five Big Five traits, namely *Openness, Conscientiousness* and *Extroversion*, while the Huber regressor and SVR proved to be better choices for predicting the remaining two Big Five traits.

Analysing the performance of the Lasso regression, which ended up being the worst performing model overall, we note that, despite the L1 regularization employed in an effort to remove noise from the feature set, it is possible that this actually led to a loss of several important features that were indicative of finer differences between personalities. As personality is a complex concept, it often tends to be both affected and manifested through the smallest differences present between individuals. As such, we speculate that L1 regularization caused the model to be less effective in efforts to capture these small differences, in turn, leading to poor performances on the task of Big Five personality trait prediction.

KerasRegressor resulted in predictions that correlate slightly worse than the baseline approach across all of the personality traits. While these results indicate worse performances than any of the other models included in Table 13, it should be noted that this approach outperformed the BERT-based method that Gjurković et al. [7] experimented on. While we tested different architectures of the KerasRegressor for this task, it is possible that a more complex deep-learning model would be capable of better capturing the relationship between the features used and personality traits.

Similarly to the KerasRegressor model, XGBoost also demonstrated less than satisfactory results. Despite improving on predictions of the *Neuroticism* trait over the baseline, the results for the other four personality traits saw a significant decrease in comparison to the baseline. We propose that, due to the small size, the data-boosting algorithm struggled to correctly predict the right value for each of the personality traits. Additionally, while the LIWC dimensions differed between feature sets, XGBoost showed almost identical results for each of the experiments. This leads us to believe that XGBoost is less capable of leveraging this language-related information and instead prioritizes features, such as *n*-grams and MBTI/Enneagram predictions from the feature set.

Table 17. Scores for predicting the gold-standard Big Five labels using combinations of *n*-grams, MBTI/Enneagram predictions and LIWC dimensions selected through process outlined in the previous subsection (Table 14) evaluated on different models.

	Features	Openness	Conscient.	Extroversion	Agreeable.	Neuroticism
Ridge Reg.	E-I	0.256	0.270	0.407	0.263	0.296
	S-N	0.265	0.275	0.392	0.273	0.290
	T-F	0.250	0.272	0.381	0.289	0.283
	J-P	0.246	0.283	0.384	0.269	0.272
Lasso Reg.	E-I	0.167	0.266	0.358	0.264	0.281
	S-N	0.181	0.268	0.347	0.268	0.270
	T-F	0.170	0.267	0.320	0.256	0.247
	J-P	0.168	0.270	0.327	0.263	0.259
Elastic-Net	E-I	0.269	0.270	**0.408**	0.264	0.310
	S-N	0.283	0.283	0.397	0.274	0.298
	T-F	0.263	0.272	0.388	0.289	0.296
	J-P	0.267	**0.285**	0.391	0.265	0.292
Huber Reg.	E-I	0.255	0.269	0.396	0.260	**0.312**
	S-N	0.263	0.276	0.384	0.272	0.288
	T-F	0.245	0.272	0.375	0.284	0.274
	J-P	0.254	**0.285**	0.378	0.268	0.266
SVR	E-I	0.230	0.274	0.370	0.282	0.291
	S-N	0.232	0.267	0.361	0.289	0.286
	T-F	0.242	0.274	0.359	**0.298**	0.279
	J-P	0.242	0.282	0.358	0.294	0.279
Keras Reg.	E-I	0.235	0.179	0.368	0.223	0.228
	S-N	0.234	0.181	0.369	0.220	0.231
	T-F	0.239	0.178	0.359	0.231	0.230
	J-P	0.249	0.171	0.359	0.210	0.227
XGBoost	E-I	0.224	0.219	0.337	0.249	0.285
	S-N	0.216	0.219	0.349	0.253	0.284
	T-F	0.222	0.224	0.335	0.250	0.287
	J-P	0.221	0.217	0.337	0.256	0.286

Underlined numbers outperform the baseline; **Bold** numbers mark the best performing result.

The best performing solution for the *Agreeableness* trait was when using the SVR model that included a subset of LIWC dimensions correlating with the *T-F* MBTI type (e.g., *you*, *article* and *sad*) in the feature set. While not the best performing model, SVR still managed to outperform the baseline approach using several different feature sets, especially when predicting the *Agreeableness* trait, for which it outperformed the baseline on every single experiment conducted.

We propose that these results are largely due to the SVR's nature to work well with smaller sets of data, as well as the error function on which it relies. However, we must remark that SVR had worse results than the baseline on both the *Openness* and *Extroversion* traits—both of which have shown the overall highest correlations with *n*-gram features, indicating that SVR places less importance on this particular set of features when making predictions.

Another well-performing model is the Huber regressor, which demonstrated overall exemplary performances when predicting the *Conscientiousness* and *Neuroticism* personality traits, with the results for the other three traits also showing promise. When it comes to predictions for the *Neuroticism* trait, we suggest that good results are due to the Huber regressors capability of working well with outliers, possibly being more capable of working with the *Neuroticism* trait, which demonstrated the presence of bimodal distribution.

Out of all the models, the overall best performing one seems to be Elastic-Net, which performed best when predicting three of the five Big Five traits, namely *Openness, Conscientiousness* and *Extroversion*. The effectiveness of Elastic-Net can be attributed largely to the good balance of both the L1 and L2 regularization norms, which eliminated noisy features while simultaneously keeping those that influenced personality prediction, thus, utilizing them to capture subtle differences in personality.

We indicate that, despite the SVR and Huber regressor slightly outperforming Elastic-Net when predicting the *Agreeableness* and *Neuroticism* dimensions, the consistent improvements in scores for many different feature sets point towards Elastic-Net being the best overall choice for the task of predicting Big Five traits with the MBTI/Enneagram predictions present in the feature set.

6. Discussion

In the previous section, we outlined and briefly discussed the results from a variety of different feature and model-selection approaches. While experiments were conducted on several different algorithms, including deep-learning and ensemble methods, experimentation on linear regression models was more prevalent for the sake of highlighting the effectiveness of features, while also offering better interpretability comparability with the baseline approach. In the end, the set of features included MBTI/Enneagram prediction, a set of n-grams stemming from the work of Gjurković et al. [7] and a set of LIWC-based features created through method described in Section 5.1.4.

While the other features that were experimented with led to limited or no improvement at all for certain personality traits, they offered valuable insight into the nature of personality traits. When considering the experiments involving the conversion of Big Five personality prediction into a classification issue rather than a regression one, followed by using these predictions as features in a regression model, we observed that, except for *Openness*, none of the other personality traits saw improvement in the results (shown in Tables 7 and 8).

Although it is possible that these improvements for the *Openness* trait can be attributed to data following a negatively skewed distribution (Figure 6), we suggest that this is a result of the classification predictions for *Openness* being more statistically independent from other 4 personality traits. Despite the previous research reporting promising results when using binning strategies [76], we propose that the possibility of information loss largely outweighs the positives of this approach [114,115] when predicting personality traits in this manner.

Another set of features that we experimented with was Reddit participation and the way in which this reflects on personality. As this particular set of features demonstrated overall little to no improvement, this suggests that personality has little effect on topical interests and the way they are expressed through Reddit (Table 16). However, as the majority of the most popular subreddits were those that centre around broader topics, it is possible that grouping certain interests into larger classes (e.g., hobbies, music-related and news), and using them as features could lead to a higher correlation with certain personality traits. Due to the breadth of the issue and the overall experimental complexity such study would warrant, we decided to leave it for future works.

The LIWC-based features introduced in our experiments led to improvements in predicting Big Five traits; however, they also introduced additionally complexity in the feature space. To best handle this and bring the most out of these features, an adequate regression model was necessary. Algorithms that achieved the best results on each of the Big Five traits did so when they used LIWC features selected through the methodology described in Section 5.1.4. The trait that saw the highest increase was *Openness*, correlating by 3.3% more with the actual trait values than with the baseline approach.

These results were achieved when using combination of Elastic-Net model as predictor and a feature set consisting of *n*-grams, MBTI/Enneagram predictions and a set of LIWC features selected for *S-N*—an MBTI type with which *Openness* demonstrated a statistically significant correlation in the past. The same model yielded the best results for *Conscientiousness* when the LIWC features were selected for the *J-P* type, and *Extroversion* when LIWC features that correlated with for the *E-I* MBTI type were used.

In the case of these traits, the increase was 1.2% for *Conscientiousness* and 2.1% for *Extroversion* with a final correlation of 40.8% for the *Extroversion* trait being the highest correlation value for an individual trait. The Huber regressor has achieved the exact same correlation percentage on the *Conscientiousness* dimension as the Elastic-Net model did when using the same features.

Additionally, the Huber regressor yielded the best results when predicting *Neuroticism* at 31.2% correlation, scoring 2.9% higher than the baseline approach. Finally, SVR model achieved an increase of 2.8% in correlation over the baseline when predicting the *Agreeableness* trait, using *n*-grams, MBTI/Enneagram predictions and LIWC features selected for the *T-F* type, with which it demonstrated a statistically significant correlation.

When analysing the results achieved using the deep-learning model, it is a bit surprising to see it not perform as well as the other options, especially considering the popularity of deep-learning approaches for tasks of automatic personality recognition in recent years [58]. However, as Gjurković et al. [7] also reported similar results when applying a deep-learning algorithm, we can conclude that linear regression models tend to be the better choice in leveraging this particular set of features for prediction.

This is possibly due to the high linearity of features, as linear regression models tend to be designed to work best in these situations. Another possible contributing factor to ensemble and deep-learning approaches performing worse than expected could be that KerasRegressor and XGBoost require more data to be efficient.

One final experiment that we conducted was the analysis of the effect Enneagram predictions on the task of predicting the Big Five traits (Table 14). Our analysis indicated that, despite a considerable increase in prediction scores for the *Conscientiousness* and *Agreeableness* traits on several different feature sets, the scores for *Neuroticism* were worse every single time when Enneagram predictions were removed from the feature set. These results signal a possibility of a relationship existing between the Enneagram types and Big Five's *Neuroticism* trait. This is significant due to the fact that *Neuroticism* has not been previously found to correlate with any of the MBTI types [13,17,18].

7. Conclusions

In this paper, we analysed the effectiveness of different features and algorithms when paired with the MBTI and Enneagram labels on the task of automatic personality recognition. We conducted multiple experiments, testing the performance of each feature set and prediction model in order to further explore the relationship between type-based models and the Big Five.

For our experiments, we looked into the effectiveness of standardising the domain of personality models by introducing the classification results of the Big Five prediction to the feature set. In addition to this, we also looked into the effect that language features extracted using the LIWC tool have on personality as well as the effect of social media participation. The best performing set of features included MBTI/Enneagram labels, a list of *n*-grams coming from previous work and a set of LIWC features that were selected based on their relationship with the MBTI types. The best performing feature set was then used as input for multiple different regression algorithms as well as a deep-learning and a boosting approach.

Our experiments suggest that an algorithm that utilised L1 and L2 normalisation led to the best performance, causing an improvement of up 3% for the Pearson correlation coefficient metric on a per-trait level. In addition to these findings, we examined the effectiveness of labels for the Enneagram model and their effectiveness on the prediction of Big Five traits. Our analysis points towards a possible relationship between the Enneagram types and the Big Five's Neuroticism trait, as the correlation scores saw a decrease of almost 3% when Enneagram predictions were taken out of the feature set.

Possible directions in which this research can be taken in the future involve taking a closer look into the effects of interests and topics on the personality prediction. While we examined the possibility of subreddit popularity having an effect on the prediction of traits, such as Extroversion, it is possible that specific hobbies and involvement in subreddits centred around them could be an indicator of one's personality. Additional directions in which this research can be expanded include applying the methods on different sets of data. While datasets that include information for multiple personality models are still scarce, other social media platforms, such as Twitter, could prove useful in collecting data for future experiments [60].

Finally, we propose that the results of this study can be helpful in further understanding personality as they indicate how well it can be captured when translating from one set of personality measures to another. The findings of our study can also be beneficial when seeking to create more believable dialogue agents, as it allows for inputs in the form of MBTI personality.

Author Contributions: Conceptualization, D.R. and R.R.; methodology, D.R.; software, D.R.; validation, D.R., R.R. and K.A.; formal analysis, D.R.; investigation, D.R.; resources, D.R.; writing—original draft preparation, D.R.; writing—review and editing, D.R., R.R. and K.A; visualization, D.R.; supervision, R.R. and K.A.; project administration, R.R. All authors have read and agreed to the published version of the manuscript.

Funding: This research received no external funding.

Institutional Review Board Statement: Not applicable.

Informed Consent Statement: Not applicable.

Data Availability Statement: This study utilizes the PANDORA dataset available at: https://psy.takelab.fer.hr/datasets/all/ (accessed on 28 March 2023), with the code for this research available at: https://github.com/hokudai-LML/personality-types-to-traits (accessed on 28 March 2023).

Acknowledgments: We would like to thank doctor Bojan Batalo from the University of Tsukuba for their continuous support in works leading up to this research.

Conflicts of Interest: The authors declare no conflict of interest.

Abbreviations

The following abbreviations are used in this manuscript:

Agreeable.	Agreeableness
Conscient.	Conscientiousness
Enne.	Enneagram
LIWC	Linguistic Inquiry and Word Count
MBTI	Myers–Briggs Type Indicator
Preds.	Predictions
Reg.	Regression
SVM	Support Vector Machine
SVR	Support Vector Regression

References

1. Vinciarelli, A.; Mohammadi, G. A survey of personality computing. *IEEE Trans. Affect. Comput.* **2014**, *5*, 273–291. [CrossRef]
2. Uleman, J.S.; Adil Saribay, S.; Gonzalez, C.M. Spontaneous inferences, implicit impressions, and implicit theories. *Annu. Rev. Psychol.* **2008**, *59*, 329–360. [CrossRef] [PubMed]
3. Funder, D.C. Accurate personality judgment. *Curr. Dir. Psychol. Sci.* **2012**, *21*, 177–182. [CrossRef]

4. Jhangiani, R.; Tarry, H.; Stangor, C. *Principles of Social Psychology-1st International Edition*; BCcampus: Victoria, BC, Canada, 2014.
5. Engler, B. *Personality Theories*, 9th ed.; Cengage Learning: Boston, MA, USA, 2013.
6. Phan, L.V.; Rauthmann, J.F. Personality computing: New frontiers in personality assessment. *Soc. Personal. Psychol. Compass* **2021**, *15*, e12624. [CrossRef]
7. Gjurković, M.; Karan, M.; Vukojević, I.; Bošnjak, M.; Snajder, J. PANDORA Talks: Personality and Demographics on Reddit. In Proceedings of the Ninth International Workshop on Natural Language Processing for Social Media, Online, 10 June 2021; pp. 138–152. [CrossRef]
8. Worrell, F.C.; Cross, W.E., Jr. The reliability and validity of Big Five Inventory scores with African American college students. *J. Multicult. Couns. Dev.* **2004**, *32*, 18–32. [CrossRef]
9. Fossati, A.; Borroni, S.; Marchione, D.; Maffei, C. The Big Five Inventory (BFI): Reliability and validity of its Italian translation in three independent nonclinical samples. *Eur. J. Psychol. Assess.* **2011**, *27*, 50. [CrossRef]
10. Morsunbul, U. The validity and reliability study of the Turkish version of quick big five personality test. *Dusunen Adam J. Psychiatry Neurol. Sci.* **2014**, *27*, 316. [CrossRef]
11. Mount, M.K.; Barrick, M.R.; Strauss, J.P. Validity of observer ratings of the big five personality factors. *J. Appl. Psychol.* **1994**, *79*, 272. [CrossRef]
12. Van der Linden, D.; te Nijenhuis, J.; Bakker, A.B. The general factor of personality: A meta-analysis of Big Five intercorrelations and a criterion-related validity study. *J. Res. Personal.* **2010**, *44*, 315–327. [CrossRef]
13. McCrae, R.R.; Costa, P.T., Jr. Reinterpreting the Myers-Briggs type indicator from the perspective of the five-factor model of personality. *J. Personal.* **1989**, *57*, 17–40. [CrossRef]
14. Pittenger, D.J. Measuring the MBTI . . . and coming up short. *J. Career Plan. Employ.* **1993**, *54*, 48–52.
15. Bess, T.L.; Harvey, R.J. Bimodal score distributions and the Myers-Briggs Type Indicator: Fact or artifact? *J. Personal. Assess.* **2002**, *78*, 176–186. [CrossRef] [PubMed]
16. Thyer, B.A.; Pignotti, M. *Science and Pseudoscience in Social Work Practice*; Springer: New York, NY, USA, 2015. [CrossRef]
17. MacDonald, D.A.; Anderson, P.E.; Tsagarakis, C.I.; Holland, C.J. Examination of the relationship between the Myers-Briggs Type Indicator and the NEO Personality Inventory. *Psychol. Rep.* **1994**, *74*, 339–344. [CrossRef]
18. Furnham, A. The big five versus the big four: The relationship between the Myers-Briggs Type Indicator (MBTI) and NEO-PI five factor model of personality. *Personal. Individ. Differ.* **1996**, *21*, 303–307. [CrossRef]
19. Fang, Q.; Giachanou, A.; Bagheri, A.; Boeschoten, L.; van Kesteren, E.J.; Kamalabad, M.S.; Oberski, D.L. On Text-based Personality Computing: Challenges and Future Directions. *arXiv* **2022**, arXiv:2212.06711. [CrossRef]
20. Digman, J.M. Personality structure: Emergence of the five-factor model. *Annu. Rev. Psychol.* **1990**, *41*, 417–440. [CrossRef]
21. Diggle, J. *Theophrastus: Characters*; Cambridge University Press: Cambridge, UK, 2022.
22. Singer, P.N.; Van der Eijk, P.J.; Tassinari, P. *Galen: Works on Human Nature-Volume 1: Mixtures (De Temperamentis)*; Cambridge University Press: Cambridge, UK, 2019.
23. Strelau, J.; Zawadzki, B. Temperament from a psychometric perspective: Theory and measurement. *SAGE Handb. Personal. Theory Assess.* **2008**, *2*, 352–373. [CrossRef]
24. Kant, I. *Kant: Anthropology from a Pragmatic Point of View*; Cambridge University Press: Cambridge, UK, 2006.
25. Stelmack, R.M.; Stalikas, A. Galen and the humour theory of temperament. *Personal. Individ. Differ.* **1991**, *12*, 255–263. [CrossRef]
26. Leary, D.E. *Immanuel Kant and the Development of Modern Psychology*; University of Richmond: Richmond, VA, USA, 1982.
27. Eysenck, H.; Eysenck, S. *Eysenck Personality Questionnaire-Revised (EPQ-R)*; EdITS LLC/Educational and Industrial Testing Service: San Diego, CA, USA, 1984. [CrossRef]
28. Cattell, H.E.P. The sixteen personality factor (16PF) questionnaire. In *Understanding Psychological Assessment*; Springer US: New York, NY, USA, 2001; Chapter 10; pp. 187–215. [CrossRef]
29. Ashton, M.C.; Lee, K. Empirical, theoretical, and practical advantages of the HEXACO model of personality structure. *Personal. Soc. Psychol. Rev.* **2007**, *11*, 150–166. [CrossRef]
30. McCrae, R.R.; John, O.P. An introduction to the five-factor model and its applications. *J. Personal.* **1992**, *60*, 175–215. [CrossRef]
31. McCrae, R.R. *The Five-Factor Model of Personality Traits: Consensus and Controversy*; Cambridge Handbooks in Psychology; Cambridge University Press: Cambridge, UK, 2009; pp. 148–161. [CrossRef]
32. Allport, G.W.; Odbert, H.S. Trait-names: A psycho-lexical study. *Psychol. Monogr.* **1936**, *47*, 22–38. [CrossRef]
33. Myers, I.B. *The Myers-Briggs Type Indicator: Manual (1962)*; Consulting Psychologists Press: Palo Alto, CA, USA, 1962. [CrossRef]
34. Jung, C.G. *Psychological Types: Volume 6*; Princeton University: Princeton, NJ, USA, 1921.
35. Štajner, S.; Yenikent, S. Why Is MBTI Personality Detection from Texts a Difficult Task? In Proceedings of the 16th Conference of the European Chapter of the Association for Computational Linguistics, Main Volume, Online, 19–23 April 2021; pp. 3580–3589. [CrossRef]
36. Myers, I.B. *Introduction to Type: A Description of the Theory and Applications of the Myers-Briggs Type Indicator*; Consulting Psychologists Press: Palo Alto, CA, USA, 1997.
37. Riso, D.R.; Hudson, R. *Personality Types: Using the Enneagram for Self-discovery*; Houghton Mifflin Harcourt: Boston, MA, USA, 1996.
38. Kale, S.H.; Shrivastava, S. The enneagram system for enhancing workplace spirituality. *J. Manag. Dev.* **2003**, *22*, 308–328. [CrossRef]

39. Kemboi, R.J.K.; Kindiki, N.; Misigo, B. Relationship between personality types and career choices of undergraduate students: A case of Moi University, Kenya. *J. Educ. Pract.* **2016**, *7*, 102–112.
40. Lounsbury, J.W.; Park, S.H.; Sundstrom, E.; Williamson, J.M.; Pemberton, A.E. Personality, career satisfaction, and life satisfaction: Test of a directional model. *J. Career Assess.* **2004**, *12*, 395–406. [CrossRef]
41. Seibert, S.E.; Crant, J.M.; Kraimer, M.L. Proactive personality and career success. *J. Appl. Psychol.* **1999**, *84*, 416–427. [CrossRef] [PubMed]
42. Gerber, A.S.; Huber, G.A.; Doherty, D.; Dowling, C.M.; Raso, C.; Ha, S.E. Personality traits and participation in political processes. *J. Politics* **2011**, *73*, 692–706. [CrossRef]
43. Caprara, G.V.; Schwartz, S.; Capanna, C.; Vecchione, M.; Barbaranelli, C. Personality and politics: Values, traits, and political choice. *Political Psychol.* **2006**, *27*, 1–28. [CrossRef]
44. Blais, A.; St-Vincent, S.L. Personality traits, political attitudes and the propensity to vote. *Eur. J. Political Res.* **2011**, *50*, 395–417. [CrossRef]
45. Taylor, A.; MacDonald, D.A. Religion and the five factor model of personality: An exploratory investigation using a Canadian university sample. *Personal. Individ. Differ.* **1999**, *27*, 1243–1259. [CrossRef]
46. Lodi-Smith, J.; Roberts, B.W. Social investment and personality: A meta-analysis of the relationship of personality traits to investment in work, family, religion, and volunteerism. *Personal. Soc. Psychol. Rev.* **2007**, *11*, 68–86. [CrossRef]
47. Roberts, B.W.; Kuncel, N.R.; Shiner, R.; Caspi, A.; Goldberg, L.R. The power of personality: The comparative validity of personality traits, socioeconomic status, and cognitive ability for predicting important life outcomes. *Perspect. Psychol. Sci.* **2007**, *2*, 313–345. [CrossRef]
48. Picard, R.W. *Affective Computing*; MIT press: Cambridge, MA, USA, 2000.
49. Pennebaker, J.W.; King, L.A. Linguistic styles: Language use as an individual difference. *J. Personal. Soc. Psychol.* **1999**, *77*, 1296–1312. [CrossRef] [PubMed]
50. Argamon, S.; Dhawle, S.; Koppel, M.; Pennebaker, J.W. Lexical predictors of personality type. In Proceedings of the 2005 Joint Annual Meeting of the Interface and the Classification Society of North America, St. Louis, MS, USA, 8–12 June 2005; pp. 1–16.
51. Mairesse, F.; Walker, M.A.; Mehl, M.R.; Moore, R.K. Using linguistic cues for the automatic recognition of personality in conversation and text. *J. Artif. Intell. Res.* **2007**, *30*, 457–500. [CrossRef]
52. Oberlander, J.; Gill, A.J. Language with character: A stratified corpus comparison of individual differences in e-mail communication. *Discourse Process.* **2006**, *42*, 239–270. [CrossRef]
53. Oberlander, J.; Nowson, S. Whose thumb is it anyway? Classifying author personality from weblog text. In Proceedings of the COLING/ACL 2006 Main Conference Poster Sessions, Sydney, Australia, 17–18 July 2006; pp. 627–634.
54. Gill, A.; Nowson, S.; Oberlander, J. What are they blogging about? Personality, topic and motivation in blogs. In Proceedings of the International AAAI Conference on Web and Social Media, San Jose, CA, USA, 17–20 May 2009; Volume 3, pp. 18–25.
55. Kosinski, M.; Matz, S.C.; Gosling, S.D.; Popov, V.; Stillwell, D. Facebook as a research tool for the social sciences: Opportunities, challenges, ethical considerations, and practical guidelines. *Am. Psychol.* **2015**, *70*, 543–556. [CrossRef] [PubMed]
56. Kosinski, M.; Stillwell, D.; Graepel, T. Private traits and attributes are predictable from digital records of human behavior. *Proc. Natl. Acad. Sci. USA* **2013**, *110*, 5802–5805. [CrossRef]
57. Wiegmann, M.; Stein, B.; Potthast, M. Celebrity profiling. In Proceedings of the 57th annual meeting of the Association for Computational Linguistics, Florence, Italy, 28 July–August 2 2019; pp. 2611–2618. [CrossRef]
58. Mehta, Y.; Majumder, N.; Gelbukh, A.; Cambria, E. Recent trends in deep learning based personality detection. *Artif. Intell. Rev.* **2020**, *53*, 2313–2339. [CrossRef]
59. Mehl, M.R.; Gosling, S.D.; Pennebaker, J.W. Personality in its natural habitat: Manifestations and implicit folk theories of personality in daily life. *J. Personal. Soc. Psychol.* **2006**, *90*, 862–877. [CrossRef]
60. Celli, F.; Lepri, B. Is big five better than MBTI? A personality computing challenge using Twitter data. In Proceedings of the Fifth Italian Conference on Computational Linguistics CLiC-it 2018, Torino, Italy, 10–12 December 2018; pp. 93–98. [CrossRef]
61. Thornton, C.; Hutter, F.; Hoos, H.H.; Leyton-Brown, K. Auto-WEKA: Combined selection and hyperparameter optimization of classification algorithms. In Proceedings of the 19th ACM SIGKDD International Conference on Knowledge Discovery and Data Mining, Chicago, IL, USA, 11–14 August 2013; pp. 847–855. [CrossRef]
62. Sun, X.; Liu, B.; Cao, J.; Luo, J.; Shen, X. Who am I? Personality detection based on deep learning for texts. In Proceedings of the 2018 IEEE International Conference on Communications (ICC), Kansas City, KS, USA, 20–24 May 2018; pp. 1–6. [CrossRef]
63. Biel, J.I.; Gatica-Perez, D. The youtube lens: Crowdsourced personality impressions and audiovisual analysis of vlogs. *IEEE Trans. Multimed.* **2012**, *15*, 41–55. [CrossRef]
64. Kazameini, A.; Fatehi, S.; Mehta, Y.; Eetemadi, S.; Cambria, E. Personality trait detection using bagged svm over bert word embedding ensembles. *arXiv* **2020**, arXiv:2010.01309. [CrossRef]
65. Devlin, J.; Chang, M.W.; Lee, K.; Toutanova, K. BERT: Pre-training of Deep Bidirectional Transformers for Language Understanding. In Proceedings of the 2019 Conference of the North American Chapter of the Association for Computational Linguistics: Human Language Technologies, Volume 1 (Long and Short Papers), Minneapolis, MN, USA, 2–7 June 2019; pp. 4171–4186. [CrossRef]
66. Kerz, E.; Qiao, Y.; Zanwar, S.; Wiechmann, D. Pushing on Personality Detection from Verbal Behavior: A Transformer Meets Text Contours of Psycholinguistic Features. *arXiv* **2022**, arXiv:2204.04629. [CrossRef]

67. Li, C.; Hancock, M.; Bowles, B.; Hancock, O.; Perg, L.; Brown, P.; Burrell, A.; Frank, G.; Stiers, F.; Marshall, S.; et al. Feature extraction from social media posts for psychometric typing of participants. In Proceedings of the International Conference on Augmented Cognition, Las Vegas, NV, USA, 3 June 2018; pp. 267–286. [CrossRef]
68. Ahmad, H.; Asghar, M.U.; Asghar, M.Z.; Khan, A.; Mosavi, A.H. A hybrid deep learning technique for personality trait classification from text. *IEEE Access* **2021**, *9*, 146214–146232. [CrossRef]
69. Li, Y.; Kazemeini, A.; Mehta, Y.; Cambria, E. Multitask learning for emotion and personality traits detection. *Neurocomputing* **2022**, *493*, 340–350. [CrossRef]
70. Schwartz, H.A.; Eichstaedt, J.C.; Kern, M.L.; Dziurzynski, L.; Ramones, S.M.; Agrawal, M.; Shah, A.; Kosinski, M.; Stillwell, D.; Seligman, M.E.; et al. Personality, gender, and age in the language of social media: The open-vocabulary approach. *PLoS ONE* **2013**, *8*, e73791. [CrossRef]
71. Farnadi, G.; Zoghbi, S.; Moens, M.F.; De Cock, M. Recognising personality traits using facebook status updates. In Proceedings of the International AAAI Conference on Web and Social Media, Cambridge, MA, USA, 8–11 June 2013; Volume 7, pp. 14–18.
72. Verhoeven, B.; Daelemans, W.; De Smedt, T. Ensemble methods for personality recognition. In Proceedings of the International AAAI Conference on Web and Social Media, Cambridge, MA, USA, 8–11 July 2013; Volume 7, pp. 35–38.
73. Celli, F.; Pianesi, F.; Stillwell, D.; Kosinski, M. Workshop on computational personality recognition: Shared task. In Proceedings of the International AAAI Conference on Web and Social Media, Cambridge, MA, USA, 8–11 July 2013; Volume 7, pp. 2–5.
74. Park, G.; Schwartz, H.A.; Eichstaedt, J.C.; Kern, M.L.; Kosinski, M.; Stillwell, D.J.; Ungar, L.H.; Seligman, M.E. Automatic personality assessment through social media language. *J. Personal. Soc. Psychol.* **2015**, *108*, 934. [CrossRef]
75. Youyou, W.; Kosinski, M.; Stillwell, D. Computer-based personality judgments are more accurate than those made by humans. *Proc. Natl. Acad. Sci. USA* **2015**, *112*, 1036–1040. [CrossRef]
76. Segalin, C.; Celli, F.; Polonio, L.; Kosinski, M.; Stillwell, D.; Sebe, N.; Cristani, M.; Lepri, B. What your Facebook profile picture reveals about your personality. In Proceedings of the 25th ACM International Conference on Multimedia, Mountain View, CA, USA, 23–27 October 2017; pp. 460–468. [CrossRef]
77. Tandera, T.; Suhartono, D.; Wongso, R.; Prasetio, Y.L. Personality prediction system from facebook users. *Procedia Comput. Sci.* **2017**, *116*, 604–611. [CrossRef]
78. Kulkarni, V.; Kern, M.L.; Stillwell, D.; Kosinski, M.; Matz, S.; Ungar, L.; Skiena, S.; Schwartz, H.A. Latent human traits in the language of social media: An open-vocabulary approach. *PLoS ONE* **2018**, *13*, e0201703. [CrossRef]
79. Ramos, R.; Neto, G.; Silva, B.; Monteiro, D.; Paraboni, I.; Dias, R. Building a corpus for personality-dependent natural language understanding and generation. In Proceedings of the Eleventh International Conference on Language Resources and Evaluation (LREC 2018), Miyazaki, Japan, 7–12 May 2018; pp. 1138–1145.
80. Xue, D.; Wu, L.; Hong, Z.; Guo, S.; Gao, L.; Wu, Z.; Zhong, X.; Sun, J. Deep learning-based personality recognition from text posts of online social networks. *Appl. Intell.* **2018**, *48*, 4232–4246. [CrossRef]
81. Marengo, D.; Sindermann, C.; Elhai, J.D.; Montag, C. One social media company to rule them all: Associations between use of Facebook-owned social media platforms, sociodemographic characteristics, and the big five personality traits. *Front. Psychol.* **2020**, *11*, 936. [CrossRef]
82. Cristani, M.; Vinciarelli, A.; Segalin, C.; Perina, A. Unveiling the multimedia unconscious: Implicit cognitive processes and multimedia content analysis. In Proceedings of the 21st ACM International Conference on Multimedia, Barcelona, Spain, 21–25 October 2013; pp. 213–222. [CrossRef]
83. Osterholz, S.; Mosel, E.I.; Egloff, B. #Insta personality: Personality expression in Instagram accounts, impression formation, and accuracy of personality judgments at zero acquaintance. *J. Personal.* **2022**, 1–17. [CrossRef]
84. Gjurković, M.; Šnajder, J. Reddit: A gold mine for personality prediction. In Proceedings of the Second Workshop on Computational Modeling of People's Opinions, Personality, and Emotions in Social Media, New Orleans, LA, USA, 6 June 2018; pp. 87–97. [CrossRef]
85. Wu, X.; Lin, W.; Wang, Z.; Rastorgueva, E. Author2Vec: A Framework for Generating User Embedding. *CoRR* **2020**. [CrossRef]
86. Radisavljević, D.; Batalo, B.; Rzepka, R.; Araki, K. Myers-Briggs Type Indicator and the Big Five Model—How Our Personality Affects Language Use. In Proceedings of the IEEE CSDE 2022, Gold Coast, Australia, 18–20 December 2022.
87. Zhou, Z.; Xu, K.; Zhao, J. Extroverts tweet differently from introverts in Weibo. *EPJ Data Sci.* **2018**, *7*, 1–22. [CrossRef]
88. Meng, K.S.; Leung, L. Factors influencing TikTok engagement behaviors in China: An examination of gratifications sought, narcissism, and the Big Five personality traits. *Telecommun. Policy* **2021**, *45*, 102172. [CrossRef]
89. Plank, B.; Hovy, D. Personality traits on twitter—or—how to get 1500 personality tests in a week. In Proceedings of the sixth Workshop on Computational Approaches to Subjectivity, Sentiment and Social Media Analysis, Lisbon, Portugal, 17 September 2015; pp. 92–98. [CrossRef]
90. Verhoeven, B.; Daelemans, W.; Plank, B. Twisty: A multilingual twitter stylometry corpus for gender and personality profiling. In Proceedings of the tenth Annual Conference on Language Resources and Evaluation (LREC 2016), Portorož, Slovenia, 23–28 May 2016; pp. 1–6.
91. Tighe, E.; Cheng, C. Modeling personality traits of filipino twitter users. In Proceedings of the Second Workshop on Computational Modeling of People's Opinions, Personality, and Emotions in Social Media, New Orleans, LA, USA, 6 June 2018; pp. 112–122. [CrossRef]

92. Balakrishnan, V.; Khan, S.; Fernandez, T.; Arabnia, H.R. Cyberbullying detection on twitter using Big Five and Dark Triad features. *Personal. Individ. Differ.* **2019**, *141*, 252–257. [CrossRef]
93. Cahyani, D.E.; Faishal, A.F. Classification of Big Five Personality Behavior Tendencies Based on Study Field with Twitter Analysis Using Support Vector Machine. In Proceedings of the 2020 seventh International Conference on Information Technology, Computer, and Electrical Engineering (ICITACEE), Online, 24–25 September 2020; pp. 140–145. [CrossRef]
94. Bassignana, E.; Nissim, M.; Patti, V. Matching Theory and Data with Personal–ITY: What a Corpus of Italian YouTube Comments Reveals About Personality. In Proceedings of the Third Workshop on Computational Modeling of People's Opinions, Personality, and Emotion's in Social Media. Association for Computational Linguistics, Barcelona, Spain, 13 December 2020; pp. 11–22. [CrossRef]
95. Mehta, Y.; Fatehi, S.; Kazameini, A.; Stachl, C.; Cambria, E.; Eetemadi, S. Bottom-up and top-down: Predicting personality with psycholinguistic and language model features. In Proceedings of the 2020 IEEE International Conference on Data Mining (ICDM), Sorrento, Italy, 17–20 November 2020; pp. 1184–1189. [CrossRef]
96. Uysal, S.D.; Pohlmeier, W. Unemployment duration and personality. *J. Econ. Psychol.* **2011**, *32*, 980–992. [CrossRef]
97. Daumé, H., III. Frustratingly Easy Domain Adaptation. In Proceedings of the 45th Annual Meeting of the Association of Computational Linguistics, Prague, Czech Republic, 23–30 June 2007; Association for Computational Linguistics: Prague, Czech Republic, 2007; pp. 256–263. [CrossRef]
98. Peng, C.Y.J.; Lee, K.L.; Ingersoll, G.M. An introduction to logistic regression analysis and reporting. *J. Educ. Res.* **2002**, *96*, 3–14. [CrossRef]
99. Hoerl, A.E.; Kennard, R.W. Ridge regression: Biased estimation for nonorthogonal problems. *Technometrics* **1970**, *12*, 55–67. [CrossRef]
100. Majumder, N.; Poria, S.; Gelbukh, A.; Cambria, E. Deep learning-based document modeling for personality detection from text. *IEEE Intell. Syst.* **2017**, *32*, 74–79. [CrossRef]
101. Ramírez-de-la Rosa, G.; Jiménez-Salazar, H.; Villatoro-Tello, E.; Reyes-Meza, V.; Rojas-Avila, J. A lexical–availability–based framework from short communications for automatic personality identification. *Cogn. Syst. Res.* **2023**, *79*, 126–137. [CrossRef]
102. Boyd, R.L.; Pennebaker, J.W. Language-based personality: A new approach to personality in a digital world. *Curr. Opin. Behav. Sci.* **2017**, *18*, 63–68. [CrossRef]
103. Boyd, R.L.; Schwartz, H.A. Natural language analysis and the psychology of verbal behavior: The past, present, and future states of the field. *J. Lang. Soc. Psychol.* **2021**, *40*, 21–41. [CrossRef] [PubMed]
104. Holtgraves, T. Text messaging, personality, and the social context. *JOurnal Res. Personal.* **2011**, *45*, 92–99. [CrossRef]
105. Pennebaker, J.W.; Francis, M.E.; Booth, R.J. Linguistic inquiry and word count: LIWC 2001. *Mahway Lawrence Erlbaum Assoc.* **2001**, *71*, 1–23.
106. Vinciarelli, A.; Pantic, M.; Heylen, D.; Pelachaud, C.; Poggi, I.; D'Errico, F.; Schroeder, M. Bridging the gap between social animal and unsocial machine: A survey of social signal processing. *IEEE Trans. Affect. Comput.* **2011**, *3*, 69–87. [CrossRef]
107. Baumgartner, J.; Zannettou, S.; Keegan, B.; Squire, M.; Blackburn, J. The pushshift reddit dataset. In Proceedings of the International AAAI Conference on Web and Social Media, Atlanta, GE, USA, 8–11 June 2020; Volume 14, pp. 830–839. [CrossRef]
108. Tibshirani, R. Regression shrinkage and selection via the lasso. *JOurnal R. Stat. Soc. Ser. B (Methodol.)* **1996**, *58*, 267–288. [CrossRef]
109. Huber, P.J. Robust statistics. In *International Encyclopedia of Statistical Science*; Springer: Berlin/Heidelberg, Germany, 2011; pp. 1248–1251. [CrossRef]
110. Awad, M.; Khanna, R. *Efficient Learning Machines*; Apress Open: New York, NY, USA, 2015; Chapter 4; pp. 67–80. [CrossRef]
111. Amirhosseini, M.H.; Kazemian, H. Machine learning approach to personality type prediction based on the myers–briggs type indicator®. *Multimodal Technol. Interact.* **2020**, *4*, 9. [CrossRef]
112. Prechelt, L. Early stopping-but when? In *Neural Networks: Tricks of the Trade*; Springer: Berlin/Heidelberg, Germany, 1998; pp. 55–69. [CrossRef]
113. Schober, P.; Boer, C.; Schwarte, L.A. Correlation coefficients: Appropriate use and interpretation. *Anesth. Analg.* **2018**, *126*, 1763–1768. [CrossRef]
114. Irwin, J.R.; McClelland, G.H. Negative consequences of dichotomizing continuous predictor variables. *J. Mark. Res.* **2003**, *40*, 366–371. [CrossRef]
115. Royston, P.; Altman, D.G.; Sauerbrei, W. Dichotomizing continuous predictors in multiple regression: A bad idea. *Stat. Med.* **2006**, *25*, 127–141. [CrossRef]

Disclaimer/Publisher's Note: The statements, opinions and data contained in all publications are solely those of the individual author(s) and contributor(s) and not of MDPI and/or the editor(s). MDPI and/or the editor(s) disclaim responsibility for any injury to people or property resulting from any ideas, methods, instructions or products referred to in the content.

Article

Understanding of Customer Decision-Making Behaviors Depending on Online Reviews

Yeo-Gyeong Noh [1], Junryeol Jeon [1] and Jin-Hyuk Hong [1,2,*]

1. School of Integrated Technology, Gwangju Institute of Science and Technology, Gwangju 61005, Republic of Korea
2. Artificial Intelligence Graduate School, Gwangju Institute of Science and Technology, Gwangju 61005, Republic of Korea
* Correspondence: jh7.hong@gist.ac.kr

Abstract: With a never-ending stream of reviews propagating online, consumers encounter countless good and bad reviews. Depending on which reviews consumers read, they get a different impression of the product. In this paper, we focused on the relationship between the text and numerical information of reviews to gain a better understanding of the decision-making process of consumers affected by the reviews. We evaluated the decisions that consumers made when encountering the review structure of star ratings paired with comments, with respect to three research questions: (1) how consumers compare two products with reviews, (2) how they individually perceive a product based on the corresponding reviews, and (3) how they interpret star ratings and comments. Through the user study, we confirmed that consumers consider reviews differently according to product presentation conditions. When consumers were comparing products, they were more influenced by star ratings, whereas when they were evaluating individual products, they were more influenced by comments. Additionally, consumers planning to buy a product examined star ratings by more stringent criteria than those who had already purchased the product.

Keywords: consumer decision-making; star rating; review comment; sentiment analysis

Citation: Noh, Y.-G.; Jeon, J.; Hong, J.-H. Understanding of Customer Decision-Making Behaviors Depending on Online Reviews. *Appl. Sci.* 2023, *13*, 3949. https://doi.org/10.3390/app13063949

Academic Editor: Christos Bouras

Received: 13 February 2023
Revised: 13 March 2023
Accepted: 18 March 2023
Published: 20 March 2023

Copyright: © 2023 by the authors. Licensee MDPI, Basel, Switzerland. This article is an open access article distributed under the terms and conditions of the Creative Commons Attribution (CC BY) license (https://creativecommons.org/licenses/by/4.0/).

1. Introduction

Numerous reviews are being made and uploaded through the Internet today. This applies not only to tangible products, such as fans and vacuum cleaners, but also to intangible consumer goods, such as concerts and movies. These reviews are generated by consumers and have a huge impact on consumers' purchasing decisions [1,2]. Consumers perceive reviews as more reliable and engaging than information provided by vendors because they are created by fellow consumers who have already purchased and have been using the products [3]. From these reviews, companies obtain valuable information from the consumers' perspectives of their products [4]. In addition, online reviewing has been integrated by many companies because it is an effective way to not only engage consumers directly with the products, but also influence consumers [5].

There is a large diversity of reviews due to reflecting consumers' perceptions, and there is usually a variety of consumer perspectives even for a single product. In a related study [6], it was confirmed that the tendency to be affected by review data varies depending on the consumer type of how many expectations are made for the product. Depending on the type of consumer, there are differences, such as being more vulnerable to negative reviews or being affected by the selection. In the case of user types in which the reliability of reviews may be seen as a measure, factors, such as sources, are also affected [7]. From a seller's point of view, naturally, they want their products to have a positive perception. However, it is not advantageous to consumers for only the large number of good reviews to have high visibility, which is often the case. Reviews should be transparent and unbiased.

User reviews implemented by companies have two main characteristics. The first is the numerical review, such as a one to five-star rating, and the second is the text review, such as a comment. In most cases, star ratings and comments are shown together. Company sites that provide consumer goods with reviews, such as Amazon, Walmart, and Movietickets provide star ratings and comments together. Yet, it is unclear which review component, either star ratings or comments, holds more weight to the consumer. Do consumers prefer consumer goods with high star ratings but poor comments or vice versa? What do consumers think when they look at the star rating to make a purchase decision? This paper explores the decisions that consumers make when presented with both review components simultaneously. Thus, three tasks were devised through our user study to simulate consumers' reading and interpretation of reviews. As a result of the user study, we had the following observations; (1) Differences in the importance of the review components according to the purpose of reading them, and (2) arbitrary interpretation of reviews by consumers that were different from the intention of the reviewers.

2. Background

2.1. Attempts to Understand Consumers through Online Reviews

Online reviews provide valuable data that reflect the interactions between consumers and products. They comprise feedback and viewpoints of consumers with respect to products, serving as a form of electronic word-of-mouth [8], based on users' willingness to share their opinions. Understanding consumers through online reviews can help researchers and practitioners in the product-related field to enhance consumer purchasing intentions and loyalty [9,10]. Based on the better understanding of consumers, recommendation systems have potential to not only augment consumers' purchasing experience, but also boost sales revenue [11].

For this, the foremost step is to gain insight into the consumer experience. Consumer experience refers to an individual's sensory, emotional, cognitive, behavioral, and relational experiences as a consumer with a product or service [12]. Depending on the situation, this experience can be remembered positively or negatively, and if it is remembered positively, the consumer will feel satisfaction [13]. By identifying the key features that contribute to consumer experience, product and service providers can identify areas where they should focus their efforts [14,15]. Text-based review data analysis is a common method used to analyze key features. Latent Semantic Analysis [16,17] can be utilized to extract hidden semantic patterns of words and phrases that make up the document corpus [18].

Based on user experience and preference information that have been identified in previous research, the recommendation system has the ability to suggest more appropriate products [19]. The recommendation system incorporates a content-based filtering technique that assesses the similarity between a product and consumer preference [20]. However, in recent times, the collaborative filtering technique has gained prominence [21]. This is largely due to the abundance of online reviews, which facilitates the learning of a model that can predict an individual's interests. The most representative methods for predicting user interests include the neighborhood method and latent-factor model [22]. The neighborhood method involves finding the group of users and items that are the most similar to the individual's interests using algorithms such as the K-nearest neighbors [22,23]. The latent-factor model method involves vectorizing users and items and modeling the relationship between them using neural networks in the latent space [24,25].

2.2. Studies on Helpfulness of Reviews

It is important to understand the helpfulness of reviews to allow consumers to be exposed to more holistic and better summarized review information, despite the excess of review information in the modern age [26]. The helpfulness of reviews can be identified by analyzing which characteristics of reviews are essential to consumers [27–29]. Quantitative characteristics, such as the numerical rating of reviews or the length of a review [30], or qualitative characteristics, such as sentiment or review complexity [31], affect the ability to infer the quality of a consumer product. In addition, factors affecting review helpfulness include reviewer-related factors (e.g., writer reputation), reader-related factors (e.g., reader identification), and environmental factors (e.g., website reputation) [27]. Through techniques such as natural language processing (NLP), machine learning, and deep learning, review helpfulness is regressed or classified by the factors mentioned above [32–36]. According to past studies, negative reviews tend to be more impactful than positive reviews [37,38] and reputation deterioration caused by negative reviews has an adverse effect on profits [39]. For both the product providers and consumers, it is necessary to present negative reviews, positive reviews, and an overall picture of consumer reviews all together, appropriately sorted.

Talton et al. studied a method of rating sorting and observed a user's choice when presented with two different rating scenarios [40]. Rating scenarios were a combination of a thumbs-up/thumbs-down ratio and a sample size. For example, there are comparisons where one distribution has a higher positive ratio but a lower number of total votes. To compare the two distributions, they calculated inter-rater reliability, a percentage value obtained by dividing the absolute value of the difference in the number of selections between two combinations by the total number of selections. Additionally, by measuring the time-to-rank, the time taken to decide which combination is more preferred, observations were made on the difficulty of ranking these scenarios.

Koji et al. developed Review Spotlight which summarizes text reviews and presents them as adjective-noun pairs [41]. Review Spotlight uses sentiment analysis to select adjacent adjectives and nouns in the texts and matches them as pairs. The pairs shown to the user have different colors depending on the three cases, where their sentiment is interpreted as positive, neutral, or negative. If the user clicks on the adjective-noun pair, the original text review of the pair is shown.

2.3. Two Types of Reviews: Numeric and Text

User reviews typically have a numerical and text portion. A rating scale is a method that requires the rater to assign a value to a measure for an assessed attribution. Several rating scales have been developed for numerical review systems. According to Chen's study, high traffic websites mainly adopt one of two types of rating schemes [42]. One is a binary thumbs-up/thumbs- down system and the other is the one to five-star rating system. Binary thumbs- up/thumbs-down rating is a rating scale that doesn't allow for neutrality. This forces the rater to assign a good or bad designation, but does not provide insight into the degree of how good or bad [43]. The one to five-star is widely used in commercial online reviews. It is scored on a scale of one to five, with more stars equated to a more positive rating. Numerical reviews are advantageous in assigning a scaled system of scoring, providing greater insight [44]. The standard of each score in the star ratings may be different for each individual rater, but since the deviation of the difference is relatively small, the distribution of scores roughly follows a normal distribution. Therefore, a star rating system is regularly averaged to provide a numeric rating.

A text review includes qualitative information that is basically impossible to provide through a rating scale review. Being subjective, variation is high for text reviews [45]. Raters will often leave the same numerical ratings, but rarely have the same text reviews. Text reviews are also more subject to subjective variation. A rater leaving a four-star review may leave a comment, "this product was very good", while another may leave the same four-star review with the comment, "this product was somewhat disappointing." Therefore, a diversified number of text reviews for a product cannot be easily summarized or averaged, so the method currently employed by most companies is to list text reviews by the number of likes or by recency. However, listing by number of likes or recency can be too random or biased.

2.4. Research Question

Based on the importance and availability of the aforementioned online reviews, we studied user selection patterns considering multiple combinations of a star rating and comments. This paper aims to provide a better understanding of the consumer's decision-making process in an environment where they need to introduce attractive reviews to consumers or give them trust when providing reviews. To this end, three research questions were specifically designed according to the purpose: (1) First, in the choice between two things, we investigated if the star rating or the comment had a stronger impact on consumers. (2) When a single product is given with a star rating and comments, how the user makes choices was also studied. (3) Finally, when a star rating was given, the reliability interval of credible comments was investigated; for example, how a review with a three-point star rating affects the trustworthiness of the comment. We formulated two specific hypotheses with the aim of understanding how consumers interpret the combination of horoscopes and reviews, based on the research questions at hand.

Hypothesis H1. *In a binary selection situation, when star rating and reviews are oppositely given, the option with a high star rating will be chosen to prevail.*

Hypothesis H2. *In a single selection situation, deciding on a single subject alone, review data will be considered to be more important than star rating.*

3. Review Data Construction

3.1. Why Movie Review

We used movie reviews as our dataset in this study. Among numerous consumer product reviews, movie reviews were selected because the price of movies is not variable and emotional language is usually used in movie reviews.

The pricing of a product is considered to be one of the primary factors that significantly impacts the outcomes of consumer reviews. The price–quality inference posits that consumers perceive price and quality as strongly correlated [46]. Moreover, a product's quality level is often objectively reflected in its performance, as perceived by consumers [47,48]. Consequently, as the price of a product increases, customers are more likely to be satisfied [49]. In light of the fact that review data itself is indicative of consumer satisfaction, the various ways in which prices are formed in the review data may act as a confounding factor in our data design. However, movie review data is relatively free from this factor since all movies are equally priced.

Sentiment analysis is a technique utilized for information extraction [50] in which the objective is to gather and categorize emotions or attitudes towards a specific topic or entity [51,52]. In our study, we opted to apply sentiment analysis to review comments as a means of classification. We specifically chose to analyze movie reviews, as they are known to be more informal and emotionally charged, often displaying personal tendencies [53]. We deemed this strategy to be effective for our research purposes. We crawled information from a search engine that provides free reviews and scores written by users on movies.

3.2. Experimental Data Preparation

The crawled movie reviews of individual users consist of a comment with a star rating mapped to it. Since each individual has different criteria for grading comments and stars, we could not immediately use all the collected comments for the experiment. Therefore, we selected comments to represent a specific star rating score range. This was done by first mapping the sentiment frequently used in comments to each specific star rating section. Then, we selected comments containing a sentiment with high frequency of use as a representative review of the star rating section. The overall process is shown in Figure 1. The GoEmotions dataset is the largest manually annotated dataset of 58,000 English Reddit comments, labeled for 27 emotional categories [54]. To perform sentiment analysis on the collected review comments, we translated the GoEmotions dataset to Korean and then trained it through KoELECTRA [55] to create a sentiment analysis model. We applied the collected Korean movie review comment data to this model to analyze and identify sentiments.

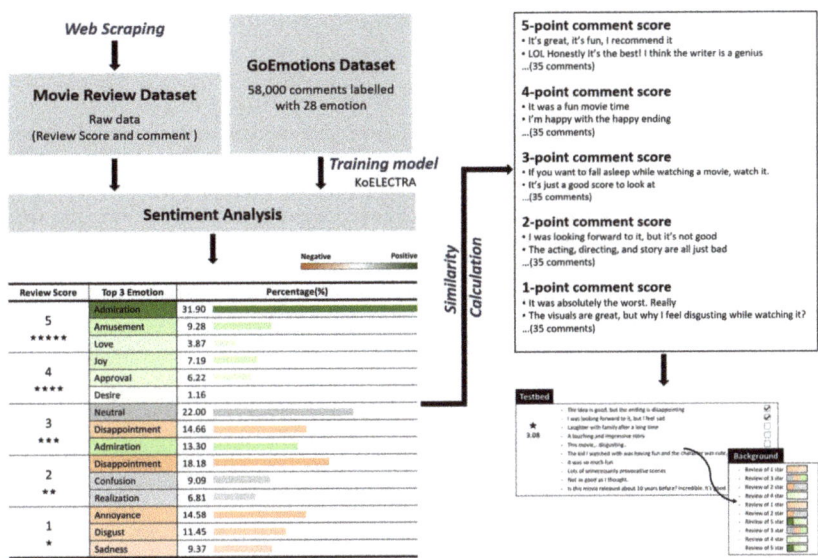

Figure 1. Overall process of experimental data preparation.

As a result of extracting emotional information from movie review scores, we correlated specific emotions to scores, as shown in Figure 1, marked with the three highest percentage emotions per each star rating. In the five-star range, admiration (31.9%) was the highest by far, followed by amusement (9.28%) and love (3.87%). The emotions that appear in the four-star range were joy (7.19%), approval (6.22%), and desire (1.16%). The three most common emotions in the three-star range were neutral (22%), disappointment (14.7%), and admiration (13.3%). Based on the general acceptance of the three-star range, we decided to only consider neutral sentiments in the three-star range. (Note that admiration and disappointment were included in the five-star and two-star ranges.) Emotions, such as disappointment (18.18%), disapproval (11.36%), and confusion (9.09%) were the most frequent in the two-star range, with emotions such as curiosity (4.54%) and optimism (2.27%) being rare. Curiosity and optimism were emotions extracted from mostly sarcastic reviewers. Lastly, annoyance (14.58%), disgust (11.45%), and sadness (9.37%) were the most common in the one-star range. We calculated the similarity of the Euclidean distance, as shown in formula (1), for the three selected emotions from the one to five-star ratings and selected the 165 comments closest to the five-star ranges (165 comments = 35 comments × 5 star ranges) as follows:

$$\text{similarity} = \sqrt{(x_2 - x_1)^2 + (y_2 - y_1)^2 + (z_2 - z_1)^2} \tag{1}$$

Emotion Type (x, y, z), Representative Emotion Value (x_2, y_2, z_2), Review Emotion Value (x_1, y_1, z_1)

Finally, we constructed a pool of comments, consisting of 165 comments for five-star ranges, and redefined the ranges as comment scores (CSs) from one to five, i.e., each CS corresponded with 35 comments.

3.3. Star Rating Score Presentation

Unlike comments, star ratings are presented as the average value of all ratings. However, for the average star rating, the collected scores ranged from 2.5 to 5 points. We divided the range of collected star ratings into five and applied it to the user study. We decided to call these five ranges the star rating ranges (SRRs) as follows. (SRR #1: $2.5 < x \leq 3.0$, SRR #2: $3.0 < x \leq 3.5$, SRR #3: $3.5 < x \leq 4.0$, SRR #4: $4.0 < x \leq 4.5$, SRR #5: $4.5 < x \leq 5.0$, x is average star rating score).

4. User Study
4.1. Overview

We designed three tasks corresponding to each research question (shown in Figure 2). Participants were selected by non-probabilistic sampling (chain-referral sampling), considering the difficulty of finding subjects representative of the experiment population. We recruited a total of 72 participants by posting announcements on social networking services and through the university community. The experiment took on average about 23 min for each participant, 6.80 USD was paid as compensation, and the experiment was conducted online. Participants' demographic information are summarized in Appendix A. In addition, a survey was conducted to understand the background of the user's movie consumption, as shown in Appendix B. This study did not include personal preferences or traits related to products. In order to exclude preferences according to age, gender, and social status, information such as genres or actors, which are characteristics of the movie, were not provided in the experiment. The development of testbed and clinical study for the progress of the experiment were conducted in the first half of 2021.

Task1. Choosing between two options

Instruction : Please choose your preferred movie from the following two suggested movies

★ 4.59
- It was a rough ride
- Except for great acting...I don't think so...
- It's just a little boring.
It's just a normal fantasy movie.

★ 2.89
- Great! I enjoyed it!
- I watched the movie with my heart fluttering from start to finish. Anyway, I recommend it.
- It's really fun.

Task2. Decision on a single option

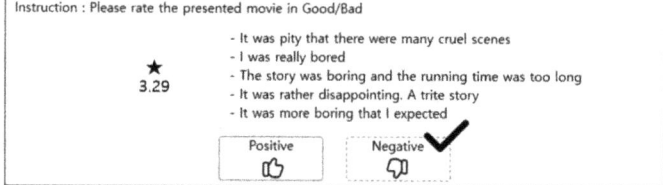

Task3. Acceptable comments to star rating

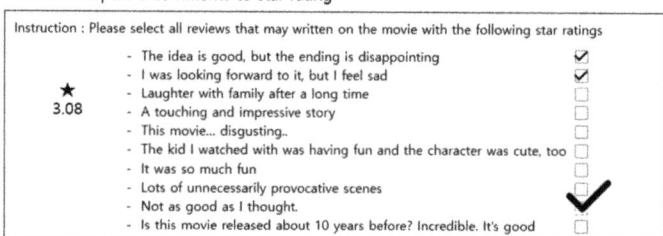

Figure 2. The interface and instruction of the testbed used in the user study; Task1 (**top**): Choosing between two options; Task2 (**middle**): Decision on a single option; Task3 (**bottom**): Acceptable comments to star rating.

4.2. Task Design

4.2.1. TASK1: Choosing between Two Things

Every day, consumers compare products by star rating and comments. The selection between two things is an insightful, simplified version of the actual selection process for multiple options. The experiment task was to decide between two movies, each with a review (one one-to-five-star rating, plus three comments), as shown in Figure 2-Task1. We compared Type A (a low star rating, high scoring comments) and Type B (a high star rating, low scoring comments), as shown in Figure 3 (left). Selection of Type A can be interpreted as indicating that comments have a stronger influence than star rating, and conversely, for selection of Type B, vice versa. To account for all combinations of Type A and Type B scenarios, a total of 100 sample spaces were compared, as shown in Figure 3 (right). Look again at the example in Figure 2-Task1. The Type B (left) combination consists of a SRR #5 ($4.5 < x \leq 5.0$) and comments of CS 3, and the Type A (right) combination consists of a SRR #1 ($2.5 < x \leq 3.0$) and comments of CS 5. This example scenario corresponds with the red cell shown in Figure 3 (right).

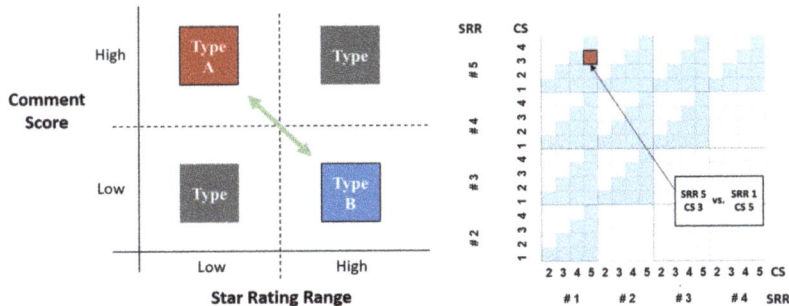

Figure 3. Subject type description (**left**); Sample space description (**right**): Red-colored cell represents the example comparison of Task1.

The Type A and Type B scenarios were randomly generated for specified ranges. For the 100 sample spaces, participants were divided evenly and randomly into two groups, and each performed selections for a half of the entire sample spaces to maintain a high level of attention. For counterbalance, a sequence of 50 selections was randomly presented to the participants. Participants chose one of two reviews/movies given this query. We computed the inter-rater reliability (I), as shown in formula (2), to measure 'contention' for each comparison, an absolute value calculated by the following expression.

$$I = |Na - Nb|/n \qquad (2)$$

Na = number of participants who selected Type A, Nb = number of participants who selected Type B, n = total number of participants who selected either Type A or Type B (n = 36).

This value gets closer to zero when people have higher contention (participants are equally split between the two choices) and closer to one with lower contention (when participants completely agree on one choice). To indicate bias for the calculated inter-rater reliability, we used the positive and negative scale. When Type B is more preferred, inter-rater reliability is a positive value; when Type A is more preferred, inter-rater reliability is a negative value.

4.2.2. TASK2: Decision on a Single Product

Unlike the choice between two things, this task is designed to determine whether a product is preferred or not when given a star rating and comments for a standalone product, mirroring the like/dislike rating system. This reproduces the situation that a consumer reads the reviews on a single product presented on a single page and takes a closer look at the product. To implement the task, we provided the participants with a one-star rating and five comments on an arbitrary movie, as shown in Figure 2-Task2. After examining the reviews, the participants had to decide if they had a desire to get it or not and click the corresponding button. For Figure 2-Task2, the SRR is #2 and the CS of the comments corresponds with 2. The sample space of the experiment consisted of all combinations of SRRs and CSs. The total sample spaces were 25 (5 possible SRRs × 5 possible CSs = 25) and each participant performed the selections for the entire sample space. For counterbalance, the sequence of 25 selections was randomly presented to the participants. During this task, we asked participants to select 'thumbs-up' or 'thumbs-down' for the movie, based on the movie's star rating and comments.

4.2.3. TASK3: Acceptable Comments to Star Rating

This task was conducted to determine the scoring range of 'acceptable' user comments for a star rating. We presented a star rating score along with 10 comments of varying scores, as shown in Figure 2-Task3. The comments presented in each task (scenario) were randomly selected, two from each possible CS (1 to 5). Therefore, 10 comments (2 comments × 5 CSs) were given for each star rating. The star rating score was randomly generated, twice for each possible SSR (#1 to #5). The entire sample space consisted of 10 total scenarios per participant due to running twice for each SRR. During this task, we asked participants to check which comments might match the star rating, with the query "Choose all comments that you think were written for the movie with the following star rating." The task provided insight into the perception of star ratings by consumers and the relationship between star ratings and comments.

5. Result

5.1. Result of TASK1

We collected 3600 selections (3600 selections = 36 participants × 2 groups × 50 selections) for 100 unique comparison scenarios. Each scenario was performed by 36 participants. Type A was selected 1745 times and Type B 1855 times, indicating that Type B was selected slightly more often. The overall trend is represented as a heatmap, shown in Figure 4. Each cell (comparison scenario) is composed of signed inter-rater reliability and indicates which type (Type A or Type B) of bias of the users. The darker the color, the more biased the selection is, and the lighter the color, the more contentious (evenly split) the selection is. The darker the red color, the more dominant Type A was for the comparison scenario; the darker the blue color, the more dominant Type B was for the comparison scenario. In general, a darker blue color appeared in the upper left quadrant and a darker red color appeared for the lower right quadrant. For example, when comparing a scenario with a SRR #5 and a CS of 3 (Type B) with a SRR #1 and a CS of 5 (Type A) (same as compared in Figure 2-Task1.), the participants chose Type B more and the inter-rater reliability at this time was +44.4%. In Figure 4, consider Type A (SRR #3, CS of 5) vs. Type B (SRR #4, CS of 1): the color is the darkest red, indicating a −100% inter-rater reliability score and that all participants chose Type A. There were 45 cells each biased for Type A and Type B. However, the median value of inter-rater reliability values was skewed for Type B (−22.2% for Type A, 33.3% for Type B). Therefore, Type B was chosen more predominantly. Both in terms of the number of choices and in terms of median values of inter-rater reliability, Type B has a slight predominance over Type A in general. Considering that Type B is a combination with a high star rating, when participants performed this task, they prioritized the star rating when making their decision. Therefore, when interpreting the results compared with the established assumption (H1), although the assumption prevailed, it was not acceptable in all cases. This result means that, when comparing two or more products and choosing one, the participants were more influenced by the quantitative review of a star rating.

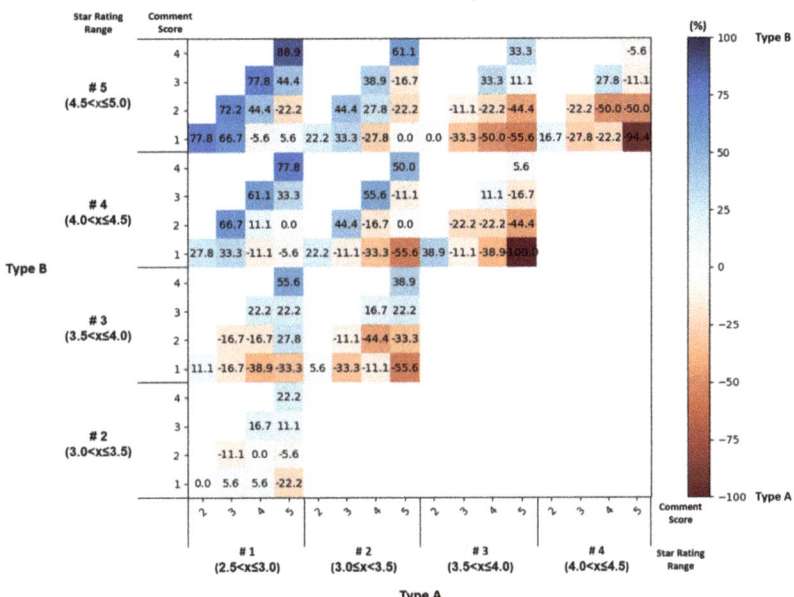

Figure 4. Heatmap of Task1 (Choosing between two things); A darker color corresponds to a higher percentage (%); Type A: red, Type B: blue.

5.2. Result of TASK2

A total of 1800 selections (1800 selections = 72 participants × 25 selections) were collected. Preference value refers to the percentage of how many thumbs-up buttons are pressed for that selection. Preference value for each cell is shown in Figure 5 (left). The darker the color of the cells in this heatmap, the higher the preference. For example, for the CS of 4 vs. SRR #2, the preference value is 69.4%. That means 69.4% of participants chose the 'thumbs-up' button and 30.6% of participants chose the 'thumbs-down' button.

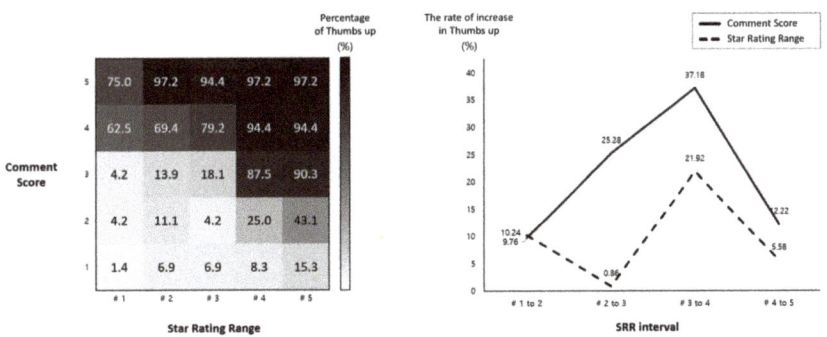

Figure 5. Heatmap of Task2 (**left**); Comparison of Thumbs up growth rates by SRR interval of comment score and star rating range (**right**).

Preference values of 50% or higher were found for: SRRs of #1, #2, #3 with comments of 3 or higher and SRRs of #4 and #5 with CSs of 3 or higher. All cells with CSs of 4 and 5 showed a preference value of 50% or higher, regardless of the star rating. For SRRs of #4 or higher with CSs of 2 or lower, the preference value was lower than 50%. Comparing these findings, CSs seem to have more influence for this type of single product evaluation.

Preference values increased more dramatically when the CS increased than when star rating increased. Figure 5 (right) shows how people's desire to get increased each time that the SRR or CS increased by one step. Preference value change was calculated by taking the difference between the averaged preference values for the designated range (#)1 to (#)2, etc. Each average preference score value was calculated by averaging each column (SRR) or row (CS), shown on the heatmap of Figure 5 (left). The preference score change when the SRR changed from #1 to #2 was 10.24%. The preference score change when CS changed from 1 to 2 was 9.76%. Except for this range, the CS change had a higher preference score change for all ranges. Thus, for evaluating a single product, participants were influenced more by comments than star ratings, contrary to the choice between two products. For the (#)3 to (#)4 score change, the preference change amount of SRR and CS showed their highest values, 21.92% and 37.18%, respectively. For the #2 to #3 change, the SRR had its lowest change of 0.86%. This result indicates that the participants barely changed their behavior, despite the star rating changing from #2 to #3 when they considered the selection on a single product.

Comments had a greater effect than star rating for consumer preference in single product evaluation. Despite a high star rating, the movie was not 'liked' if the CS was low; despite a low star rating, if the CS was high, the movie had more chances to be 'liked.' Additionally, the preference increase due to the increase of CS was higher than the preference increase due to the increase of the star rating. As a result, the hypothesis proposed for a single selection (H2) has been supported, as the review data was found to have a greater impact on decision-making. It seems that preference (liking or disliking) of a single product is more influenced by subjective information in the form of comments, rather than star ratings.

5.3. Result of TASK3

A total of 720 data points for Task 3 were collected (72 participants × 10 scenarios). Figure 6 shows the resulting range of CSs suitable for each SRR. We conducted the Mann–Whitney U test to check whether the ranges of CSs assigned to each SRR were significantly different from each other. In the Mann–Whitney U test, there was a significant difference (*** $p < 0.001$) in all cases, except when comparing SRR #4 and #5 s ($p = 0.056$). The difference between SRRs #4 and #5 was the smallest, and the difference between SRRs #2 and #3 was the largest.

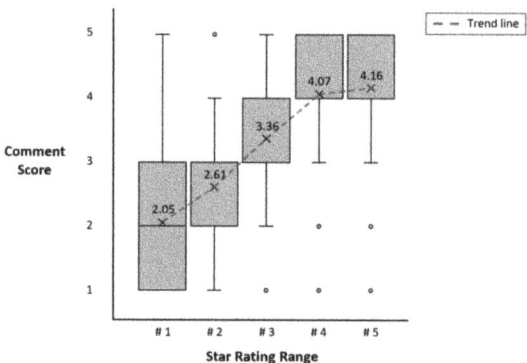

Figure 6. Acceptable comment score according to star rating range.

The average calculated CS for each SRR was 2.05, 2.61, 3.36, 4.07, and 4.16 for #1 to #5, respectively. When we substitute the representative sentiment, according to the CS that we obtained through sentiment analysis into the result of Task 3, the participants selected a negative sentiment for SRRs #1 and #2, a neutral sentiment for SRR #3, and a positive sentiment for SRRs #4 and #5.

Interestingly, when converting SRR #3 to its actual rating range, it corresponds to 3.5 to 4.0 stars for a one-to-five-star scale. The sentiment collected for this SRR #3 was neutral. Thus, despite a one-to-five-star scale indicating mathematical neutrality at the three-star rating, our sentiment analysis shows that consumers view 3.5 to 4.0 stars as a neutral rating. Thus, consumers who are planning to purchase a product view the star rating with a stricter standard than reviewers who have already made a purchase and given a review. Both SRRs #4 and #5 were viewed with positive sentiment, with no significant difference in consumer sentiment between the two. SRRs #4 and #5 are accepted by consumers in a positive way, but the difference in the positive level is not that big compared with the difference in star rating. It seems that people do not notice the difference between the high star ratings.

CSs changed most significantly when SRRs changed from #2 to #3. The average CS increased by 0.56, 0.75, 0.71, and 0.09 when the SRR changed from #1 to #5, respectively. Thus, for SRRs #2 to #3, CS increased by the largest margin. From a sentiment point of view, consumers seem to have a jump from the low 3.0 values to the high 4.0 values as the largest increase in perceived value.

5.4. Star Rating Inflation

From the results of Task 1, we determined that consumers were more influenced by star ratings when making decisions between two products. Based on Task 3, consumers were highly critical of star ratings, viewing a 3.5 to 4.0 star rating with a neutral sentiment, despite 3.0 being the mathematically neutral rating. Additionally, consumers did not demonstrate a difference in sentiment for star ratings between 4.0 and 5.0. Thus, ratings in the 4.0 to 5.0 range lose discriminating power. When the rating scale loses discrimination, consumers are less able to differentiate between good and bad products. Maintaining an exceedingly high star rating, such as one higher than 4.9, becomes a necessity for sellers who are trying to differentiate their products. As a result, sellers may create some fake reviews to increase their star rating, as revealed in the study by Luca et al. [56]. This behavior leads to a vicious cycle in which star ratings lose their discriminating power and the star rating scale becomes inflated. This cycle is detrimental to both consumers and sellers. Consumers cannot discriminate between products and sellers compete for a volatile and often inaccurate rating. To ameliorate this issue, the sentiment of consumers reading reviews and the sentiment of reviewers writing them need to be better aligned. In addition, comments can be better integrated and structured to give insight into products and guide consumers.

6. Discussion

6.1. Limitation

In this paper, we selected star ratings and comments as the main influencing factors to understand a consumer's decision-making behaviors towards movies. For this reason, we inevitably excluded information, such as genre, story, and directing or visual style, and consistently designed review datasets representing each score. Additionally, analysis without considering the preference for numerical and text review according to the user's characteristics and user's state data (e.g., gaze and biometric) can be pointed out as a limitation. For instance, it appears that the gaze data of the participant executing the task can be utilized for secondary analysis. By analyzing the area of star ratings or comments through gaze analysis, it becomes feasible to scrutinize which component garners greater attention from the user [57,58]. Furthermore, when inspecting the relevant component, the user's psychological state may also be inferred [59,60]. We analyzed representative emotions in each score range through sentiment analysis and constructed a dataset with comments that represent those emotions well. However, constructing comment data based on emotional similarity does not sufficiently eliminate all confounding factors. For example, from the review comment, "I watched a movie with my children, and they enjoyed it a lot", the participant might assume that the movie was of the animation genre. One participant commented that he chose movies that were similar to his tastes by inferring genres. Consequently, given the availability of data that can be used to infer cinematic characteristics, it is challenging to generalize this finding to all other products.

In addition, the diversity of expressions was not accounted for. More specifically, reviews using sarcasm or expressing emotions other than the representative emotions were excluded from our dataset. This situation often occurs when some common idioms are used in comments. As has been discussed in other studies [61–63], it is very challenging to accurately read the implied intentions of the user. This is because emotions can be expressed in reverse through paradoxical expressions or irony. For example, the comment "I'm okay with not eating sweet potatoes for a while", was mapped to a one-star rating. In Korea, the expression 'eating sweet potatoes' is negative. However, this comment could not be used because it was analyzed as a positive emotion of 'caring' through sentiment analysis. It seems possible to experiment with more abstract comments if the natural language processing technique, such as sarcasm detection [64], that analyzes various expressions beyond the superficial expression of emotions, is applied. Therefore, detecting an unusual emotion or sarcasm is a very challenging issue.

6.2. Future Work

In this study, the movie review dataset was used because of zero price variability for movies and the frequent use of emotional language in movie reviews. These characteristics reduce the variables that we need to control, but at the same time, there are a number of areas that we need further analysis on in the future. Since most consumer goods do not have the same price, it seems necessary to study how consumers behave when reviews and prices are provided at the same time in future research. Additionally, we set a comment score in this study and selected three representative sentiments for it through sentiment analysis. However, this sentiment is generated from movie review data. If sentiment analysis is applied to review data for other consumer goods, it is expected that different representative sentiments will be selected even with the same comment score. To solve this problem, it seems necessary to extend the sentiment analysis of the movie domain to other domains. Transfer learning techniques can be used to extend domains [65]. After collecting review data of the domain to be applied, additional training is performed on the pre-trained model. This enables sentiment analysis for other review domains. It seems possible to find sentiment expressions that are similarly used for each domain through word embedding [66]. Through this, sentiment analysis can be applied to multiple domains with only one model without transfer learning.

7. Conclusions

This paper aimed to identify how people perceive and behave against reviews of star ratings paired with comments. A choice between two products, single product evaluation, and the relationship between comments and star ratings were investigated by user behavioral and sentiment analysis. First, we confirmed the usability of movie review data through exploratory data analysis and designed three viable user studies with 72 participants. Consumers value star ratings and comments differently, based on the scenario (experiment 1 and 2) in which they are provided. For the first study, a choice problem between two products, star rating had more influence than comments when choosing a more preferred movie. Yet, if movie ratings and comments are presented for a single movie, a 'like' or 'dislike' decision was more influenced by comments. Additionally, consumers were strict when assigning positive sentiment towards star ratings when viewing them as consumers, rather than when reviewing products themselves. Our study aims to contribute insights into identifying consumer perception patterns concerning star ratings and comment reviews. These findings could potentially aid in the development of more effective review structures.

Author Contributions: Conceptualization, Y.-G.N. and J.-H.H.; methodology, Y.-G.N.; software, Y.-G.N.; validation, Y.-G.N. and J.J.; formal analysis, Y.-G.N.; writing—original draft preparation, Y.-G.N.; writing—review and editing, Y.-G.N. and J.J.; visualization, Y.-G.N. and J.J; supervision, J.-H.H.; project administration, J.-H.H. All authors have read and agreed to the published version of the manuscript.

Funding: This work was supported by the National Research Foundation of Korea (NRF) grant funded by the Korea government(MSIT) (2021R1A4A1030075).

Institutional Review Board Statement: Not applicable.

Informed Consent Statement: Not applicable.

Data Availability Statement: The data presented in this study are available on request from the corresponding author. The data are not publicly available due to privacy.

Conflicts of Interest: The authors declare no conflict of interest.

Appendix A

Table A1. Demographic Information.

		N (n = 72)	Percentage (%)
Gender	Male	42	58.33
	Female	30	41.67
Age	18–23	17	23.61
	24–29	23	31.94
	30–35	13	18.06
	36–41	10	13.89
	42+	9	12.50
Occupation	High school student	4	5.56
	University student	17	23.61
	Graduate student	23	31.94
	Office worker	12	16.67
	Practitioner	9	12.50
	Freelance	4	5.56
	Others	3	4.17
Major	Engineering	34	47.22
	Nature	13	18.06
	Liberal arts	6	8.33
	Education	11	15.28
	Society	3	4.167
	Entertainments and sports	1	1.39
	others	34	47.22

Appendix B

Table A2. Background of Participants' Movie Consumption.

		N (n = 72)	Percentage (%)
Frequency of watching movies	4 or more per month	14	19.44
	2 or more per month	16	22.22
	1 or more per month	20	27.78
	1 movie 2 month	19	26.39
	Rarely watch	3	4.17

Table A2. *Cont.*

		N (n = 72)	Percentage (%)
Major genre of watching (Multiple choice)	SF/Fantasy/Adventure	43	59.72
	Thriller/Mystery	34	47.22
	Romantic comedy	29	40.28
	Action	27	37.50
	Drama	24	33.33
	Comedy	21	29.17
	Animation	18	25.00
	Regardless of genre	18	25.00
	Others	3	4.167
Criteria for selecting movies	Story	62	86.11
	Evaluation of online rating	43	59.72
	Genre	41	56.94
	Recommendation of acquaintance	33	45.83
	Companion's taste	24	33.33
	Actor	22	30.56
	Director	21	29.17
	Film festival entry or award	19	26.39
	Others	4	5.56
How to get information (Multiple choice)	Movie information tv program	48	66.67
	OTT service	31	43.06
	Customer review information	30	41.67
	Social network service (SNS)	23	31.94
	Advertisement	21	29.17
	others	19	26.39

References

1. Chen, A.; Lu, Y.; Wang, B. Customers' purchase decision-making process in social commerce: A social learning perspective. *Int. J. Inf. Manag.* **2017**, *37*, 627–638. [CrossRef]
2. Lee, J.; Park, D.-H.; Han, I. The effect of negative online consumer reviews on product attitude: An information processing view. *Electron. Commer. Res. Appl.* **2008**, *7*, 341–352. [CrossRef]
3. Blazevic, V.; Hammedi, W.; Garnefeld, I.; Rust, R.T.; Keiningham, T.; Andreassen, T.W.; Donthu, N.; Carl, W. Beyond traditional word-of-mouth: An expanded model of customer-driven influence. *J. Serv. Manag.* **2013**, *24*, 294–313. [CrossRef]
4. Dellarocas, C.; Zhang, X.M.; Awad, N.F. Exploring the value of online product reviews in forecasting sales: The case of motion pictures. *J. Interact. Mark.* **2007**, *21*, 23–45. [CrossRef]
5. Maslowska, E.; Malthouse, E.C.; Viswanathan, V. Do customer reviews drive purchase decisions? The moderating roles of review exposure and price. *Decis. Support Syst.* **2017**, *98*, 1–9. [CrossRef]
6. Tsao, W.-C. Which type of online review is more persuasive? The influence of consumer reviews and critic ratings on moviegoers. *Electron. Commer. Res.* **2014**, *14*, 559–583. [CrossRef]
7. Bae, S.; Lee, T. Product type and consumers' perception of online consumer reviews. *Electron. Mark.* **2011**, *21*, 255–266. [CrossRef]
8. Donthu, N.; Kumar, S.; Pandey, N.; Pandey, N.; Mishra, A. Mapping the electronic word-of-mouth (eWOM) research: A systematic review and bibliometric analysis. *J. Bus. Res.* **2021**, *135*, 758–773. [CrossRef]
9. Ismagilova, E.; Rana, N.P.; Slade, E.L.; Dwivedi, Y.K. A meta-analysis of the factors affecting eWOM providing behaviour. *Eur. J. Mark.* **2021**, *55*, 1067–1102. [CrossRef]
10. Dessart, L.; Pitardi, V. How stories generate consumer engagement: An exploratory study. *J. Bus. Res.* **2019**, *104*, 183–195. [CrossRef]
11. Oestreicher-Singer, G.; Sundararajan, A. The visible hand? Demand effects of recommendation networks in electronic markets. *Manag. Sci.* **2012**, *58*, 1963–1981. [CrossRef]
12. Ye, Q.; Law, R.; Gu, B. The impact of online user reviews on hotel room sales. *Int. J. Hosp. Manag.* **2009**, *28*, 180–182. [CrossRef]
13. Yang, Y.; Pan, B.; Song, H. Predicting hotel demand using destination marketing organization's web traffic data. *J. Travel Res.* **2014**, *53*, 433–447. [CrossRef]
14. Zhou, L.; Ye, S.; Pearce, P.L.; Wu, M.-Y. Refreshing hotel satisfaction studies by reconfiguring customer review data. *Int. J. Hosp. Manag.* **2014**, *38*, 1–10. [CrossRef]
15. Dong, J.; Li, H.; Zhang, X. Classification of customer satisfaction attributes: An application of online hotel review analysis. In Proceedings of the Digital Services and Information Intelligence: 13th IFIP WG 6.11 Conference on e-Business, e-Services, and e-Society, Sanya, China, 28–30 November 2014; pp. 238–250.

16. Evangelopoulos, N. Citing Taylor: Tracing Taylorism's technical and sociotechnical duality through latent semantic analysis. *J. Bus. Manag.* **2011**, *17*, 57–74.
17. Visinescu, L.L.; Evangelopoulos, N. Orthogonal rotations in latent semantic analysis: An empirical study. *Decis. Support. Syst.* **2014**, *62*, 131–143. [CrossRef]
18. Kitsios, F.; Kamariotou, M.; Karanikolas, P.; Grigoroudis, E. Digital marketing platforms and customer satisfaction: Identifying eWOM using big data and text mining. *Appl. sci.* **2021**, *11*, 8032. [CrossRef]
19. Gartrell, M.; Alanezi, K.; Tian, L.; Han, R.; Lv, Q.; Mishra, S. SocialDining: Design and analysis of a group recommendation application in a mobile context. *Comput. Sci. Tech. Rep.* **2014**, *10*, 1034.
20. Vanetti, M.; Binaghi, E.; Carminati, B.; Carullo, M.; Ferrari, E. Content-based filtering in on-line social networks. In Proceedings of the International Workshop on Privacy and Security Issues in Data Mining and Machine Learning, Barcelona, Spain, 24 September 2010; pp. 127–140.
21. Srifi, M.; Oussous, A.; Ait Lahcen, A.; Mouline, S. Recommender systems based on collaborative filtering using review texts—A survey. *Information* **2020**, *11*, 317. [CrossRef]
22. Sánchez, C.N.; Domínguez-Soberanes, J.; Arreola, A.; Graff, M. Recommendation System for a Delivery Food Application Based on Number of Orders. *Appl. Sci.* **2023**, *13*, 2299. [CrossRef]
23. Roy, A.; Banerjee, S.; Sarkar, M.; Darwish, A.; Elhoseny, M.; Hassanien, A.E. Exploring New Vista of intelligent collaborative filtering: A restaurant recommendation paradigm. *J. Comput. Sci.* **2018**, *27*, 168–182. [CrossRef]
24. Da'u, A.; Salim, N.; Idris, R. An adaptive deep learning method for item recommendation system. *Knowl.-Based Syst.* **2021**, *213*, 106681. [CrossRef]
25. Wu, H.; Zhang, Z.; Yue, K.; Zhang, B.; He, J.; Sun, L. Dual-regularized matrix factorization with deep neural networks for recommender systems. *Knowl.-Based Syst.* **2018**, *145*, 46–58. [CrossRef]
26. Karimi, S.; Wang, F. Online review helpfulness: Impact of reviewer profile image. *Decis. Support Syst.* **2017**, *96*, 39–48. [CrossRef]
27. Rietsche, R.; Frei, D.; Stöckli, E.; Söllner, M. Not all Reviews are Equal-a Literature Review on Online Review Helpfulness. In Proceedings of the 27th European Conference on Information Systems (ECIS), Stockholm/Uppsala, Sweden, 8–14 June 2019; p. 58.
28. Otterbacher, J. 'Helpfulness' in online communities: A measure of message quality. In Proceedings of the SIGCHI Conference on Human Factors in Computing Systems (CHI '09), Boston, MA, USA, 4–9 April 2009; pp. 955–964.
29. Huang, A.H.; Yen, D.C. Predicting the helpfulness of online reviews—A replication. *Int. J. Hum.-Comput. Interact.* **2013**, *29*, 129–138. [CrossRef]
30. Huang, A.H.; Chen, K.; Yen, D.C.; Tran, T.P. A study of factors that contribute to online review helpfulness. *Comput. Hum. Behav.* **2015**, *48*, 17–27. [CrossRef]
31. Salehan, M.; Kim, D.J. Predicting the performance of online consumer reviews: A sentiment mining approach to big data analytics. *Decis. Support Syst.* **2016**, *81*, 30–40. [CrossRef]
32. Chua, A.Y.; Banerjee, S. Helpfulness of user-generated reviews as a function of review sentiment, product type and information quality. *Comput. Hum. Behav.* **2016**, *54*, 547–554. [CrossRef]
33. Singh, J.P.; Irani, S.; Rana, N.P.; Dwivedi, Y.K.; Saumya, S.; Roy, P.K. Predicting the "helpfulness" of online consumer reviews. *J. Bus. Res.* **2017**, *70*, 346–355. [CrossRef]
34. Ma, Y.; Xiang, Z.; Du, Q.; Fan, W. Effects of user-provided photos on hotel review helpfulness: An analytical approach with deep leaning. *Int. J. Hosp. Manag.* **2018**, *71*, 120–131. [CrossRef]
35. Chen, C.; Qiu, M.; Yang, Y.; Zhou, J.; Huang, J.; Li, X.; Bao, F.S. Multi-domain gated CNN for review helpfulness prediction. In Proceedings of the WWW '19: The World Wide Web Conference, San Francisco, CA, USA, 13–17 May 2019; pp. 2630–2636.
36. Mitra, S.; Jenamani, M. Helpfulness of online consumer reviews: A multi-perspective approach. *Inf. Process. Manag.* **2021**, *58*, 102538. [CrossRef]
37. Filieri, R. What makes an online consumer review trustworthy? *Ann. Tour. Res.* **2016**, *58*, 46–64. [CrossRef]
38. Filieri, R.; Raguseo, E.; Vitari, C. Extremely negative ratings and online consumer review helpfulness: The moderating role of product quality signals. *J. Travel Res.* **2021**, *60*, 699–717. [CrossRef]
39. Chen, P.-Y.; Wu, S.-Y.; Yoon, J. The impact of online recommendations and consumer feedback on sales. In Proceedings of the ICIS 2004, Washington, DC, USA, 12–15 December 2004; p. 58.
40. Talton, J.O., III; Dusad, K.; Koiliaris, K.; Kumar, R.S. How do people sort by ratings? In Proceedings of the CHI Conference on Human Factors in Computing Systems (CHI '19), Glasgow, UK, 4–9 May 2019; pp. 1–10.
41. Yatani, K.; Novati, M.; Trusty, A.; Truong, K.N. Review spotlight: A user interface for summarizing user-generated reviews using adjective-noun word pairs. In Proceedings of the SIGCHI Conference on Human Factors in Computing Systems (CHI '11), Vancouver, BC, Canada, 7–13 May 2011; pp. 1541–1550.
42. Chen, C.-W. Five-star or thumbs-up? The influence of rating system types on users' perceptions of information quality, cognitive effort, enjoyment and continuance intention. *Internet Res.* **2017**, *27*, 478–494. [CrossRef]
43. Maharani, W.; Widyantoro, D.H.; Khodra, M.L. Discovering Users' Perceptions on Rating Visualizations. In Proceedings of the 2nd International Conference in HCI and UX Indonesia 2016 (CHIuXiD '16), Jakarta, Indonesia, 13–15 April 2016; pp. 31–38.
44. Zhao, Y.; Xu, X.; Wang, M. Predicting overall customer satisfaction: Big data evidence from hotel online textual reviews. *Int. J. Hosp. Manag.* **2019**, *76*, 111–121. [CrossRef]

45. Hu, Y.; Kim, H.J. Positive and negative eWOM motivations and hotel customers' eWOM behavior: Does personality matter? *Int. J. Hosp. Manag.* **2018**, *75*, 27–37. [CrossRef]
46. Lichtenstein, D.R.; Burton, S. The relationship between perceived and objective price-quality. *J. Mark. Res.* **1989**, *26*, 429–443. [CrossRef]
47. Anderson, E.W.; Sullivan, M.W. The antecedents and consequences of customer satisfaction for firms. *Mark. Sci.* **1993**, *12*, 125–143. [CrossRef]
48. Fornell, C.; Johnson, M.D.; Anderson, E.W.; Cha, J.; Bryant, B.E. The American customer satisfaction index: Nature, purpose, and findings. *J. Mark.* **1996**, *60*, 7–18. [CrossRef]
49. Wang, J.-N.; Du, J.; Chiu, Y.-L.; Li, J. Dynamic effects of customer experience levels on durable product satisfaction: Price and popularity moderation. *Electron. Commer. Res. Appl.* **2018**, *28*, 16–29. [CrossRef]
50. Turney, P.D. Thumbs up or thumbs down?: Semantic orientation applied to unsupervised classification of reviews. In Proceedings of the 40th Annual Meeting on Association for Computational Linguistics: ACL '02, Stroudsburg, PA, USA, 7–12 July 2002; pp. 417–424.
51. Ptaszynski, M.; Dybala, P.; Rzepka, R.; Araki, K. An automatic evaluation method for conversational agents based on affect-as-information theory. *J. Jpn. Soc. Fuzzy Theory Intell. Inform.* **2010**, *22*, 73–89. [CrossRef]
52. Wankhade, M.; Rao, A.C.S.; Kulkarni, C. A survey on sentiment analysis methods, applications, and challenges. *Artif. Intell. Rev.* **2022**, *55*, 5731–5780. [CrossRef]
53. Na, J.C.; Thura Thet, T.; Khoo, C.S. Comparing sentiment expression in movie reviews from four online genres. *Online Inf. Rev.* **2010**, *34*, 317–338. [CrossRef]
54. Demszky, D.; Movshovitz-Attias, D.; Ko, J.; Cowen, A.; Nemade, G.; Ravi, S. GoEmotions: A dataset of fine-grained emotions. In Proceedings of the 58th Annual Meeting of the Association for Computational Linguistics, Online, 5–10 July 2020; pp. 4040–4054.
55. KoELECTRA: Pretrained ELECTRA Model for Korean Home Page. Available online: https://github.com/monologg/KoELECTRA (accessed on 18 March 2023).
56. Luca, M.; Zervas, G. Fake it till you make it: Reputation, competition, and Yelp review fraud. *Manag. Sci.* **2016**, *62*, 3412–3427. [CrossRef]
57. Noh, Y.-G.; Hong, J.-H. Designing Reenacted Chatbots to Enhance Museum Experience. *Appl. Sci.* **2021**, *11*, 7420. [CrossRef]
58. Kong, H.-K.; Zhu, W.; Liu, Z.; Karahalios, K. Understanding Visual Cues in Visualizations Accompanied by Audio Narrations. In Proceedings of the 2019 CHI Conference on Human Factors in Computing Systems, Glasgow, UK, 4–9 May 2019; p. 50.
59. Choi, Y.; Kim, J.; Hong, J.-H. Immersion Measurement in Watching Videos Using Eye-tracking Data. *IEEE Trans. Affect. Comput.* **2022**, *13*, 1759–1770. [CrossRef]
60. Soleymani, M.; Mortillaro, M. Behavioral and Physiological Responses to Visual Interest and Appraisals: Multimodal Analysis and Automatic Recognition. *Front. ICT* **2018**, *5*, 17. [CrossRef]
61. Farias, D.H.; Rosso, P. Irony, sarcasm, and sentiment analysis. In *Sentiment Analysis in Social Networks*; Pozzi, F., Fersini, E., Messina, E., Liu, B., Eds.; Morgan Kaufmann: Burlington, MA, USA, 2017; pp. 113–128.
62. Agrawal, A.; Hamling, T. Sentiment analysis of tweets to gain insights into the 2016 US election. *Columbia Undergraduate Sci. J.* **2017**, *11*, 34–42. [CrossRef]
63. Maynard, D.; Greenwood, M. Who cares about Sarcastic Tweets? Investigating the Impact of Sarcasm on Sentiment Analysis. In Proceedings of the Ninth International Conference on Language Resources and Evaluation (LREC'14), Reykjavik, Iceland, 26–31 May 2014; pp. 4238–4243.
64. Verma, P.; Shukla, N.; Shukla, A. Techniques of sarcasm detection: A review. In Proceedings of the 2021 International Conference on Advance Computing and Innovative Technologies in Engineering (ICACITE), Greater Noida, India, 4–5 March 2021; pp. 968–972.
65. Yoshida, Y.; Hirao, T.; Iwata, T.; Nagata, M.; Matsumoto, Y. Transfer learning for multiple-domain sentiment analysis—Identifying domain dependent/independent word polarity. In Proceedings of the Twenty-Fifth AAAI Conference on Artificial Intelligence (AAAI-11), San Francisco, CA, USA, 7–11 August 2011; pp. 1286–1291.
66. Giatsoglou, M.; Vozalis, M.G.; Diamantaras, K.; Vakali, A.; Sarigiannidis, G.; Chatzisavvas, K.C. Sentiment analysis leveraging emotions and word embeddings. *Expert Syst. Appl.* **2017**, *69*, 214–224. [CrossRef]

Disclaimer/Publisher's Note: The statements, opinions and data contained in all publications are solely those of the individual author(s) and contributor(s) and not of MDPI and/or the editor(s). MDPI and/or the editor(s) disclaim responsibility for any injury to people or property resulting from any ideas, methods, instructions or products referred to in the content.

Review

Fake News Detection on Social Networks: A Survey

Yanping Shen [1,*], Qingjie Liu [1], Na Guo [1], Jing Yuan [1] and Yanqing Yang [2]

[1] School of Information Engineering, Institute of Disaster Prevention, Beijing 101601, China; liuqingjie@cidp.edu.cn (Q.L.); guona@cidp.edu.cn (N.G.); yuanjing@cidp.edu.cn (J.Y.)
[2] College of Information Science and Engineering, Xinjiang University, Urumqi 830046, China; qing0991@163.com
* Correspondence: shenyanping@cidp.edu.cn

Abstract: In recent years, social networks have developed rapidly and have become the main platform for the release and dissemination of fake news. The research on fake news detection has attracted extensive attention in the field of computer science. Fake news detection technology has made many breakthroughs recently, but many challenges remain. Although there are some review papers on fake news detection, a more detailed picture for carrying out a comprehensive review is presented in this paper. The concepts related to fake news detection, including fundamental theory, feature type, detection technique and detection approach, are introduced. Specifically, through extensive investigation and complex organization, a classification method for fake news detection is proposed. The datasets of fake news detection in different fields are also compared and analyzed. In addition, the tables and pictures summarized here help researchers easily grasp the full picture of fake news detection.

Keywords: fake news; detection; dataset; social networks

1. Introduction

With the rapid development of network technology and the widespread application of social networks, platforms such as Facebook, Twitter, YouTube, Weibo and Zhihu (the last two are based in China) have become important sources and main venues for users to publish, obtain, and share information. However, they also provide a hotbed for the proliferation of fake news on the internet. After the 2016 US presidential election, social networks have been facing increasing pressure to crack down on fake news [1].

The term "fake news" has evolved over time, and there is still no exact definition for the term. Fake news was initially defined as intentional and verifiable false news that may mislead readers [2]. Later, other definitions were established, such as in the Collins English Dictionary, where the term was defined as false and often sensational information spread under the cover of news reports. For all definitions of fake news, all agreed that the authenticity of fake news is false. However, they did not reach an agreement on whether fake news included satire, rumors, conspiracy theories, misinformation, and hoaxes [3–5]. Recently, some scholars reported that politicians define the news that does not support them as fake news. It is known that this term cannot express the authenticity of the information [6]. Therefore, the term "fake news" has been rejected because it has meanings of "unstable" and "absurd" [7].

Meanwhile, many terms and concepts related to fake news have been found in the literature. The elementary terms related to fake news generally include misinformation, disinformation and misinformation. There are also general terms including false information, false news, information disorder, etc. [3]. The above terms are some concepts that attempt to define false information in the fields of news and politics. In fact, the phenomenon of fluid terminology brings semantic confusion, making it difficult to study and detect fake news [8]. It may be more accurate to use more objective, comprehensive and verifiable general terms [8,9].

However, the attention of "fake news" has become relatively high recently. The term "fake news" was defined by the Macquarie Dictionary in 2021 and by the American Dialect Association and Collins English Dictionary in 2017 [10]. The Google Trends service shows the search popularity of the common terms mentioned above in the past ten years, as shown in Figure 1. Trends shows that the search popularity of these common terms did not significantly differ before June 2016. However, starting from around the second half of 2016, the popularity of the term fake news exceeded that of several other terms. This confirms that the term fake news has been a topic of public debate since the 2016 US presidential election. With the prevalence of COVID-19 and the US presidential election in 2020, the search popularity of the term fake news reached its peak by March 2020. In the past three years, the search index for fake news has generally been higher than other terms. Given the popularity of the term fake news, further analysis needs to be conducted using the term fake news instead of other general terms.

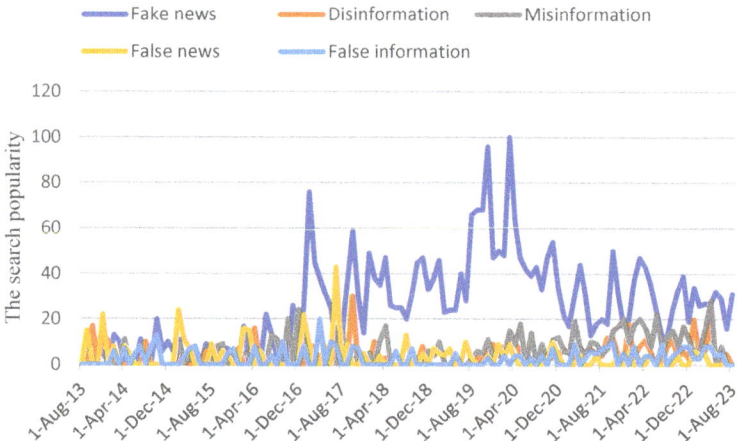

Figure 1. The Google Trends analysis of common terms related to fake news in the past decade.

For financial and ideological purposes, a significant amount of fake news has been produced and distributed through social networks [11]. Zhang et al. [12,13] showed that fake news has three basic characteristics as follows: volume, velocity, and veracity (3V). This means that fake news is massive, uncertain, and in spread in real time. Most people define fake news as "deliberate and verifiable false news" [14,15]. Authenticity and intention are two important characteristic dimensions of fake news [3,16]. Table 1 shows the similarities and differences of basic terms related to fake news based on these two important characteristics [3,17].

Table 1. Comparison of terms related to fake news.

Terms	Characteristics	Authenticity	Intent to Harm
Fake news	Commonly used, false and often sensational information	N	Y
Disinformation	International organizations like to use, referring to fake and harmful information	N	Y
Misinformation	Refers to false content but lacks potential intent to cause harm	N	N

In order to cope with the chaos of information dissemination in the new media era, some online fact-checking systems have emerged. Fact checking is the task of evaluating whether news content is true. It was initially developed in journalism and is an important task that is usually manually executed by specialized organizations, such as PolitiFact [18]. Even though the automatic fact-checking systems have been developed to assist human fact-checkers, they heavily rely on the knowledge base and require regular updates [19]. Additionally, it is difficult to deal with sudden topics and knowledge, such as COVID-19 symptoms in the early stages of the pandemic. In addition, most online fact-checking systems mainly focus on verifying political news [12]. With the large amount of real-time information being created, commented on, and shared through social networks, automatic fake news detection technology is urgently needed [13,20–22].

The relatively new and representative literature in the domain of fake news detection has been analyzed in this paper. In particular, we focused on how to derive detection methods based on the characteristics of fake news. The references were obtained by searching for keywords "fake news" or terms equivalent to the fake news declared by the authors, such as "false news" through Google Scholar. In addition, the above keywords were also combined with "detection", "detection model", "overview", "survey", "review", "dataset", etc., for search and selection. Subsequently, we studied the papers and systemically sorted the fake news detection models and datasets. Finally, considering factors such as the citation frequency and the publication time, the methods with a greater impact on the existing technologies were listed.

The contributions of this article are as follows: Firstly, the characteristics of fake news and its related terms are analyzed. Secondly, the classification of different fake news detection methods are compared from the perspective of characterization to detection. Thirdly, the existing approaches for detecting fake news are reviewed and divided in three categories, including content-based, propagation-based and source-based methods. The advantages and disadvantages of the three detection methods are also given; moreover, the datasets of fake news detection in different fields are analyzed. Finally, some insights are provided to provide a direction for further research.

2. Development and Classification of Fake News Detection Technology

Many existing studies have adopted an intuitive method: that is, to detect fake news based on the news content. Capuano et al. [23] showed that the manual fact-checking method is insufficient when faced with a massive amount of fake news information. They reviewed content-based fake news detection methods adopting machine learning. In addition to content features, many other types of features can be added to the detection model. Shu et al. [14] pointed out that fake news is deliberately written to misguide readers, making it difficult to detect fake news based on its content. It is necessary to include other auxiliary information to assist in making decisions. Sahoo et al. [13] also indicated that detecting fake news based on shared content is difficult and requires the addition of new information related to the users' profiles. Specifically, multiple features related to each user's Facebook account and some news content features were combined. Sheikhi et al. [24] reported that fake news detection systems are generally divided into news content and social context methods based on their data sources. The social context method focuses on social features such as user interactions and participation based on the given news.

In summary, the fake news detection technology has evolved from expert systems to automatic detection, as shown in Figure 2. For the automatic detection technology, the above literature indicates that the news content alone may be not enough and requires additional auxiliary information, such as social context, for fake news detection.

There have been various classification methods for detecting fake news in recent years. Guo et al. [15] classified false information and other related terms based on the intention, dividing the existing false information detection methods into content-based, social context-based, feature fusion-based, and deep learning-based methods based on the type of features used. Most of the literature categorizes fake news detection into broader categories, including content-based and contextual aspects [3].

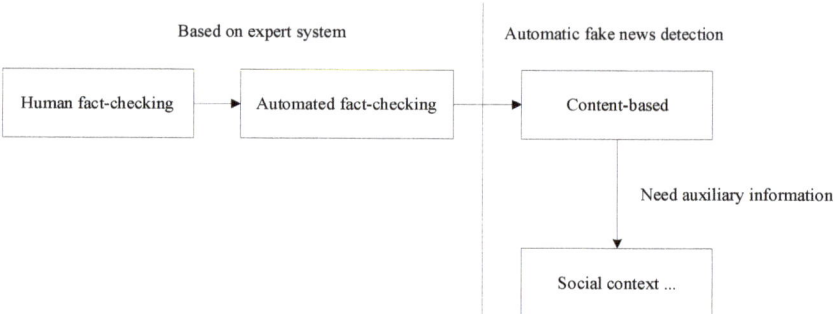

Figure 2. Fake news detection has evolved from expert systems to automatic detection.

To facilitate the research of fake news detection on social networks, it is necessary to first study the characterizations of fake news and then propose the fake news detection methods. This paper sorts the classification of the detection methods from this perspective, as shown in Figure 3. Shu et al. [14] first used the theory of psychology and sociology to describe the background of fake news detection. Fake news can be categorized into traditional media fake news and social network fake news. The detection of fake news on traditional news media mainly depends on the content of the news, and social contextual information can serve as additional information for detecting fake news on social networks. Then, fake news detection methods can be classified into news content methods and social context methods. Furthermore, the news content model can be divided into knowledge-based and style-based methods. The social context model involves the social participation of relevant users and can be categorized into stance analysis and propagation analysis.

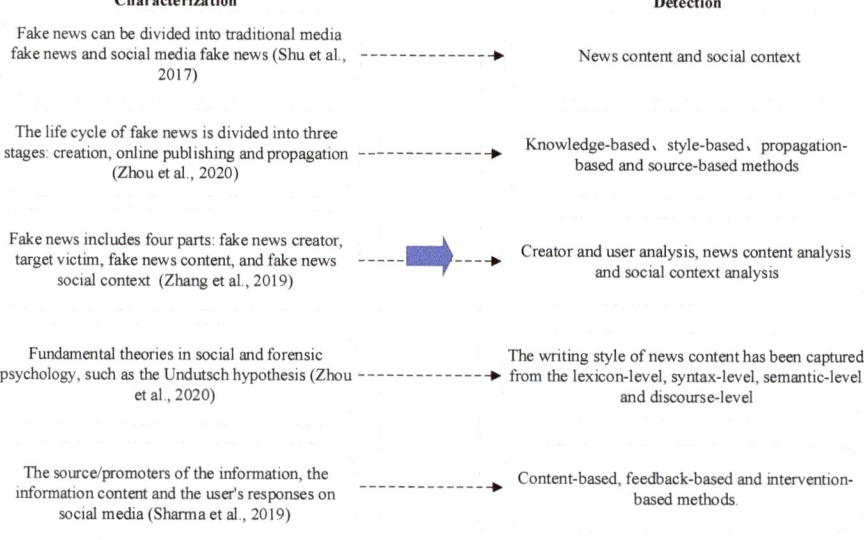

Figure 3. Fake news from characterization to detection [12,14,17,25,26].

To better understand the fake news detection task, fake news can be investigated from both temporal and spatial dimensions. From the time angle, Zhou et al. [17] divided the life cycle of fake news into three stages: creation, publication, and propagation. According to the life cycle of fake news, they detailed the methods for detecting fake news from four aspects: knowledge, writing style, communication mode, and source. From the spatial dimension, Zhang et al. [12] divided fake news and its related content into four parts: creator, target victim, content, and social context. Therefore, fake news detection can be categorized into creator and user-based, content-based, and social context-based detection.

As the Undeutsch hypothesis implies, fake news may differ from real news in writing style. Therefore, Zhou et al. [25] studied the writing style of fake news based on vocabulary, syntax, semantics and discourse for early detection of fake news. Sharma et al. [26] studied different characteristics of fake news for detection, including the source, promoters, content, and responses on social networks. Consequently, they divided the fake news detection methods into three types, including content-based, feedback-based and intervention-based methods.

3. Fake News Detection

3.1. Problem Formulation

Let $N = \{n_1, n_2, \ldots, n_M\}$ represent M news items. Each news n_j has at least one publisher P and a maximum of K users $U = \{u_1, u_2, \ldots, u_K\}$ forwarding the news. For the fake news detection task, the objective is to obtain the learning function $p(c \mid n_j, N, P, U, \theta)$ to detect fake news. c represents the class label of the news, and θ denotes the hyper-parameters. The learning function can be described as follows:

$$p(n_j) = \begin{cases} 0, & \text{if } n_j \text{ is fake} \\ 1, & \text{otherwise} \end{cases} \quad (1)$$

Some fake news detection approaches are only based on the news itself, while in other detection methods, the information of the news publishers or forwarders is also considered. The following section will describe the fake news detection methods in detail.

3.2. Fundamental Theories of Fake News Detection

The cognitive and behavioral theories in social sciences and economics, as the fundamental theories of fake news, can promote the establishment of reasonable and interpretable fake news detection models.

To illustrate the role of fundamental theories, such as cognitive and behavioral theories, in fake news detection, a comparison was made between fake news detection and traditional distributed denial of service (DDoS) attack detection, as shown in Figure 4.

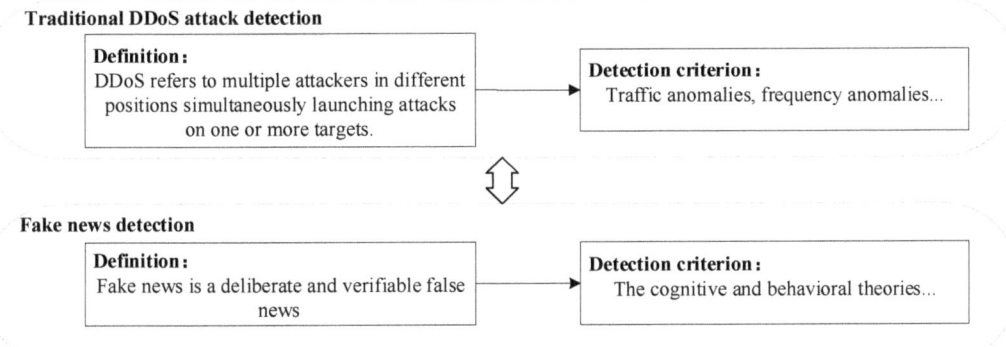

Figure 4. The role of the fundamental theories in fake news detection.

Why are these two detection models compared? Fake news and distributed denial of service attacks are both very important issues in the field of information security. The research on fake news detection and distributed denial of service attack detection models plays an important role in protecting the network ecological environment. Before detection, the unique features of fake news and distributed denial of service attacks should be identified first. Therefore, it is necessary to first study the definitions of fake news and DDoS and then further summarize their characteristics, namely the detection criterion, based on the definitions. For example, traffic anomalies and frequency anomalies can serve as the detection criteria for distributed denial of service attacks. Meanwhile, relevant theories have revealed that fake news and true news are different in terms of presentation form and propagation patterns. Cognitive and behavioral theories can serve as the criteria for detecting fake news. For example, the four-factor theory suggests that lies are expressed differently in terms of emotions, behavioral control, and other aspects [17]. The latest research also shows that fake news spreads faster and further than true news [12,13]. Thus, the content or propagation features can be extracted based on the above theories for fake news detection.

3.3. Fake News Detection Approaches

News involves multiple factors in the propagation process. Chi et al. [27] thought that most online data involve four aspects: people, relationship, content, and time. Ruchansky et al. [28] showed fake news has the following three characteristics: the text of the article, the user response after receiving the text, and the user source for promoting the article. In order to summarize and unify relevant concepts, scholars have proposed the term "social context" to describe how news is disseminated online [12].

As the classification of fake news detection approaches into content-based and context-based is too broad, this article attempts to describe different fake news detection methods based on detection characteristics. Meel et al. [29] detected false information based on text/content, visual, user, message, propagation, temporal, structural, and linguistic features. It is worth noting that some of these features can be grouped together. For example, text/content, visual, message, and linguistic features all belong to content-based features. The propagation, temporal, and structural features can be referred to as propagation features. Moreover, source detection, which refers to identifying individuals or locations in the network where false information begins to spread, plays a crucial role in reducing misinformation [29].

Consequently, to highlight the characteristics of the detection method, this paper divides fake news detection methods into the following three types: content-based, propagation-based, and source-based methods. The pros and cons of the three detection approaches are shown in Table 2 [17,30]. Based on this viewpoint, the current fake news detection methods have been carefully organized, as shown in Table 3.

Table 2. The pros and cons of fake news detection methodologies.

	Content-Based	Propagation-Based	Source-Based
Pros	• Easy to obtain data; • Can evaluate the authenticity and the intention of the news; • Suitable for early detection of fake news.	• Language independence; • More robust and better resistance to adversarial attacks.	• Sounds simple but effective; • Quickly finds fake news and is suitable for early detection of fake news.
Cons	• Requires linguistic knowledge, time-consuming, laborious; • Unable to resist adversarial attacks; • Due to many types of fake news, difficult to be transferred.	• Need a large amount of news dissemination information; • Not suitable for the early detection of fake news.	• Difficult to obtain real user data; • Need to overcome the problem of adversarial user camouflage.

3.3.1. The Content-Based Methods

(1) Knowledge-based

For this method, a knowledge base or knowledge graph must first be established. Knowledge can be expressed in the form of triples [17]. Then, the extracted knowledge from a piece of news is compared with the facts in the knowledge base to check its authenticity. The above is actually the implementation process of an automatic fact-checking system which can be categorized into fact extraction and fact checking.

Some studies use knowledge-based methods to detect fake news. For example, researchers have been exploring how to realize the automation of fact-checking, using technologies based on natural language processing (NLP) and machine learning to automatically predict the authenticity of reports [31]. Mayank et al. [32] proposed a knowledge graph-based (KG) framework which includes the news encoder, the entity encoder, and the final classification layer to detect fake news articles. Firstly, the NLP-based technology is used to encode the headline of the news. Then, named entity recognition (NER) is adopted to recognize and extract named entities from the texts, and then named entity disambiguation (NED) is adopted to map the entities to the KG. The ComplEx model is adopted to obtain the representation of the entities. Finally, the embedding of the two parts is combined to detect fake news. Hu et al. [33] designed a new graph neural model, which compares the news with an external knowledge base. Firstly, a directed heterogeneous document graph containing topics and entities for each type of news is constructed. Then, through a carefully designed entity comparison network, different entity representations are compared. Finally, the above comparison features are fed into the fake news classifiers. Pan et al. [34] proposed a fake news detection method based on incomplete and imprecise knowledge graphs using the TransE and B-TransE methods. Firstly, three different knowledge graphs are created to generate background knowledge. Then, the B-TransE method is used to establish the entity and relation embedding and check whether news articles are authentic. The results show that even an incomplete knowledge graph can be used for fake news detection.

(2) Style-based

The style-based approach can evaluate news intent [17]. The pre-trained language model based on neural networks and deep learning has brought about breakthrough development in natural language processing technology. These provide better ideas for fake news detection.

Nasir et al. [35] proposed a combination of convolutional neural networks and recurrent neural networks for fake news classification. Zhou et al. [25] proposed a theory-driven fake news detection method. The news content is analyzed based on vocabulary, syntax, semantics, and discourse. Furthermore, the supervised machine learning method is used to detect fake news. Choudhary et al. [36] proposed a linguistic model to extract the syntactic, grammatical, sentimental, and readability features of specific news. Moreover, the sequential learning method based on a neural network is adopted to detect fake news. Alonso et al. [10] pointed out that the authors of fake news adopt various stylistic tricks, such as stimulating readers' emotions, to promote their creative success. The role of emotional analysis in detecting fake news has been studied. Verma et al. [37] detected fake news based on linguistic features combined with the word-embedding technology. The extracted linguistic features include syntactic and semantic features. Umer et al. [38] designed a fake news stance detection framework based on titles and news text. Firstly, principal component analysis (PCA) and the Chi-square test are used to extract features, and then the extracted features are fed into the CNN-LSTM classifier.

With the development of multimedia technology, fake news tries to attract and mislead consumers by using the multimedia content of images or videos, making visual features become an integral part to fake news. Cao et al. [39] showed that although visual content is very important, our understanding of the visual features in fake news detection is still limited. Qi et al. [40] designed a CNN-based neural network framework to fuse the visual information of the frequency domain and pixel domain to detect fake news. Uppada et al. [41] proposed the credibility neural network to evaluate the credibility of images on

OSNs. They applied the spatial characteristics of CNNs to find physical changes in images and analyze whether images reflect negative emotions.

In recent years, the issue of facial manipulation videos has received widespread attention, especially deepfake techniques that use deep learning tools to manipulate images and videos. The deepfake algorithm can use the generative model to replace the face in the target video [42]. As people become increasingly interested in deepfake technology, more and more deepfake detection technologies are underway. Early attempts mainly focused on the inconsistency of features caused by the facial synthesis process, while recently proposed methods center on basic features, where camera fingerprinting and biological signal-based schemes perform better [42].

(3) Multimodal-based

Currently, most fake news detection schemes focus on a single modality (such as text or visual features only). Due to the combination of multimodal information in social media in recent years, fake news detection based on multimodal information has gradually attracted extensive research. How to effectively fuse the data from different modalities is a challenge.

Song et al. [43] designed a multimodal fake news detection method that combines multimodal attention residuals and multi-channel convolutional neural networks (CNNs). Singh et al. [44] proposed a multimodal approach for detecting fake news using the text and visual features of news stories. Khattar et al. [45] proposed a fake news detection model consisting of a bimodal variational autoencoder and a binary classifier. The bimodal variational autoencoder is used for the multimodal representation, with the input being the text of the post and the accompanying images. Segura-Bedmar et al. [46] used unimodal and multimodal methods to classify fake news in a fine-grained way. Experiments showed that the multimodal CNN method that combines text and image data has the best performance. Wang et al. [47] extracted features based on text, images, and user attributes for Sina Weibo as well as image–text correlation features. Finally, a new deep neural network framework was established to detect fake news.

3.3.2. Propagation-Based Methods

Many studies have shown that fake news differs from true news in terms of dissemination methods [48]. A promising fake news detection method involves the study of propagation-based methods. The information propagation on social networks has strong temporal characteristics as follows: sudden updates, fast propagation speed, and rapid disappearance [29]. Based on the factors involved in the news propagation, the propagation-based approaches can be categorized into (1) only considering the propagation pattern of the news itself, known as news cascade, and (2) capturing other additional information, such as comments, during the news propagation process [17].

(1) News cascade

In recent years, the news cascade, a tree-like structure, has been used to describe news propagation. This structure only represents the dissemination mode of news and does not contain any additional information [17].

Graph neural networks (GNNs) can better simulate the news propagation on social networks, and thus, there are numerous fake news detection methods based on the GNNs. Monti et al. [48] applied the geometric deep learning model to describe the propagation mode of fake news. Four types of features, including user profile, user activity, news cascade, and content, are extracted to describe news, users, and their activity. Silva et al. [49] showed that the news propagation model, including the corresponding tweets and retweets, can be seen as a tree. The tree is composed of multiple cascades, each consisting of a series of tweets/retweets. They proposed a propagation network model called Propagation2Vec for the early detection of fake news. A hierarchical attention mechanism is adopted to encode the propagation network, which can assign corresponding weights to different cascades. At the same time, a technique for reconstructing a complete propagation network in the early stages has been proposed.

Barnabò et al. [50] used three graph neural network algorithms, including GraphSAGE, GAT and GCN, for fake news detection, with the URL diffusion cascade as the input. An active learning method has also been proposed for sample annotation. Due to the dynamic nature of information diffusion networks in the real world, Song et al. [51] designed a fake news detection model called the Dynamic Graph Neural Network (DGNF). Specifically, the discrete time dynamic graph (DTDG) is used to model news propagation networks. Jeong et al. [52] designed a hypergraph neural network to jointly model multiple propagation trees for fake news detection. Han et al. [53] used a GNN to model information propagation patterns. Continual learning techniques are used to gradually train GNNs so that they can achieve stable performance across different datasets. Wei et al. [54] proposed the concept of a propagation forest to cluster propagation trees at the semantic level. A new framework based on Unified Propagation Forest (UniPF) was proposed to fully explore the potential correlation between propagation trees and improve the performance of fake news detection. Murayama et al. [55] designed a point process model for fake news dissemination on Twitter. In the model, the dissemination of fake news consists of two parts as follows: the cascade of original news and the cascade of assertion of news falsehoods. The experiments indicated that the proposed method is helpful in detecting and mitigating fake news.

(2) Propagation graph

Sometimes, it is also necessary to consider other additional information, such as comments and user characteristics, in the dissemination of information. Various propagation graphs can be created to model the information dissemination process.

Zhou et al. [17] showed that the propagation-based fake news detection method can be classified into news cascades and self-defined graphs. For self-defined graphs, homogeneous, heterogeneous, or hierarchical networks can be constructed to describe the propagation pattern of fake news. Ni et al. [56] proposed a new neural network framework for detecting fake news based on source tweets and their propagation structures. The text semantic attention network and the propagation structure attention network are used to obtain the semantic features and propagation structure features of tweets, respectively. Shu et al. [57] designed a hierarchical propagation network for fake news detection. Firstly, a hierarchical propagation network is constructed, based on which the corresponding features are extracted from the structural, temporal, and linguistic perspectives. Finally, the effectiveness of these propagation network features in fake news detection is verified. Davoudi et al. [58] used both the propagation tree and the stance network for early fake news detection. A new method for constructing the stance network has been proposed, and various graph-based features are extracted for sentiment analysis. The node2vec technology uses graph embedding to obtain feature representations of the two networks mentioned above. Yang et al. [30] designed a model called PostCom2DR for rumor detection. Firstly, a response graph is created between posts and comments. Secondly, a two-layer GCN and self-attention mechanism are used to obtain the features of comments. Finally, a post-comment co-attention mechanism is applied to fuse information. Shu et al. [59] described the relationship among publishers, news, and users which may improve the detection of fake news. A tri-relationship-embedding framework, which models both publisher–news relationship and user–news interactions, is proposed for fake news detection. Nguyen et al. [60] proposed a novel graph representation called the Fact News Graph (FANG), which models all major social participants and their interactions to improve representation quality.

3.3.3. Source-Based Methods

The source-based methods evaluate the authenticity of news based on the credibility of the source. The source mainly includes the news authors and social media users. This method, an indirect way to detect fake news, may seem arbitrary but is very effective [17].

(1) News author-based

Many studies have shown that considering news authors can improve the performance of fake news detection. For example, Sitaula et al. [61] reported that adding features such as the number of authors and the historical connection between authors and fake news can improve fake news detection. They showed that fake news can be effectively detected by adopting a few features based on the source's credibility. Yuan et al. [62] indicated that the content-based and propagation-based methods are not suitable for the early detection of fake news. They pointed out that the credibility of publishers and users can be adopted to quickly locate fake news in a massive amount of news. They proposed a multi-head attention network that considers publishing and forwarding relations to optimize fake news detection. Luvembe et al. [63] used the stacked BiGRU to extract dual emotional features derived from publisher emotions and social emotions. An adaptive genetic weight update-random forest (AGWu-RF) was proposed to improve the accuracy of fake news detection.

(2) Social media user-based

Fake news can also be detected by detecting malicious social media users. Social media users are the communicators of news on social media, and malicious users are more likely to spread fake news. The social robot, sybil account, fake profile, fake account, etc., are generally malicious users of social media [64]. Kudugunta et al. [65] proposed a deep neural network based on long short-term memory (LSTM) architecture. This network uses content and metadata to detect robots at the tweet level. Rostami et al. [66] pointed out that feature selection is the main process of fake account detection based on machine learning. A multi-objective hybrid feature selection method is used for feature selection. Gao et al. [67] designed an end-to-end Sybil detection model based on the content. The self-normalizing CNN and bi-SN-LSTM are simultaneously adopted to improve the performance of fake news detection. Bazmi et al. [68] pointed out that according to the fundamental theories in fake news detection, the credibility of authors and users varies across different themes. They proposed a network model named MVCAN, which jointly models the potential topic credibility of authors and users in fake news detection. Zhang et al. [69] extracted a set of explicit and implicit features from the text information, and a deep diffusion network model was proposed based on the connections among the news content, creators, and news subjects.

Table 3. Classifications and comparisons of various fake news detection methods.

	Detection Approach	Main Models	Datasets	Characteristics
	Knowledge-based	biLSTM, ComplEx [32] GNN [33] TransE [34]	Kaggle, CoAID LUN, SLN Kaggles + BBC news, etc.	The authenticity of the news is evaluated
Content-based	Style-based	CNN, RNN [35], BOW, SVM, LR, etc. [25] Sequential neural network [36] WE, LFS, KNN, Ensemble [37] PCA, CNN-LSTM [38] CNN-RNN [40] SENAD, CNN [41]	FA-KES, ISOT PolitiFact, BuzzFeed Buzzfeed, Random political WELFake FNC Weibo dataset FakeNewsNet	Evaluate news intention
	Multimodal-based	Residual network, CNN [43] LIWC, LDA, LR, KNN, etc. [44] VAE [45] CNN, BiLSTM, BERT [46] ERNIE, CNN, FNN [47]	Tweet, Weibo Kaggle Twitter, Weibo Fakeddit Weibo	Obtain better results than unimodal methods; there is significant room for improvement
Propagation-based	News cascade	Geometric DL [48] Propagation2Vec [49] GraphSAGE, GAT and GCN [50] DTDG, GNN [51] Hypergraph NN [52] GNN [53] UniPF [54] Point process model [55]	Twitter PolitiFact, GossipCop FbMultiLingMisinfo, PolitiFact Weibo, FakeNewsNet, Twitter FakeNewsNet FakeNewsNet FakeNewsNet, Twitter Twitter	Evaluate news intention; more robust; directly capture news propagation
	Propagation graph	Graph attention networks [56] Hierarchical propagation network, GNB, DT [57] Stance network, RNN [58] Tri-relationship, TriFN [59] FANG [60]	Twitter15, Twitter16 FakeNewsNet FakeNewsNet FakeNewsNet Twitter	Evaluate news intention; more robust; indirectly capture the propagation of the news

Table 3. *Cont.*

	Detection Approach	Main Models	Datasets	Characteristics
Source-based	News author-based	Seven classifiers [61] Multi-head attention network, CNN [62] BiGRU, AGWu-RF [63]	Buzzfeed news, PolitiFact Twitter15, Twitter16, Weibo RumorEval19, Pheme	Detect fake news by assessing the credibility of the authors or publishers
	Social media user-based	LSTM [65] mRMR, RF, NB, SVM [66] CNN, bi-SN-LSTM [67] Multi-view co-attention network [68] Diffusive network [69]	Presented by others Twitter MIB PolitiFact, GossipCop PolitiFact	Detect fake news by detecting malicious social media users

Figure 5 introduces the concepts related to fake news detection, including the following four aspects: fundamental theory, feature type, detection technique, and detection approach. The fundamental theories include the four-factor theory, Undeutsch hypothesis, social identity theory, confirmation bias, desirability bias, Naïve realism, etc. [17]. Furthermore, the feature type consists of content-based, i.e., "linguistic-based" or "visual-based". Propagation-based features can be comment-based and network-based. The comment-based method is a direct method for detecting fake news. Furthermore, different types of networks can be constructed, such as stance networks and diffusion networks, in the process of news dissemination. Existing network metrics (e.g., degree coefficient, clustering coefficient, etc.) can be used as network features, or embedding algorithms can be applied to extract network-embedding representations. Furthermore, user-based features can be categorized into individual level and group level. The credibility of users can be obtained through individual-level features such as registration age and the number of followers/followers. Group-level features are used to describe the characteristics of different communities formed by the spreaders of fake news and true news [14]. Moreover, it is worth mentioning that detection technologies include human-based, AI-based and blockchain-based technologies [3]. Human-based technology relies on knowledge for fake news detection through crowdsourcing and fact-checking techniques. AI-based technology applies shallow or deep machine learning approaches to detect fake news. For the blockchain-based technology, fake news can be detected by checking news sources and tracking news. Finally, as for the detection approach, some of the "content-based", "propagation-based", or "source-based" methods are adopted and illustrated in Section 3.3.

Figure 5. An overview of the concepts related to fake news detection.

3.4. Discussion

For the three fake news detection methods mentioned in this paper, the content-based detection method is generally language-related. This means that the detection model applicable to political fake news may not be applicable to fake news detection in other fields. If additional information, such as comments generated during the information dissemination, is not considered, the propagation-based method is language-independent. Therefore, this method can better resist adversarial attacks and has better robustness. The source-based method sounds simple but effective. The characteristics of users are generally collected from the user's homepage on social media, such as their personal description, gender, fan base, followers, place of residence, and hobbies. However, with the increasing awareness of privacy protection among people, many users are unwilling to disclose too much of their information. Some source-based detection methods have shifted to detecting malicious users through text published by users, and so they are also language-related.

The content-based detection method and source-based method are suitable for the early detection of fake news, as fake news can be recognized before it spreads on social media. However, the propagation-based detection is inefficient for the early detection of fake news, as it is difficult to detect fake news before it spreads. It can be seen that the content-based detection method and the source-based detection method generally have higher detection efficiency than the propagation-based detection.

The above three fake news detection methods are not independent. It is possible to jointly predict fake news from multiple perspectives; therefore, the advantages of the different approaches can be combined.

4. Datasets

For fake news detection, useful features are extracted from social media datasets, and an effective detection model is established to detect fake news in the future. Supervised learning methods are widely used in the field of fake news detection. Although these methods have shown promising results, reliable annotated datasets are needed to train the detection model [70]. Therefore, establishing a large-scale dataset with multidimensional information is very important.

Vlachos and Riedel [71] were the first to publish a dataset in the field of fake news, but only 221 statements were issued. In recent years, several fake news datasets have been proposed, which involve politics, security, health, and satire. The relatively new and representative datasets are classified based on the data sources of the news, as shown in Table 4. The CHECKED [72] and Weibo21 [73] datasets were collected from Weibo, COVID-19 [74] and FibVID [75] were collected from Twitter, and BuzzFeed [76] and Fakeddit [77] were collected from Facebook and Reddit, respectively. Datasets such as IFND [78], FakeNewsNet [79], FA-KES [80], ISOT [81], GermanFakeNC [82], BanFakeNews [83] and LIAR [84] are sourced from news websites.

Table 4. The relatively new and representative datasets in fake news detection.

Source	Dataset Name	Feature Types	Modality	Annotation Methods for True News	Annotation Methods for Fake News	News Domain	Language	Time
Weibo	CHECKED	Textual, visual, temporal, network	Text, images, video	Weibo accounts (People's Daily)	Weibo Community Management Center	COVID-19	Chinese	2019–2020
	Weibo21	Textual	Text	Verified by the NewsVerify	Weibo Community Management Center	variety	Chinese	2014–2021
Twitter	COVID-19	Textual	Text	The official government accounts	Fact-checking websites	COVID-19	English	-
	FibVID	Textual, propagation, social context	Text	Fact-checking websites	Fact-checking websites	COVID-19	English	2020

Table 4. *Cont.*

Source	Dataset Name	Feature Types	Modality	Annotation Methods for True News	Annotation Methods for Fake News	News Domain	Language	Time
Facebook	BuzzFeed	Textual	Text	Fact-checked manually	fact-checked manually	political	English	2016
News websites	IFND	Textual, visual	Text, images	Official websites	Fact-checking websites	variety	Indian	2013–2021
	FakeNewsNet	Textual, visual, social context, spatio-temporal	Text, images	trusted media websites	Fact-checking websites	variety	English	-
	FA-KES	Textual	Text	a semi-supervised fact-checking approach	a semi-supervised fact-checking approach	Syrian war	English	2011–2018
	ISOT	Textual	Text	Official websites	Unreliable websites	politics	English	2016–2017
	GermanFakeNC	Textual	Text	Well-known publishers	fact-checked manually	variety	Germany	-
	BanFakeNews	Textual	Text	Trusted news portals	Popular websites	variety	Bangladesh	-
	LIAR	Textual, network	Text	PolitiFact.com	PolitiFact.com	political	English	2007–2016
Reddit	Fakeddit	Textual, visual	Text, images	the distant supervision	the distant supervision	variety	English	2008–2019

Table 4 shows the characteristics of the datasets, including the source, dataset name, feature types, modality, news domain, language, annotation methods for true news, annotation methods for fake news, and time. The main differences between these datasets lie in the data features they contain, the granularity of classification, and the language of the news. The CHECKED and Weibo21 are Chinese datasets; IFND, GermanFakeNC, and BanFakeNews are Indian, German, and Bangladeshi datasets, respectively; and the rest are English datasets.

4.1. Annotation Methods: Manual to Automatic

The annotation of news is a major bottleneck in constructing datasets, which mainly includes manual annotation and automatic annotation. Manual annotation is labor-intensive as it requires a careful examination of news content and other additional information. Crowdsourcing methods can be used for the annotation to reduce the burden on experts. Moreover, automatic annotation based on machine learning has also received attention. The annotation methods for news are also listed in Table 4. Most datasets are usually collated from fact-checking websites and marked by human experts [80]. The fact-checking systems claim to operate on the principles of "independence", "objectivity", and "neutrality". Due to the fact that the news content may be entirely or partially true, most fact-checking websites use a multi-level method to describe the authenticity of the news when evaluating information rather than using two simple judgments, "True" and "Fake". For example, PolitiFact, a trusted fact-checking website in the United States, provides six rating levels through a scale called the Trust-O-Meter, including "True", "Mostly True", "Half True", "Mostly False", "False", and "Pants on Fire". Figure 6 shows a Facebook post that is rated as "False" by the Truth-O-Meter. Table 5 lists the rating levels of common fact-checking websites, including PolitiFact [85], Snopes [86], TruthOrFiction [87], CheckYourFact [88], FactCheck [89], Gossip [90], and Ruijianshiyao [91].

Most fake news detection methods are generally defined as binary classification problems. This means that news is classified in a coarse-grained manner. Thus, it is necessary to classify the news labels into two categories, "True" and "Fake" [75,92], when using fact-checking websites to label news. For example, Khan et al. [92] used the LIAR dataset collated from PolitiFact.com to carry out the fake news detection experiment. Statements labeled as "Half True", "Mostly True" and "True" were classified as "True", while statements with "Pants on Fire", "False", and "Mostly False" were classified as "Fake". Nakamura et al. [77] introduced the Fakeddit dataset, where samples have two-way, three-way, and six-way category labels. In addition to the label of "True", fake labels in the six-way classification labels include "Manipulated Content", "False Connection", "Satire/Parody", "Misleading Content" and "Imposter Content". The three-way classification model was created to determine whether a sample is true, the sample is fake with real text, or the sample is fake and contains false text. The two-way classification labels only include fake or true. Segura-Bedmar et al. [46] classified fake news based on texts and images in a fine-grained way on the Fakeddit dataset. Results showed that the detection rate of some fake news categories increased after using images. The detection rate of other categories has also slightly improved after using images.

Figure 6. Illustrations of the fact-checking website PolitiFact.com.

Table 5. The rating levels of common fact-checking websites.

Fact-Checking Websites	Rating	Number of Rating Label Types
PolitiFact	True, Mostly True, Half True, Mostly False, False, Pants on Fire	6
Snopes	Research In Progress, True, Mostly True, Mixture, Mostly False, False, Unproven, Unfounded, Outdated, Miscaptioned, Correct Attribution, Misattributed, Legend, Scam, Legit, Labeled Satire, Originated as Satire, Recall, Lost Legend, Fake	20
TruthOrFiction	Truth, Fiction, Reported to be Truth, Unproven, Truth and Fiction, Previously Truth, Disputed, Pending Investigation	8
CheckYourFact	True, False, Misleading, Unsubstantiated	4
FactCheck	False, misleading, Missing content	3
Gossip	Scale from 0 to 10	-
Ruijianshiyao	High, Medium, Low	3

The semi-supervised fact-checking approach and distant supervision methods are also used for automatic annotation. Salem et al. [80] used a semi-supervised fact-checking method to label news articles in a dataset. The crowdsourcing method is used to obtain information from news articles and then to check whether it matches the information in the VDC database, which represents the truth of the facts. Finally, unsupervised machine learning is used to cluster the articles into two sets based on the extracted information. Nakamura et al. [77] did not manually label each sample. Instead, distant supervision was used to generate the final tag.

4.2. Feature Types

The feature types include textual and visual features as well as social contextual features. Patwa et al. [74] published a manually annotated dataset containing 10,700 real and fake news articles about COVID-19. Only English text contents were considered. The TF-IDF technology was used for feature extraction, and SVM and other machine learning methods were used to detect fake news. Shu et al. [79] proposed the FakeNewsNet dataset which contains two comprehensive datasets with different characteristics in terms of news content, social context, and spatio-temporal information. The FakeNewsNet contains 23,921, texts which are obtained from the fact-checking websites PolitiFact and GossipCop. The main feature of the dataset is that it includes which users have forwarded the news in addition to the original news. Thus, the propagation pattern of the news can be plotted into a network graph for further processing. Sharma et al. [78] proposed the Indian fake news dataset IFND, which consists of text and images. A multimodal method was also proposed, which considers both text and visual features for fake news detection. In the Fakeddit dataset, Nakamura et al. [77] showed that multimodal features perform best, followed by text-only and image-only features. Finally, the authors point out that some auxiliary information, such as the submission metadata, is useful for future research.

4.3. Discussion

The existing fake news datasets have significant differences in labeling categories, modalities, topic domains, and other aspects. Most datasets are related to politics and economics with limited coverage of topic areas. Additionally, some samples in some multi-category datasets were not annotated. Furthermore, most datasets only contain linguistic features. Few datasets contain both linguistic and social contextual features. From Table 3, it can be seen that most propagation-based detection methods use the FakeNewsNet dataset for experiments, and some source-based detection methods also use this dataset. The content-based detection method uses text-based or image-based datasets for experimentation. The dataset plays a very important role in training fake news detection models. As fake news takes on different forms, it is necessary to continuously iterate and update the datasets in terms of multimodality, dataset size, topic domains, and other aspects.

5. Conclusion

In the post-truth era, the public pays attention to the truth. Researchers have sought to improve performance in detecting fake news. About 40% of the research focuses on fake news detection by adopting machine learning [29]. Nevertheless, some key areas remain unresolved. The following highlights the current research gaps and future work directions.

(1) Fake news detection is an interdisciplinary study, involving graph mining, NLP, information retrieval (IR), and other fields [17]. We need to have a deeper understanding of what fake news is and what the nature of fake news is. More importantly, the cooperation between experts in different fields should be strengthened to study fake news.

(2) Due to the high cost of manual tagging, semi-supervised or unsupervised fake news detection methods should be studied. Automatic annotation methods can also be sought to reduce annotation costs.

(3) At present, most detection models roughly classify news into the following two categories: "true" and "fake". Experiments have shown that applying multimodal information can improve detection performance, but there is room for improvement in terms of aligning and fusing information from different modalities.

(4) The current fake news detection models are usually limited to specific topics. A deeper understanding of fake news is needed to identify its unique invariant features. Cross-topic fake news detection models should be studied.

Author Contributions: Methodology, Y.S.; validation, Y.S., Q.L., N.G., J.Y. and Y.Y.; investigation, Y.S. and Q.L.; resources, Y.S. and N.G.; data curation, Y.S. and Y.Y.; writing—original draft preparation, Y.S.; writing—review and editing, Y.S.; visualization, Y.S. and J.Y.; supervision, Q.L.; project administration, Q.L.; funding acquisition, Y.S. All authors have read and agreed to the published version of the manuscript.

Funding: This research was funded by the Fundamental Research Funds for the Central Universities, grant number ZY20215151; and the Natural Science Project of Xinjiang University Scientific Research Program, grant number XJEDU2021Y003.

Institutional Review Board Statement: Not applicable.

Informed Consent Statement: Not applicable.

Data Availability Statement: Not applicable.

Conflicts of Interest: The authors declare no conflict of interest.

References

1. Papanastasiou, Y. Fake News Propagation and Detection: A Sequential Model. *Manag. Sci.* **2020**, *66*, 1826–1846. [CrossRef]
2. Allcott, H.; Gentzkow, M. Social media and fake news in the 2016 election. *J. Econ. Perspect.* **2017**, *31*, 211–236. [CrossRef]
3. Aïmeur, E.; Amri, S.; Brassard, G. Fake news, disinformation and misinformation in social media: A review. *Soc. Netw. Anal. Min.* **2023**, *13*, 1–36. [CrossRef]
4. Islam, M.R.; Liu, S.; Wang, X.; Xu, G. Deep learning for misinformation detection on online social networks: A survey and new perspectives. *Soc. Netw. Anal. Min.* **2020**, *10*, 1–20. [CrossRef]
5. Alam, F.; Cresci, S.; Chakraborty, T.; Silvestri, F.; Dimitrov, D.; Martino, G.D.S.; Shaar, S.; Firooz, H.; Nakov, P. A survey on multimodal disinformation detection. *arXiv* **2021**, arXiv:2103.12541.
6. Vosoughi, S.; Roy, D.; Aral, S. The spread of true and false news online. *Science* **2018**, *359*, 1146–1151. [CrossRef]
7. Habgood-Coote, J. Stop talking about fake news! *Inquiry* **2019**, *62*, 1033–1065. [CrossRef]
8. Baptista, J.P.; Gradim, A. A Working Definition of Fake News. *Encyclopedia* **2022**, *2*, 632–645. [CrossRef]
9. Pennycook, G.; Rand, D.G. The psychology of fake news. *Trends Cogn. Sci.* **2021**, *25*, 388–402. [CrossRef]
10. Alonso, M.A.; Vilares, D.; Gómez-Rodríguez, C.; Vilares, J. Sentiment Analysis for Fake News Detection. *Electronics* **2021**, *10*, 1348. [CrossRef]
11. Au, C.H.; Ho, K.K.W.; Chiu, D.K.W. The role of online misinformation and fake news in ideological polarization: Barriers, catalysts, and implications. *Inf. Syst. Front.* **2022**, 1331–1354. [CrossRef]
12. Zhang, X.; Ghorbani, A.A. An overview of online fake news: Characterization, detection, and discussion. *Inf. Process. Manag.* **2019**, *57*, 102025. [CrossRef]
13. Sahoo, S.R.; Gupta, B. Multiple features based approach for automatic fake news detection on social networks using deep learning. *Appl. Soft Comput.* **2020**, *100*, 106983. [CrossRef]
14. Shu, K.; Sliva, A.; Wang, S.; Tang, J.; Liu, H. Fake news detection on social media: A data mining perspective. *ACM SIGKDD Explor. Newsl.* **2017**, *19*, 22–36. [CrossRef]
15. Guo, B.; Ding, Y.; Yao, L.; Liang, Y.; Yu, Z. The future of false information detection on social media: New perspectives and trends. *ACM Comput. Surv. (CSUR)* **2020**, *53*, 1–36. [CrossRef]
16. Tandoc, E.C., Jr.; Lim, Z.W.; Ling, R. Defining "fake news" A typology of scholarly definitions. *Digit. J.* **2018**, *6*, 137–153.
17. Zhou, X.; Zafarani, R. A survey of fake news: Fundamental theories, detection methods, and opportunities. *ACM Comput. Surv. (CSUR)* **2020**, *53*, 1–40. [CrossRef]
18. Guo, Z.; Schlichtkrull, M.; Vlachos, A. A survey on automated fact-checking. *Trans. Assoc. Comput. Linguist.* **2022**, *10*, 178–206. [CrossRef]
19. Nakov, P.; Corney, D.; Hasanain, M.; Alam, F.; Elsayed, T.; Barrón-Cedeño, A.; Papotti, P.; Shaar, S.; Martino, G.D.S. Automated Fact-Checking for Assisting Human Fact-Checkers. In Proceedings of the Thirtieth International Joint Conference on Artificial Intelligence, Montreal, QC, Canada, 19–27 August 2021; pp. 4551–4558.
20. Mridha, M.F.; Keya, A.J.; Hamid, A.; Monowar, M.M.; Rahman, S. A Comprehensive Review on Fake News Detection with Deep Learning. *IEEE Access* **2021**, *9*, 156151–156170. [CrossRef]
21. Paka, W.S.; Bansal, R.; Kaushik, A.; Sengupta, S.; Chakraborty, T. Cross-SEAN: A cross-stitch semi-supervised neural attention model for COVID-19 fake news detection. *Appl. Soft Comput.* **2021**, *107*, 107393. [CrossRef]
22. Garg, S.; Sharma, D.K. Linguistic features based framework for automatic fake news detection. *Comput. Ind. Eng.* **2022**, *172*, 108432. [CrossRef]
23. Capuano, N.; Fenza, G.; Loia, V.; Nota, F.D. Content-Based Fake News Detection with Machine and Deep Learning: A Systematic Review. *Neurocomputing* **2023**, *530*, 91–103. [CrossRef]
24. Sheikhi, S. An effective fake news detection method using WOA-xgbTree algorithm and content-based features. *Appl. Soft Comput.* **2021**, *109*, 107559. [CrossRef]
25. Zhou, X.; Jain, A.; Phoha, V.V.; Zafarani, R. Fake news early detection: A theory-driven model. *Digit. Threat. Res. Pract.* **2020**, *1*, 1–25. [CrossRef]
26. Sharma, K.; Qian, F.; Jiang, H.; Ruchansky, N.; Zhang, M.; Liu, Y. Combating fake news: A survey on identification and mitigation techniques. *ACM Trans. Intell. Syst. Technol. (TIST)* **2019**, *10*, 1–42. [CrossRef]
27. Chi, Y.; Zhu, S.; Hino, K.; Gong, Y.; Zhang, Y. iOLAP: A framework for analyzing the internet, social networks, and other networked data. *IEEE Trans. Multimed.* **2009**, *11*, 372–382.
28. Ruchansky, N.; Seo, S.; Liu, Y. CSI. In Proceedings of theCIKM'17: ACM Conference on Information and Knowledge Management, Singapore, 6–10 November 2017; pp. 797–806.
29. Meel, P.; Vishwakarma, D.K. Fake news, rumor, information pollution in social media and web: A contemporary survey of state-of-the-arts, challenges and opportunities. *Expert Syst. Appl.* **2019**, *153*, 112986. [CrossRef]
30. Yang, Y.; Wang, Y.; Wang, L.; Meng, J. PostCom2DR: Utilizing information from post and comments to detect rumors. *Expert Syst. Appl.* **2022**, *189*, 116071. [CrossRef]
31. Zeng, X.; Abumansour, A.S.; Zubiaga, A. Automated fact-checking: A survey. *Lang. Linguist. Compass* **2021**, *15*, e12438. [CrossRef]

32. Mayank, M.; Sharma, S.; Sharma, R. DEAP-FAKED: Knowledge graph based approach for fake news detection. In Proceedings of the 2022 IEEE/ACM International Conference on Advances in Social Networks Analysis and Mining (ASONAM), Istanbul, Turkey, 10–13 November 2022; pp. 47–51.
33. Hu, L.; Yang, T.; Zhang, L.; Zhong, W.; Tang, D.; Shi, C.; Duan, N.; Zhou, M. Compare to the knowledge: Graph neural fake news detection with external knowledge. In Proceedings of the 59th Annual Meeting of the Association for Computational Linguistics and the 11th International Joint Conference on Natural Language Processing (Volume 1: Long Papers), Online, 1–6 August 2021; pp. 754–763.
34. Pan, J.Z.; Pavlova, S.; Li, C.; Li, N.; Li, Y.; Liu, J. Content based fake news detection using knowledge graphs. In Proceedings of the International Semantic Web Conference, Monterey, CA, USA, 8–12 October 2018; pp. 669–683.
35. Nasir, J.A.; Khan, O.S.; Varlamis, I. Fake news detection: A hybrid CNN-RNN based deep learning approach. *Int. J. Inf. Manag. Data Insights* **2021**, *1*, 100007. [CrossRef]
36. Choudhary, A.; Arora, A. Linguistic feature based learning model for fake news detection and classification. *Expert Syst. Appl.* **2020**, *169*, 114171. [CrossRef]
37. Verma, P.K.; Agrawal, P.; Amorim, I.; Prodan, R. WELFake: Word Embedding Over Linguistic Features for Fake News Detection. *IEEE Trans. Comput. Soc. Syst.* **2021**, *8*, 881–893. [CrossRef]
38. Umer, M.; Imtiaz, Z.; Ullah, S.; Mehmood, A.; Choi, G.S.; On, B.-W. Fake News Stance Detection Using Deep Learning Architecture (CNN-LSTM). *IEEE Access* **2020**, *8*, 156695–156706. [CrossRef]
39. Cao, J.; Qi, P.; Sheng, Q.; Yang, T.; Guo, J.; Li, J. Exploring the role of visual content in fake news detection. In *Disinformation, Misinformation, and Fake News in Social Media: Emerging Research Challenges and Opportunities*; Springer: Cham, Switzerland, 2020; pp. 141–161.
40. Qi, P.; Cao, J.; Yang, T.; Guo, J.; Li, J. Exploiting multi-domain visual information for fake news detection. In Proceedings of the 2019 IEEE international conference on data mining (ICDM), Beijing, China, 8–11 November 2019; pp. 518–527.
41. Uppada, S.K.; Manasa, K.; Vidhathri, B.; Harini, R.; Sivaselvan, B. Novel approaches to fake news and fake account detection in OSNs: User social engagement and visual content centric model. *Soc. Netw. Anal. Min.* **2022**, *12*, 1–19. [CrossRef] [PubMed]
42. Yu, P.; Xia, Z.; Fei, J.; Lu, Y. A Survey on Deepfake Video Detection. *IET Biom.* **2021**, *10*, 607–624. [CrossRef]
43. Song, C.; Ning, N.; Zhang, Y.; Wu, B. A multimodal fake news detection model based on crossmodal attention residual and multichannel convolutional neural networks. *Inf. Process. Manag.* **2020**, *58*, 102437. [CrossRef]
44. Singh, V.K.; Ghosh, I.; Sonagara, D. Detecting fake news stories via multimodal analysis. *J. Assoc. Inf. Sci. Technol.* **2021**, *72*, 3–17. [CrossRef]
45. Khattar, D.; Goud, J.S.; Gupta, M.; Varma, V. Mvae: Multimodal variational autoencoder for fake news detection. In *The World Wide Web Conference*; Association for Computing Machinery: New York, NY, USA, 2019; pp. 2915–2921.
46. Segura-Bedmar, I.; Alonso-Bartolome, S. Multimodal Fake News Detection. *Information* **2022**, *13*, 284. [CrossRef]
47. Wang, H.; Wang, S.; Han, Y. Detecting fake news on Chinese social media based on hybrid feature fusion method. *Expert Syst. Appl.* **2022**, *208*, 118111. [CrossRef]
48. Monti, F.; Frasca, F.; Eynard, D.; Mannion, D.; Bronstein, M.M. Fake news detection on social media using geometric deep learning. *arXiv* **2019**, arXiv:1902.06673, 2019.
49. Silva, A.; Han, Y.; Luo, L.; Karunasekera, S.; Leckie, C. Propagation2Vec: Embedding partial propagation networks for explainable fake news early detection. *Inf. Process. Manag.* **2021**, *58*, 102618. [CrossRef]
50. Barnabò, G.; Siciliano, F.; Castillo, C.; Leonardi, S.; Nakov, P.; Martino, G.D.S.; Silvestri, F. Deep active learning for misinformation detection using geometric deep learning. *Online Soc. Netw. Media* **2023**, *33*, 100244. [CrossRef]
51. Song, C.; Teng, Y.; Zhu, Y.; Wei, S.; Wu, B. Dynamic graph neural network for fake news detection. *Neurocomputing* **2022**, *505*, 362–374. [CrossRef]
52. Jeong, U.; Ding, K.; Cheng, L.; Guo, R.; Shu, K.; Liu, H. Nothing stands alone: Relational fake news detection with hypergraph neural networks. In Proceedings of the 2022 IEEE International Conference on Big Data (Big Data), Osaka, Japan, 17–20 December 2022; pp. 596–605.
53. Han, Y.; Karunasekera, S.; Leckie, C. Graph neural networks with continual learning for fake news detection from social media. *arXiv* **2020**, arXiv:2007.03316.
54. Wei, L.; Hu, D.; Lai, Y.; Zhou, W.; Hu, S. A Unified Propagation Forest-based Framework for Fake News Detection. In Proceedings of the 29th International Conference on Computational Linguistics, Gyeongju, South Korea, 12–17 October 2022; pp. 2769–2779.
55. Murayama, T.; Wakamiya, S.; Aramaki, E.; Kobayashi, R. Modeling the spread of fake news on Twitter. *PLoS ONE* **2021**, *16*, e0250419. [CrossRef]
56. Ni, S.; Li, J.; Kao, H.-Y. MVAN: Multi-View Attention Networks for Fake News Detection on Social Media. *IEEE Access* **2021**, *9*, 106907–106917. [CrossRef]
57. Shu, K.; Mahudeswaran, D.; Wang, S.; Liu, H. Hierarchical propagation networks for fake news detection: Investigation and exploitation. In Proceedings of the International AAAI Conference on Web and Social Media, Limassol, Cyprus, 5–8 June 2020; Volume 14, pp. 626–637.
58. Davoudi, M.; Moosavi, M.R.; Sadreddini, M.H. DSS: A hybrid deep model for fake news detection using propagation tree and stance network. *Expert Syst. Appl.* **2022**, *198*, 116635. [CrossRef]

59. Shu, K.; Wang, S.; Liu, H. Beyond news contents: The role of social context for fake news detection. In Proceedings of the twelfth ACM International Conference on Web Search and Data Mining, Melbourne, Australia, 11–15 February 2019; pp. 312–320.
60. Nguyen, V.H.; Sugiyama, K.; Nakov, P.; Kan, M.-Y. Fang: Leveraging social context for fake news detection using graph representation. In Proceedings of the 29th ACM International Conference on Information & Knowledge Management, San Francisco, CA, USA, 27 October–1 November 2020; pp. 1165–1174.
61. Sitaula, N.; Mohan, C.K.; Grygiel, J.; Zhou, X. Credibility-based fake news detection. In *Disinformation, Misinformation, and Fake News in Social Media: Emerging Research Challenges and Opportunities*; Springer: Cham, Switzerland, 2020; pp. 163–182.
62. Yuan, C.; Ma, Q.; Zhou, W.; Han, J.; Hu, S. Early Detection of Fake News by Utilizing the Credibility of News, Publishers, and Users based on Weakly Supervised Learning. In Proceedings of the 28th International Conference on Computational Linguistics, Barcelona, Spain, 8–13 December 2020; pp. 5444–5454.
63. Luvembe, A.M.; Li, W.; Li, S.; Liu, F.; Xu, G. Dual emotion based fake news detection: A deep attention-weight update approach. *Inf. Process. Manag.* **2023**, *60*, 103354. [CrossRef]
64. Ramalingam, D.; Chinnaiah, V. Fake profile detection techniques in large-scale online social networks: A comprehensive review. *Comput. Electr. Eng.* **2018**, *65*, 165–177. [CrossRef]
65. Kudugunta, S.; Ferrara, E. Deep neural networks for bot detection. *Inf. Sci.* **2018**, *467*, 312–322. [CrossRef]
66. Rostami, R.R.; Karbasi, S. Detecting Fake Accounts on Twitter Social Network Using Multi-Objective Hybrid Feature Selection Approach. *Webology* **2020**, *17*, 1–18. [CrossRef]
67. Gao, T.; Yang, J.; Peng, W.; Jiang, L.; Sun, Y.; Li, F. A Content-Based Method for Sybil Detection in Online Social Networks via Deep Learning. *IEEE Access* **2020**, *8*, 38753–38766. [CrossRef]
68. Bazmi, P.; Asadpour, M.; Shakery, A. Multi-view co-attention network for fake news detection by modeling topic-specific user and news source credibility. *Inf. Process. Manag.* **2023**, *60*, 103146. [CrossRef]
69. Zhang, J.; Dong, B.; Yu, P.S. Fake News Detection with Deep Diffusive Network Model. *arXiv* **2018**, arXiv:1805.08751.
70. Yang, S.; Shu, K.; Wang, S.; Gu, R.; Wu, F.; Liu, H. Unsupervised fake news detection on social media: A generative approach. In Proceedings of the AAAI conference on artificial intelligence, Hilton, HI, USA, 27 January–1 February 2019; Volume 33, pp. 5644–5651.
71. Vlachos, A.; Riedel, S. Fact checking: Task definition and dataset construction. In Proceedings of the ACL 2014 Workshop on Language Technology and Computational Social Science, Baltimore, MD, USA, 26 June 2014.
72. Yang, C.; Zhou, X.; Zafarani, R. CHECKED: Chinese COVID-19 fake news dataset. *Soc. Netw. Anal. Min.* **2021**, *11*, 1–8. [CrossRef] [PubMed]
73. Nan, Q.; Cao, J.; Zhu, Y.; Wang, Y.; Li, J. MDFEND: Multi-domain fake news detection. In Proceedings of the 30th ACM International Conference on Information & Knowledge Management, Queensland, Australia, 1–5 November 2021; pp. 3343–3347.
74. Patwa, P.; Sharma, S.; Pykl, S.; Guptha, V.; Kumari, G.; Akhtar, M.S.; Ekbal, A.; Das, A. Fighting an infodemic: COVID-19 fake news dataset. In Proceedings of the Combating Online Hostile Posts in Regional Languages during Emergency Situation: First International Workshop, CONSTRAINT 2021, Collocated with AAAI 2021, Virtual Event, 8 February 2021; Springer International Publishing: Berlin/Heidelberg, Germany, 2021; pp. 21–29.
75. Kim, J.; Aum, J.; Lee, S.; Jang, Y.; Park, E.; Choi, D. FibVID: Comprehensive fake news diffusion dataset during the COVID-19 period. *Telemat. Inform.* **2021**, *64*, 101688. [CrossRef]
76. BuzzFeed. Available online: https://github.com/BuzzFeedNews/2016-10-facebook-fact-check/tree/master/data (accessed on 10 August 2023).
77. Nakamura, K.; Levy, S.; Wang, W.Y. r/fakeddit: A new multimodal benchmark dataset for fine-grained fake news detection. *arXiv* **2019**, arXiv:1911.03854, 2019.
78. Sharma, D.K.; Garg, S. IFND: A benchmark dataset for fake news detection. *Complex Intell. Syst.* **2023**, 2843–2863. [CrossRef]
79. Shu, K.; Mahudeswaran, D.; Wang, S.; Lee, D.; Liu, H. FakeNewsNet: A Data Repository with News Content, Social Context, and Spatiotemporal Information for Studying Fake News on Social Media. *Big Data* **2020**, *8*, 171–188. [CrossRef]
80. Salem, F.K.A.; Al Feel, R.; Elbassuoni, S.; Jaber, M.; Farah, M. Fa-kes: A fake news dataset around the syrian war. In Proceedings of the international AAAI conference on web and social media, Limassol, Cyprus, 5–8 June 2019; Volume 13, pp. 573–582.
81. ISOT Fake News Dataset. Available online: https://www.uvic.ca/engineering/ece/isot/datasets/ (accessed on 22 August 2021).
82. Vogel, I.; Jiang, P. Fake news detection with the new German dataset "GermanFakeNC". In *International Conference on Theory and Practice of Digital Libraries*; Springer International Publishing: Cham, Switzerland, 2019; pp. 288–295.
83. Hossain, M.Z.; Rahman, M.A.; Islam, M.S.; Kar, S. Banfakenews: A dataset for detecting fake news in bangla. *arXiv* **2020**, arXiv:2004.08789.
84. Wang, W.Y. "liar, liar pants on fire": A new benchmark dataset for fake news detection. *arXiv* **2017**, arXiv:1705.00648.
85. PolitiFact. Available online: https://www.politifact.com/ (accessed on 12 August 2023).
86. Snopes. Available online: https://www.snopes.com/ (accessed on 12 August 2023).
87. TruthOrFiction. Available online: https://www.truthorfiction.com/ (accessed on 12 August 2023).
88. CheckYourFact. Available online: https://checkyourfact.com/ (accessed on 12 August 2023).
89. FactCheck. Available online: https://www.factcheck.org/ (accessed on 12 August 2023).
90. Gossip. Available online: https://www.gossipcop.com (accessed on 12 August 2023).

91. Ruijianshiyao. Available online: http://www.newsverify.com (accessed on 12 August 2023).
92. Khan, J.Y.; Khondaker, M.; Islam, T.; Iqbal, A.; Afroz, S. A benchmark study on machine learning methods for fake news detection. *arXiv* **2019**, arXiv:1905.04749.

Disclaimer/Publisher's Note: The statements, opinions and data contained in all publications are solely those of the individual author(s) and contributor(s) and not of MDPI and/or the editor(s). MDPI and/or the editor(s) disclaim responsibility for any injury to people or property resulting from any ideas, methods, instructions or products referred to in the content.

MDPI AG
Grosspeteranlage 5
4052 Basel
Switzerland
Tel.: +41 61 683 77 34

Applied Sciences Editorial Office
E-mail: applsci@mdpi.com
www.mdpi.com/journal/applsci

Disclaimer/Publisher's Note: The statements, opinions and data contained in all publications are solely those of the individual author(s) and contributor(s) and not of MDPI and/or the editor(s). MDPI and/or the editor(s) disclaim responsibility for any injury to people or property resulting from any ideas, methods, instructions or products referred to in the content.